INTRODUCTION TO
SCIENTIFIC
PROGRAMMING

COMPUTATIONAL PROBLEM SOLVING
USING *MATHEMATICA*® AND C

INTRODUCTION TO
SCIENTIFIC PROGRAMMING

COMPUTATIONAL PROBLEM SOLVING
USING *MATHEMATICA*® AND C

JOSEPH L. ZACHARY

Joseph L. Zachary
Department of Computer Science
University of Utah
3190 Merrill Engineering Bldg.
Salt Lake City, UT 84112, USA

Publisher: Allan M. Wylde
Publishing Associate: Keisha Sherbecoe
Production Supervisor: Natalie Johnson
Manufacturing Supervisor: Jeffrey Taub
Production: Black Hole Publishing Service
Cover Designer: Irene Imfeld

Library of Congress Cataloging-in-Publication Data

Zachary, Joseph. L.
    Introduction to scientific programming : computational problem solving using
    Mathematica and C / Joseph L. Zachary.
      p. cm.
    Includes bibliographical references and index.
    ISBN 0-387-98250-7 (alk. paper)
      1. Electronic digital computers—Programming.   2. *Mathematica*
    3. C (Computer program language)   I. Title.
    QA76.6.Z32   1996
    510'.1'135133—dc20   96-9378

Printed on acid-free paper.

Photocomposed copy prepared by Black Hole Publishing Service, Berkeley, CA.

Printed and bound by Hamilton Printing Co., Rensselaer, NY.
Printed in the United States of America.

9 8 7 6 5 4 3 2 1

ISBN 0-387-98250-7 Springer-Verlag New York Berlin Heidelberg SPIN 10628428

TELOS, The Electronic Library of Science, is an imprint of Springer-Verlag New York with publishing facilities in Santa Clara, California. Its publishing program encompasses the natural and physical sciences, computer science, economics, mathematics, and engineering. All TELOS publications have a computational orientation to them, as TELOS' primary publishing strategy is to wed the traditional print medium with the emerging new electronic media in order to provide the reader with a truly interactive multimedia information environment. To achieve this, every TELOS publication delivered on paper has an associated electronic component. This can take the form of book/diskette combinations, book/CD-ROM packages, books delivered via networks, electronic journals, newsletters, plus a multitude of other exciting possibilities. Since TELOS is not committed to any one technology, any delivery medium can be considered. We also do not foresee the imminent demise of the paper book, or journal, as we know them. Instead we believe paper and electronic media can coexist side-by-side, since both offer valuable means by which to convey information to consumers.

The range of TELOS publications extends from research level reference works through textbook materials for the higher education audience, practical handbooks for working professionals, and broadly accessible science, computer science, and high technology general interest publications. Many TELOS publications are interdisciplinary in nature, and most are targeted for the individual buyer, which dictates that TELOS publications be affordably priced.

Of the numerous definitions of the Greek word "telos," the one most representative of our publishing philosophy is "to turn," or "turning point." We perceive the establishment of the TELOS publishing program to be a significant step towards attaining a new plateau of high quality information packaging and dissemination in the interactive learning environment of the future. TELOS welcomes you to join us in the exploration and development of this exciting frontier as a reader and user, an author, editor, consultant, strategic partner, or in whatever other capacity one might imagine.

TELOS, The Electronic Library of Science
Springer-Verlag Publishers
3600 Pruneridge Avenue, Suite 200
Santa Clara, CA 95051

THE
ELECTRONIC
LIBRARY
OF
SCIENCE

TELOS Diskettes

Unless otherwise designated, computer diskettes packaged with TELOS publications are 3.5" high-density DOS-formatted diskettes. They may be read by any IBM-compatible computer running DOS or Windows. They may also be read by computers running NEXTSTEP, by most UNIX machines, and by Macintosh computers using a file exchange utility.

In those cases where the diskettes require the availability of specific software programs in order to run them, or to take full advantage of their capabilities, then the specific requirements regarding these software packages will be indicated.

TELOS CD-ROM Discs

For buyers of TELOS publications containing CD-ROM discs, or in those cases where the product is a stand-alone CD-ROM, it is always indicated on which specific platform, or platforms, the disc is designed to run. For example Macintosh only, Windows only; cross-platform, and so forth.

TELOSpub.com (Online)

Interact with TELOS online via the Internet by setting your World-Wide-Web browser to the URL: *http://www.telospub.com*

The TELOS Web site features new product information and updates, an online catalog and ordering, samples from our publications, information about TELOS, data-files related to and enhancements of our products, and a broad selection of other unique features. Presented in hypertext format with rich graphics, it's your best way to discover what's new at TELOS.

TELOS also maintains these additional Internet resources:

gopher://gopher.telospub.com
ftp://ftp.telospub.com

For up-to-date information or questions regarding TELOS online services, send the one-line e-mail message to:

send info to: info@TELOSpub.com

To Bethy and Sammy

Preface

This book teaches beginning science and engineering students how to solve the computational problems they will likely encounter during their academic and professional careers. Although it deals with introductory concepts, the book provides a solid foundation on which students will be able to base a lifetime of learning. It also addresses two shortcomings typical of introductory programming classes taken by science and engineering students.

First, introductory programming classes often ignore the context in which programming is actually done. When the goal is to teach students how to write programs, the temptation is to tell the students exactly what programs to write. Unfortunately, this means that the person who makes up the assignments ends up doing what is often the most interesting part: reducing a real-world problem to a programming problem.

Students can become so used to having someone else do this creative work that they end up with no idea how to apply their programming skills to problems that arise elsewhere. As a result, they tend not to make use of computing in their science and engineering classes unless they are explicitly told to do so. Even then, many students are completely lost until someone tells them what program to write and how to go about writing it.

Second, introductory programming classes are most often designed for students planning to major in computer science. Even when the problems that inspire computational solutions are highlighted, they are often not particularly relevant to science and engineering students. More fundamentally, the technical slant on programming that is appropriate for computer science majors is not always appropriate for science and engineering majors.

The treatment of floating-point numbers is an excellent case-in-point. Conventional introductory courses in computer science make almost no use of floating-point numbers; scientific programming, in contrast, is all but impossible without them. Computer science students might be taught about the IEEE standard for representing floating-point numbers; science and engineering majors must be taught the computational properties of floating-point numbers and their role in representing inexact physical measurements.

With the exception of Chapters 1 and 10, which contain overview material, each of the chapters in this book is organized around the solution of a problem from science or engineering. Each chapter begins with a problem description, develops a mathematical model of the problem, devises a computational method for solving the model, creates an implementation based on the method, and finally assesses the solution yielded by the implementation.

The thread that ties this book together, of course, is its coverage of the essentials of scientific programming using *Mathematica* and C. I very deliberately develop these essentials, however, in the broader context of computational problem solving. To that end, I have selected problems that permit me to motivate, introduce, and explain the essential programming concepts in a natural order. Although the bulk of each chapter is devoted to exploring implementation issues, each chapter also contains material on problem description, model building, method development, and solution assessment that put the implementation issues squarely in context.

Programming Languages

There is no single best language for either beginning or advanced programmers to use. Programming languages are nothing more than sophisticated tools. Skilled programmers, just like master mechanics, must have a variety of tools at their disposal. The appropriate choice of programming language depends on such factors as the nature of the programming problem at hand, the availability of solutions to similar problems, the programmer's background, and the computing environment in which the programmer is working.

In recognition of this reality, I use *Mathematica* in the first half of this book and C in the second. *Mathematica* and C are representatives of two broad classes of tools that are widely used for scientific programming. *Mathematica* is a computer algebra system, and C is a conventional programming language.

Mathematica is a system for doing numerical, symbolic, and graphical mathematics, and it is also the name for the programming language that is used within that system. I chose to use *Mathematica* instead of C in the first half of the book for a number of reasons. First, *Mathematica* is easier to use and to understand than is C. Second, it is possible to do interesting things in *Mathematica* with only a modicum of training. Students can easily find immediate applications for *Mathematica* in their other classes. Finally, most of the basic concepts that a student must master to understand *Mathematica* carry over to the process of learning C. I exploit this fact in the second half of the book.

Mathematica is not the only computer algebra system. Its major competitors are Maple and Macsyma. I could have equally well used either in place of *Mathematica* in this book; in fact, I have already written a version of this book that uses Maple.

C is a conventional programming language that is widely used for all types of programming, including scientific programming. Despite the power of computer algebra systems such as *Mathematica*, proficiency in conventional languages remains an essential skill for computational scientists. Conventional languages have more than 40 years of development behind them, and a large body of software written in such languages as Fortran and C is available for reuse. Conventional languages are more flexible than computer algebra systems, and for large applications are also considerably faster.

There are a large number of conventional programming languages, but the two that are most often used for scientific programming are C and Fortran. I arbitrarily chose C over Fortran because I do most of my own programming in C, but I could have written essentially the same book using Fortran in place of C. The truth of the matter is that Fortran and C—at least when used for scientific programming—are much more alike than they are different.

Despite the fact that it features two programming languages, this book is not made up of two unrelated halves. I use *Mathematica* to examine such issues as the computational properties of numbers, the use of operators and functions to build expressions, the exploitation of function libraries, and the organization of programs around programmer-defined functions. All of this material from the *Mathematica* half carries over to the C half. This means that instead of starting over from scratch when I introduce C, I am able to pick up where I left off in my discussion of *Mathematica*.

I do not attempt to cover all of either *Mathematica* or C. *Mathematica* is far too large to even imagine covering in a book of this type, and C contains a number of aspects that are either irrelevant to scientific programming or are easily picked up from any C reference after the material in this book is covered. My idea is to show a representative and useful subset of each language in the context of computational problem solving.

As I was writing the book, I identified several extremely useful features of *Mathematica* that deserved discussion but which did not fit conveniently into the narrative. Rather than wedge a discussion of these features into the main body of the book, I explore them in Appendix A. Several of the chapters contain pointers to these topics; the reader can decide whether or not to pursue them as interest dictates.

Mathematica is a product of Wolfram Research Inc., which sells *Mathematica* systems (in both professional and student versions) for all types of computers. The programs in this book are compatible with both *Mathematica* 3.0 and (with a few exceptions) the older *Mathematica* 2.2. Appendix D details the changes you must make if you wish to use the programs with *Mathematica* 2.2.

C compilers are available from a variety of vendors. All of the C programs in this book will work with any C compiler that observes the ANSI C standard, which almost any C compiler released in this decade will do. In fact, the programs will also work with any compiler for ANSI standard C++, so long as that compiler has access to an ANSI C input/output library.

Required Background

In adopting the approach of beginning each chapter with a problem and then pursuing its solution through model, method, implementation, and assessment, I was forced to make assumptions about the math and science backgrounds of the students who would use this book. The degree of preparation that I am assuming is typical of college freshmen intending to major in science or engineering.

Seven of the 16 problems are of the kind encountered in high school or college physics classes. They variously involve mass, velocity, acceleration, force, gravity, friction, drag, springs, and heat. Whenever I present a physics problem, I include an explanation of the physical principles involved, but it helps if the student has worked with at least some of these concepts before.

The remaining nine problems are mathematical in nature. These problems, and all of the models in the book, require an appropriate mathematical background. I make extensive use of algebra, geometry, and the trigonometry of right triangles. I also make one use of the law of cosines, one use of integration, and three uses of differentiation. Although it is possible to understand the implementations without fully understanding the models that lead to them, it is best if the student has at least completed the prerequisites required to take an introductory calculus course.

Many of the methods hinge on elementary concepts from numerical analysis, and the implementations make use of *Mathematica* and C. I have assumed no particular background in any of these areas. To succeed as a programmer, however, a student must be willing to experiment. The kind of student who is reluctant to try something out until he or she completely understands it will not prosper.

Chapter Overview

I have summarized below, as is appropriate for each chapter, the problem around which it is organized, the mathematical background required to understand the model, the computational method that it introduces, and the *Mathematica* or C background that I develop in the chapter to produce the implementation that solves the problem.

- **Chapter 1** explores the differences among the experimental, theoretical, and computational approaches to science, using the work of Tycho Brahe, Isaac Newton, and Johannes Kepler as motivating examples. I present an extended discussion of Kepler's five-year struggle to computationally determine the orbit of Mars, and use it to introduce the five-step approach to computational problem solving that I employ in the remainder of the book.

- **Chapter 2** marks the beginning of my treatment of *Mathematica*. I use the problem of determining the average population density of the earth as the starting point for learning how to use arithmetic expressions to do simple calculations. I devote a major portion of the chapter to illustrating the differences between *Mathematica*'s rational and floating-point numbers.

- **Chapter 3** describes the simple experiment and geometrical argument that Eratosthenes of Cyrene used to determine the circumference of the earth in 225 B.C. This leads into a discussion of the uncertainty inherent in calculations involving physical measurements, and I explain how to use the methods of significant digits and interval arithmetic to cope with this uncertainty. Along the way, I introduce the notions of variables and assignment statements.

- **Chapter 4** introduces the amusing problem of determining how far beyond the edge of a table an overhanging stack of n identical blocks can be made to extend. This is the first problem in the book that has a non-trivial model. To develop the model, which boils down to an additive series of n terms, I appeal to the concept of center of gravity. Much of the chapter is devoted to explicitly summing the series for different values of n. This allows me to illuminate the tradeoff between exact but (sometimes extremely) slow rational number arithmetic and fast but (sometimes catastrophically) inexact floating-point number arithmetic.

- **Chapter 5** uses the Pythagorean Theorem to develop a model and two methods for the problem of determining the line-of-sight distance to the horizon from the top of a hill. The first method involves subtracting two numbers that differ only in their nonsignificant digits and leads to a catastrophic loss of significance in the result. The second method uses an algebraic simplification to avoid the difficulties of the first. The implementation based on this method exploits programmer-defined functions.

- **Chapter 6** exploits the problem of predicting the population growth of the United States as an opportunity to branch out from the numerical focus of Chapters 2–5 and examine symbolic programming. I use such symbolic *Mathematica* functions as `Factor`, `Solve`, and `Limit` to develop the models of simple, compound, and continuous interest and to apply them to the problem of modeling population growth.

- **Chapter 7** uses the problem of visualizing the ballistic trajectory of a projectile to explore *Mathematica*'s `Plot` function and its variants. I develop the model by resolving the projectile's initial velocity into its horizontal and vertical components, which means that this is the first chapter to make use of trigonometry. I explain how to produce a variety of two-dimensional visualizations, including graphs, parametric plots, and animations.

- **Chapter 8** explores the problem of determining the power required to move a ship along a specified one-dimensional trajectory through the water. This entails developing a function that takes a symbolic trajectory expression, the ship's mass, and the ship's drag coefficient as parameters and produces a power consumption expression as its result. The model and its implementation both rely on differentiation to obtain velocity and acceleration from position. I use *Mathematica*'s relational operators in conjunction with its `Which` construct to describe a trajectory for the ship.

- **Chapter 9** poses a problem that boils down to finding the radius of an arc that divides a circle into equal areas. The model is based on the geometric properties of circles and triangles and uses a bit of trigonometry. The end result is a simple transcendental equation that defies symbolic solution. This motivates the explanation of the bisection method that follows. I then develop, in measured steps, a *Mathematica* function that implements the method. The function exploits assignment, `If`, `While`, and `Module` expressions. My intention is not to conduct an in-depth treatment of imperative programming in *Mathematica*, but to foreshadow programming in C.

- **Chapter 10** marks the transition point between my treatments of *Mathematica* and C. By this point the student should have a solid grasp of, among other things, the computational properties of numbers, expression-oriented functional programming, and the use of programmer-defined functions. C, of course, has numbers, expressions, and programmer-defined functions, which means that I can build directly on this background as I begin discussing statement-oriented procedural programming in C. I devote this chapter to outlining the differences between *Mathematica* and C and to exploring, at a high level, C versions of *Mathematica* programs from previous chapters.

- **Chapter 11** describes the problem of controlling the ankle, knee, and hip joints of a robot so it can do squat exercises. By making two simplifying assumptions, I develop a trigonometric model in which the position of the robot's shoulder is determined by the length of its torso and the angle of its knee. The C implementation of this model leads into a discussion of straight-line programs that touches on numerical types, expressions, assignment statements, the C math library, and the C input/output library.

- **Chapter 12** uses the problem of calculating the motion of a block on an inclined ramp to explore the variants of the C `if` statement. I package the implementation into a programmer-defined function, which lets me explore such notions as function headers, function bodies, function prototypes, and the runtime behavior of functions. The increasingly more general models that I develop in the course of the chapter are all based on trigonometry.

- **Chapter 13** explains the problem of determining the center coordinates of each rod in a stack of cylindrical rods. The model makes extensive use of trigonometry. Unlike the other trigonometric models in the book, this model involves non-right triangles and hence the law of cosines. I do not introduce any new aspects of C in this chapter. Instead, I use the implementation of the model as an opportunity to illustrate the role of functional decomposition in program design, implementation, and testing.

- **Chapter 14** begins with the problem of determining the positioning of a beam that is laid against a cubical box. Although the model relies only on the properties of similar triangles and some algebraic manipulations, it leads to a fourth-order polynomial equation. I develop and implement Newton's method as the means of solving this equation. Along the way I explain the workings of `while` and `do` loops in C.

- **Chapter 15** begins with the problem of determining the required length of a steel sheet if it is to be 10 meters long after it is corrugated. Solving this problem involves finding the length of the curve that characterizes the corrugation, which in turn requires evaluating an integral. This leads into a discussion of the rectangular and trapezoidal methods for numerical integration, and eventually to their implementations. The implementations of these methods give me the opportunity to illustrate the use of `for` loops, the possibility of passing functions as parameters, and the advantages of dividing large programs into multiple files.

- **Chapter 16** poses the problem of determining the position, as a function of time, of a block undergoing damped harmonic oscillation. I do not attempt to derive the model for this problem, but instead point out that determining the solution entails finding the complex roots of a quadratic equation. I then show how Newton's method can be easily generalized to find complex roots. Implementing this version of Newton's method in C leads to a discussion of the use of C structures to create programmer-defined types and abstract datatypes.

- **Chapter 17** describes the problem of determining the temperature, as a function of time, at the center of an insulated silver rod after heat sources have been applied to its ends. I develop a finite-element model for this problem from first principles, beginning with three facts about heat transfer at steady state. The model is entirely algebraic. I then explain how a finite-element method can be used to solve such a model. Finally, I use C arrays to implement the model.

- **Chapter 18** addresses the problem of visualizing the temperature of the rod from Chapter 17 as it undergoes heating. My strategy is to write a C program that writes to a file a sequence of *Mathematica* commands that, when loaded into *Mathematica*, produces the desired animation. This gives closure to the book and illustrates that no single tool is appropriate for all computational problems. The resulting implementation allows me to explain how to pass arrays as parameters and how to read from and write to files.

For the most part, the models, methods, and implementations that I develop in the book do not depend on any mathematics beyond geometry, algebra, and the trigonometry of right triangles. This makes the book accessible to beginning science and engineering students without compromising my goal of illustrating the entire problem-solving process. There are, however, several places where I use slightly more advanced mathematics.

- Chapter 8 uses differentiation to determine symbolic velocity and acceleration expressions from position expressions. Differentiation is used in both the model and the implementation, which makes this chapter difficult for students who haven't begun taking a calculus course. Fortunately, this chapter can be skipped with very little impact on the continuity of the book.

- Chapter 13 applies the law of cosines to non-right triangles. Students who are unfamiliar with the law of cosines may come away unclear on the derivation of the model. This will not stand in the way of understanding the bulk of the chapter, however, which focuses on designing, implementing, and testing C programs.

- Chapter 15 hinges on the problem of definite integration, but describes it in purely geometric terms as the problem of finding the area enclosed by a curve.

- Chapters 14 and 16 deal with Newton's method. A full understanding of how the method works requires understanding a bit about first derivatives of functions. The implementations themselves, however, do not depend on differentiation.

Pedagogical Features and Supporting Material

This book contains a variety of features designed to help students learn. It is also supported by an enclosed diskette and by a Web site.

- At the end of each chapter I review the key concepts that were developed in the chapter.
- Every chapter except the first is followed by a comprehensive set of exercises.
- Except for Chapters 1 and 10, every chapter is organized in two different ways. Each chapter is divided in the usual way into titled sections and subsections. In addition, the transitions among the five steps of the computational problem-solving process are marked by icons that appear in the margins. The icons are described in Chapter 1.
- There are four appendices containing supplementary information. Appendix A discusses a handful of interesting *Mathematica* topics that were not appropriate for the body of the text; Appendix B summarizes the *Mathematica* functions that are used in the text; Appendix C summarizes the C functions that are used in the text; Appendix D details the changes you must make to the programs in this book if you are using *Mathematica* version 2.2. The book ends with a list of references to further reading.
- For each chapter that contains *Mathematica* code, I have included on the diskette a *Mathematica* notebook containing all of the code. The notebook gives students a way to use the code without having to type it back in, and includes explanatory text that complements the book's narrative. I have included both *Mathematica* 3.0 and 2.2 versions of the notebooks.
- At several points in the book I use functions from a *Mathematica* library that I created specifically for this book. The implementation of that library, along with directions for its use, is included (in both *Mathematica* 3.0 and 2.2 formats) on the diskette.
- The diskette contains the complete implementation of every C program that I present in the book. In many cases, the implementations that appear in the book consist only of an interesting fragment. In other cases, a long program is spread over several pages. The programs on the diskette are complete, consolidated, and contain explanatory comments.
- Two identical Web sites contain online material that is designed to support the book. Much of it will be useful in, for example, student laboratory sessions. The Web sites are

```
http://www.telospub.com/catalog/MATHEMATICA/IntroSciProgMma.html
http://www.cs.utah.edu/~zachary/IntroSciProgMma.html
```

The material includes Java applets, HTML-based tutorials, and *Mathematica* notebooks, and I have included several examples on the diskette. The development of the material is an ongoing project, and it will continue to accumulate over the next few years. You are welcome to use the online material over the Internet or to download it for local use.

- My electronic mail address is `zachary@cs.utah.edu`. I would enjoy hearing from you if you have questions, comments, or suggestions.

Acknowledgments

This book grew out of a collaboration with Chris Johnson that dates back to the beginning of 1993, when Chris interested me in the problem of introductory computational science education. We teamed up to design a new introductory course in scientific programming, develop the course notes and online material that were the precursors to this book, and teach the first offering of the course. The course has been taught seven times at the University of Utah since its initial offering in the winter of 1994. After the initial offering, Chris began encouraging me to convert the course notes and online material into a textbook. I would never have started this book without the benefit of his energy and vision, and I would never have completed it without his help and encouragement.

Eric Eide and Ken Parker played prominent roles in the development of the original course notes and online material in late 1993 and early 1994. The original course used Maple, C, and Fortran. Eric wrote the bulk of the material on Fortran and C that appeared in the original notes. The precision and clarity of his writing was an inspiration. Ken developed preliminary versions of the Maple material and came up with most of the problems that were featured in the notes. I have carried over many of those problems to this book. Ken has also answered many of my questions about physics, mathematics, and numerical analysis over the years.

The Undergraduate Computational Engineering and Science (UCES) project is an initiative of the Department of Energy to improve the undergraduate curriculum through the use of computation. In 1993 I became a charter member of the project, which has grown to include about 50 educators and scientists. My involvement with this group has been an important source of professional enrichment, and the members of UCES have contributed in no small way to this book. The idea of using both a symbolic algebra system and a conventional programming language in a problem-oriented course on scientific programming came out of the first meeting. The five-step approach to solving computational problems that I use throughout this book evolved in a subsequent meeting.

Tom Marchioro is the executive director of the UCES project, and he has supported and encouraged the development of this book from the beginning. Jim Corones, who founded the UCES project; Dave Martin, who preceded Tom as executive director; and Tom have cooperated to provide funding for the development of the original course notes and online material. I hope that their speculative investment has paid off.

Judy Zachary read early drafts of many of the chapters of the Maple-based precursor to this book, and made careful suggestions for improving their technical content and presentation. Her efforts complemented those of the reviewers, who commented mostly on the big picture. Judy also cheerfully served as a sounding board for many of my preliminary ideas.

Andre Weideman visited the University of Utah during the 1996–97 academic year, and twice taught the course that uses the Maple-based version of this book as its text. He suggested a number of improvements to my treatment of numerical analysis as well as to the exercises. It was a pleasure to work with him.

The reviewers—Alkis Akritas, Tom Marchioro, Vaidy Sunderam, Bob White, and John Ziebarth—collectively made a number of valuable suggestions. Their suggestions for sharpening my treatment of *Mathematica* were particularly valuable.

Frank Stenger was an early champion of the use of computer algebra systems in introductory computing classes at the University of Utah. When I had run out of ideas for problems, I visited Frank, who suggested two.

Paul Wellin of Wolfram Research answered my questions and sent me several different versions of *Mathematica*.

A Course and Curriculum Development grant from the Division of Undergraduate Education of the National Science Foundation funded the development of online material in support of undergraduate computational science education in general and this book in particular.

The students, faculty, and staff in the Department of Computer Science at the University of Utah have combined to make the first ten years of my teaching career extremely rewarding.

Joe Zachary
Salt Lake City
April 11, 1997

Contents

Computational Science

Tycho Brahe was a Danish astronomer who lived from 1546 until 1601. Under the patronage of the King of Denmark, he established Renaissance Europe's first observatory on the Danish island of Hven. Even though the telescope would not be invented until eight years after his death, Tycho was able to measure accurately the positions of celestial bodies to between one-thirtieth and one-sixtieth of a degree. To accomplish this feat he invented a succession of instruments, including the sextant in 1569, the mural quadrant in 1582, and the portable ring armillary in 1591. For more than twenty years he devoted himself to systematically measuring and recording the positions of the sun, the moon, and the planets.

Johannes Kepler was a German astronomer who lived from 1571 until 1630. He assisted Tycho for the last year and a half of Tycho's life. When Tycho died, Kepler came into possession of Tycho's lifetime accumulation of detailed astronomical observations. Kepler's goal was to find the pattern behind the motions of the planets. Over a period of years he filled his ledgers with extensive calculations as he looked for patterns in Tycho's numbers. He ultimately published the following three laws of planetary motion, the first two in 1609 and the third in 1619.

1. A planetary orbit is an ellipse with one focus at the sun.
2. The line between a planet and the sun sweeps out equal areas in equal times.
3. The square of a planet's orbital period is proportional to the cube of its average distance from the sun.

Isaac Newton was an English physicist who lived from 1642 until 1727. Among his many other accomplishments, which included the invention of calculus, he discovered his now familiar laws of motion and gravitation. He also had the insight to see that these laws apply in the same way to every object, whether it is a falling apple or an orbiting planet. While Kepler's laws match Tycho's observations, they do not explain why the planets behave as they do. Newton's laws provide this explanation, and in fact Kepler's laws are a consequence of, and can be derived from, Newton's laws.

Using modern terminology, we would say that Tycho was an *experimental* scientist, Newton was a *theoretical* scientist, and Kepler was a *computational* scientist. Many scientists today actually have these adjectives in their job descriptions; there are, for example, experimental chemists, theoretical physicists, and computational fluid dynamicists. At the foundation of modern science and engineering are the experimental, theoretical, and computational approaches to science.

1.1 Experiment, Theory, and Computation

The three approaches to science are not entirely distinct. Some degree of overlap is evident in the work of all scientists and engineers. Tycho, Newton, and Kepler each exploited elements of experiment, theory, and computation in his work.

Experimental scientists work by observing how nature behaves. Tycho experimented by designing, building, and using sextants, quadrants, and armillaries to make astronomical observations of unprecedented precision. Modern-day scientists experiment by designing, building, and using supercolliders to measure the interactions of high-energy subatomic particles, and they experiment by designing, building, and using the Cosmic Background Explorer satellite to observe the background microwave radiation of the universe.

Theoretical scientists use the language of mathematics to explain and predict the behavior of nature. Newton's laws of motion explain how objects move under the influence of external forces, and his law of gravitation explains how gravitational force depends on mass and distance. For their part, modern-day scientists have developed theories that explain how high-energy subatomic particles interact with one another, and theories that explain how the universe may have evolved following its creation in the Big Bang.

Theories are often inspired by experimental results, and experiments are often devised to validate theories. For example, the Big Bang theory holds that the universe as we know it today evolved over billions of years from its original form as an incredibly dense and energetic concentration of subatomic particles. Some of the high-energy interactions that occurred during the crucial first few hundred microseconds of the universe were the same, according to the theory, as the interactions that have been experimentally observed in high-energy supercolliders. Conversely, one requirement of the Big Bang theory is that the background radiation of the universe should exhibit minute variations, a prediction that has been verified by the Cosmic Background Explorer.

Computational scientists use theoretical and experimental knowledge to create computer-based models of aspects of nature. They then use these models to simulate the behavior of nature and relate the results that they observe back to the real world. In a sense, computational scientists create a simulated version of nature on which they perform experiments. The computational approach to science is most often applied in cases where physical experimentation is difficult, expensive, or impossible.

Work in computational science sometimes leads to new theories or suggests new experiments. Kepler's computational work on the orbits of planets was in this vein. Kepler began his career believing that the orbits of the planets were circular. By performing extensive calculations on Tycho's measurements, he discovered that this was not the case and eventually discovered his three laws of planetary motion.

Computational science is also used to test theories. A computational scientist, for example, might draw on the available theoretical and experimental knowledge to create, on a computer, a simulated version of the universe as it existed immediately after the Big Bang. The simulation could then be run and the evolution of the simulated universe observed. If the simulated universe resulting from the simulated Big Bang turned out to be unlike the real universe, this would be evidence that the theories on which the simulation was based were flawed.

Computational science is perhaps most often used to apply well-established theories in ways that would be impossible without computers. Computer-based simulations can be used, for example, to predict possible changes in the climate, estimate the potential for earthquake damage in a region, or assess the aerodynamics of a new aircraft design.

Kepler notwithstanding, the widespread exploitation of the computational approach to science is a fairly new phenomenon. Although climate simulations could in principle be done by hand, in practice the climate might change before the calculations were complete. Whereas Kepler had to play the dual roles of computational scientist and computer, the power of modern computers has made possible much more extensive computational work.

1.2 Solving Computational Problems

The process of solving a problem in computational science does not begin with writing a computer program. In fact, much creative work must be done before the first line of a program can be written. We can idealize the process of computational problem solving into the following five steps.

1. Identify the problem.
2. Pose the problem in terms of a mathematical model.
3. Identify a computational method for solving the model.
4. Implement the computational method on a computer.
5. Assess the answer in the context of the implementation, method, model, and problem.

In practice, the boundaries between these five steps can become blurred, and for specific problems one or two of the steps may be more important than the others. Nonetheless, having this five-step strategy firmly in mind will help to focus our efforts as we solve problems. In this section we will discuss the five steps and illustrate them with the work that Kepler did to establish that the orbit of Mars is elliptical.

1.2.1 Problem

The first step is to thoroughly explore the problem that is to be solved. We must be sure we know what question we are answering, what theoretical and experimental knowledge we can bring to bear, and what kind of answer we wish to obtain. All of these considerations require knowledge about the underlying science. A computational scientist is first of all a scientist; Kepler was first of all an astronomer.

We must be sure at the outset that the problem is one that can be approached computationally. The problem that Kepler took on in 1600 was to characterize the path taken by Mars as it orbits the sun (a computational question), not to measure how Mars moves (an experimental question) or to explain why Mars moves as it does (a theoretical question).

The theoretical ideas with which Kepler framed his problem were those of the Copernican model of the solar system. Nicolaus Copernicus had postulated in 1543 that the planets move around the sun in perfect circles at constant speeds. In Kepler's day this model had just begun to supplant the ancient Ptolemaic system, which held that the sun and planets revolved around the earth.

Although the Copernican system gave a simpler explanation for many of the observed motions of the planets, it didn't account very well for the fact that the rate at which a planet moves is different at different points of its orbit. Mars, which at times was five degrees away from the position in the sky predicted by the Copernican system, was particularly problematic.

The experimental facts on which Kepler drew were the observations of his mentor, Tycho Brahe. These observations gave Kepler the angle between the sun and Mars as observed from the earth on many different days over a period of many years. Kepler also knew that the orbital period of Mars is about 687 days.

Kepler sought a mathematical characterization of the Martian orbit, and to do this would ultimately require solving two subproblems. The first, ironically, was to use Tycho's observations of the sun and Mars to locate the center of the *earth's* orbit, which he assumed was circular. Once he had found the orbit of the earth, Kepler's second subproblem was to determine points through which Mars moved in its orbital plane and to find a mathematical expression describing a smooth path through those points.

1.2.2 Model

The second step in the problem-solving process is to create a mathematical model of the problem we are solving. We must reduce the problem as originally stated into one expressed in purely mathematical terms, using mathematical expertise to extract the essentials from the underlying physical description of the problem. Once the problem is expressed mathematically, it can be solved computationally.

Kepler created two different models corresponding to the two subproblems we identified in the last section. Figure 1.1 illustrates the model that Kepler used to determine the earth's orbit, which was a prerequisite to finding coordinates of the

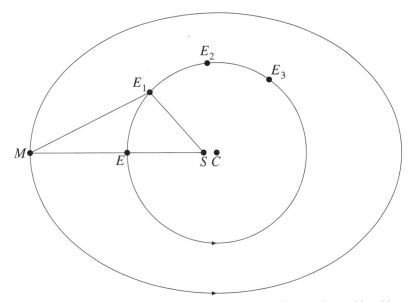

Figure 1.1 Using triangulation to determine three points on the earth's orbit

Martian orbit. S denotes the sun, E the earth, and M Mars at a time when Mars and the sun are in opposition to one another.

Kepler assumed that the orbit of the earth was circular, but that it was centered on some point C other than the sun. Fortunately, this assumption is close to the truth. The earth's orbit is in fact an almost circular ellipse with the sun at one focus. The center of this ellipse is offset from the sun, as suggested in Figure 1.1.

E_1 denotes the position of the earth in its orbit one Martian year (687 days) after the opposition at point E. During this time, the earth has completed one orbit and is 43 days shy of completing a second, while Mars has returned to its position at M. Similarly, E_2 and E_3 denote the positions of the earth two and three Martian years, respectively, after the opposition.

Observations of the sun made when the earth is at points E and E_1 establish the angle MSE_1, and observations of Mars and the sun made when the earth is at point E_1 establish the angle ME_1S. To obtain the required observations of the sun and Mars, Kepler needed only to consult Tycho's records. By taking the distance MS to be a unit of length, Kepler could then determine the coordinates of E_1 by applying geometric and trigonometric reasoning.

By repeating this process for E_2 and E_3, Kepler could locate three points on the orbit of the earth. Since three points are sufficient to fix the center of a circle, Kepler could then determine the location of C relative to the sun.

(This is a slight simplification of the model Kepler actually constructed. His model did not depend on observations made when Mars was at opposition, and in fact did not require that the earth be in position E_1 one Martian year after an opposition.)

Kepler's model of the second subproblem depended on his first. Once he had determined the orbit of the earth, he could use Tycho's observations of Mars to determine coordinates on the Martian orbit. All that remained was to determine an equation describing a smooth closed curve through those points.

1.2.3 Method

Once a problem has been described in mathematical terms, the third step is to settle on a method for actually calculating the solution. It is usually easier to set up the solution to a problem mathematically than it is to follow through and actually find the answer. Finding the answer often requires the use of numerical analysis, which lies at the intersection of applied mathematics and computer science. Numerical analysis is the science of doing numerical calculations accurately and efficiently on a computer.

Kepler had to identify a computational method for solving each of his models. The first model (Figure 1.1) led to the simpler method. Finding each of the three points on the earth's orbit required doing only well-understood trigonometric and algebraic operations.

The second model was more challenging; Kepler, in fact, was unable to find a good method for dealing with it. Given a set of points lying on the Martian orbit, Kepler had to find an equation describing a smooth closed curve through those points. To make matters more difficult, as we will see below, the points were not exact. More precisely, then, Kepler had to find an equation describing a smooth closed curve passing acceptably close to the points.

The method that Kepler ultimately used to deal with this problem was one of trial and error. He made a succession of guesses of what the equation might be, and then checked whether or not the data fit the guess. His first supposition was that the orbit was circular. When this didn't fit with the data, he next considered different sorts of oval orbits and finally a variety of elliptical orbits. In the almost 400 years since Kepler did this, numerical analysts (including Isaac Newton) have developed systematic methods for fitting curves to experimental data. A modern-day Kepler could choose one of these methods and save a lot of time and effort in the process.

1.2.4 Implementation

Once we have a computational method for dealing with our mathematical model, the fourth step is to carry out the method with a computer, whether human or silicon. It may be possible to do the required calculations by hand; it may be possible to do the calculations with a pocket calculator; or it may require writing a computer program.

If a computer program is required, a variety of programming languages, each with different properties, are available to choose from, as well as a variety of computers, ranging from the most basic home computers to the fastest parallel supercomputers. The ability to choose the proper combination of programming language and computer, and use them to create and execute a correct and efficient implementation of the method, requires knowledge in computer science.

Kepler did all of his calculations by hand, using as computational aids tables of trigonometric functions and logarithms. When Kepler began his work on the Martian orbit, he made a bet that he would solve the problem within a week. In the end he spent a total of approximately five years on the problem, and ultimately filled his notebooks with 900 pages of calculations. With a modern programming language, even his trial-and-error approach to finding an equation for the Martian orbit would have required only an afternoon's effort.

1.2.5 Assessment

Even after the problem is posed, a model created, a method developed, and an implementation crafted to obtain a solution, we are not done. The fifth and final step is to assess the solution. In many respects, assessment is the most open-ended and difficult of the five steps involved in solving computational problems.

Numerical answers can never be taken at face value. Errors and assumptions, both intentional and unintentional, inevitably creep into the problem, model, method, and implementation. We must ask ourselves how large an error is likely to be included in the solution. As we identify errors, we must try to fix the problems that caused them and come up with an improved solution. When we finally identify the best solution possible, we must find a good way to communicate it to others.

The experimental data that Kepler used in stating the problem—Tycho's observations of Mars—were not perfectly accurate. At best the measurements were accurate to within one-thirtieth of a degree, which represents an error of over 80,000 miles in the position of Mars. Kepler had to account for this error when he was trying to fit an equation to the coordinates he calculated.

Kepler's model is based on a number of assumptions. The construction in Figure 1.1, for example, assumes that the earth's orbit is circular when it is in fact slightly eccentric. An assessment of Kepler's results would have to estimate the amount of error that this assumption might have introduced into his solution. Minor assumptions sometimes have major consequences.

Kepler's method for computing coordinates of the earth's orbit depends on the values of trigonometric functions. In any calculation these values must be rounded off to a fixed number of decimal places. Although these unavoidable numerical errors are not likely to change the solution significantly, their effect must be assessed.

Kepler's implementation also offered an important potential source of error. A mistake in any of his 900 pages of by-hand calculations could have compromised his answer. Today, a mistake in a computer program could have the same effect. A full assessment must work to uncover such mistakes.

Kepler's final result was an equation for the elliptical orbit of Mars. At different points of its orbit, Mars moves at different speeds. In describing this result to others, the equation is essential but it is not the only means that might be employed. A graph of the orbit would make its elliptical eccentricity more evident; an animation of its orbit would make its speed fluctuations more evident.

The final step in Kepler's assessment of his work on the orbit of Mars was his publication, in 1627, of the Rudolphine Tables. These tables allowed astronomers to determine the positions of the planets at any point in the past or future, and were about 30 times more accurate than either the Alfonsine Tables (which were based on the Ptolemaic system) or the Prutenic Tables (which were based on the Copernican system). The correspondence of computed results with physical reality is, in the end, the most convincing form of assessment.

1.3 Onward

We have just seen that a computational scientist must have expertise in the underlying science, in mathematics, in numerical analysis, and in computer science. We will assume throughout this book that you already have an introductory-level background in science and mathematics. This will leave us free to focus on programming and, to a lesser extent, on the essentials of numerical analysis.

There are many kinds of computers, including personal computers, engineering workstations, and state-of-the-art supercomputers. For the most part, however, a program written in a high-level programming language can be made to run on just about any kind of machine. (The exceptions to this rule occur mostly with supercomputers, which often require specialized programming techniques to achieve optimal performance.) Thus, we need not be concerned in this text with what kind of computer you will be using.

There are also many kinds of programming languages. Fortunately, most languages are based on a common set of principles; once you understand these principles, it is much easier to learn a new language. Unfortunately, it is difficult to grasp these principles without first learning a programming language. As a result, we cannot write this book in the abstract. We must pick some real languages to be the object of your study.

There is no single language that is appropriate for all problems. As a result, an accomplished programmer is generally the master of more than one programming language. We will focus in this text on a pair of languages that enable two different styles of scientific programming: *Mathematica* and C.

Mathematica is a very high-level computer algebra system that supports writing programs to do symbolic, numerical, and graphical mathematics. There are other computer algebra systems—for example, Maple and Macsyma—and the choice of *Mathematica* as a representative of this category is to a large extent arbitrary. C is a lower-level conventional language that supports general-purpose computation, including scientific programming. There are many other conventional languages that we might have chosen, including Fortran and C++.

We will study *Mathematica* in the first half of the text and C in the second half. Because *Mathematica* and C share common principles, we will not have to start over when we begin our study of C. Instead, we will build on what you have learned in the context of *Mathematica*. We will have much more to say about the similarities and differences between *Mathematica* and C in Chapter 10, at which point you will be much better equipped to appreciate them.

We will make no attempt to cover all that there is to know about either *Mathematica* or C. Our primary goals are to enable you to write useful scientific programs in both languages while giving you a strong enough background in the underlying principles that you can easily learn more on your own. Neither goal would be served by spreading ourselves too thin.

With the exception of this chapter and Chapter 10, which will mark the transition from *Mathematica* to C, every chapter in this book will be organized around solving a representative problem from computational science. We will lay out a problem, develop a model, devise a computational method, create an implementation in *Mathematica* or C, and assess the solution. (In each chapter, the transitions between the steps of the problem-solving process are marked with icons. Look at the beginnings of Sections 1.2.1–1.2.5 to see the icons we will use.) It is crucial that you appreciate the context in which scientific programming is performed. Otherwise, you may end up with no idea of how to apply your programming skills.

2

Population Density: Computational Properties of Numbers

The human population of the earth is large and is increasing rapidly. Table 2.1 gives the estimates of the United States Bureau of the Census for the growth of the world's population over the last 2000 years. This table reveals that the population in 1900 was approximately 1.6 billion. This increased to around 2.5 billion in 1950 and more than doubled over the next 40 years to exceed 5.6 billion in 1995. The Census Bureau estimates that the earth's population reached 5.713 billion during September 1995.

Are there already too many people in the world, or can the earth comfortably absorb even larger numbers? This is a crucial question that can be addressed from a governmental, scientific, or religious perspective. As with most complex issues, there is considerable disagreement.

Partisans on one side or the other of public policy issues often use metaphors to express an issue in terms the average citizen can easily understand. For example, President Reagan once characterized the size of the U.S. federal budget, on national television, by giving the height of a stack of dollar bills equal in value to the budget.

A common tactic when discussing the question of global population is to calculate how much space each person would have if, say, California were evenly divided up among all the people on earth. In this chapter we will calculate the length of a side of the square of land that would be allocated to each person if the earth's entire surface were parceled out this way.

Table 2.1   Lower and upper bounds on the estimated population of the earth, in millions, from the year 1 to the present. These estimates were compiled from a variety of sources by the Census Bureau and are available from its Web site at `http://www.census.gov/`.

Year	Bounds Lower	Upper		Year	Bounds Lower	Upper
1	170	400		1600	545	579
400	190	206		1650	470	545
500	190	206		1700	600	679
600	200	206		1750	629	961
700	207	210		1800	813	1125
800	220	224		1850	1128	1402
900	226	240		1900	1550	1762
1000	254	345		1950	2400	2556
1100	301	320		1960	3039	3039
1200	360	450		1970	3706	3706
1300	360	432		1980	4458	4458
1400	350	374		1990	5282	5282
1500	425	540		1995	5691	5691

2.1 Model

Before we can use a computer, we must first find a model for our problem. A model will allow us to state the problem in mathematical terms. We will then be able to identify a computational method for producing a solution. It is well worth the time to evaluate possible models before proceeding. A bad choice will ultimately lead to less-than-useful answers no matter how careful we are in the subsequent steps of the process.

One possibility is to model the earth as a perfect sphere of radius 4000 miles whose population is 5.713 billion. Beginning from these assumptions, we can easily calculate the earth's surface area and average population density. This is a very crude model that exhibits the following problems. You can probably think of others.

• The radius of the earth is not exactly 4000 miles—that's just an approximate figure that you might remember from elementary school.

• The earth is not a perfect sphere. It has a larger radius at the equator than at the poles, and the surface itself is very irregular.

• Most of the earth's surface is water. Presumably, we're only interested in dividing up the dry land.

We will be much better off if we turn to an atlas for an improved estimate of the earth's land area. If we do this, we can model the earth's land area as 57.8 million square miles and its population as 5.713 billion people.

This improved model is far from perfect. The estimated land area includes inland water such as the Great Lakes and counts Antarctica and other uninhabitable regions. The population estimate, while it was (or will be) correct at some instant in time, only approximates a value that changes every second. Nevertheless, our model is about the best we can do. It is acceptable as a basis for our computations so long as we acknowledge its imperfections when we present our final answer. It can also serve as a good starting point in case we ever want to develop an improved model.

2.2 Method

Now that we have a model, the obvious method for computing an answer to the original problem is to divide the earth's estimated land area by its estimated population, convert the quotient to square feet, and then take the square root. It is not always so easy to find an appropriate computational method, but, as in this case, a simple model often leads to a simple method.

2.3 Implementation

The method that we have just identified could be carried out with pencil and paper, or by punching buttons on a pocket calculator, or by writing and running a C or Fortran program, to give just three examples. We will use *Mathematica* to carry out the computations because this affords a good way to start learning about it.

Our primary goal in exploring the implementation is to understand the differences between the two types of numbers, rational and floating-point, that *Mathematica* provides. We will begin in Section 2.4 by computing two different answers, first with rational numbers and then with floating-point numbers. This will raise a number of questions concerning these two types of numbers that we will deal with in Sections 2.5–2.8.

2.4 Arithmetic Expressions

To use *Mathematica*, type in a command (which you should think of as a question) and it will print out a result (which you should think of as the answer). The key to using *Mathematica* is learning what kinds of commands are available and how to pose them. The best approach to learning *Mathematica* (or any programming language) is to begin with the simple things and gradually build up to the more complicated ones. You will probably never know everything there is to know about *Mathematica* (few people do), but that is no reason to shy away from making productive use of it.

2.4.1 Simple Arithmetic Expressions

The most basic kind of command is the *arithmetic expression*. The simplest arithmetic expression is a numeral: type it in and *Mathematica* echoes its value:

```
In[1]:= 57.8

Out[1]= 57.8
```

(2.1)

Our question is, What is the value of 57.8? and *Mathematica*'s answer is, 57.8. The In[1]:= is supplied by *Mathematica* and indicates the beginning of the command. Similarly, the Out[1]= marks the beginning of the answer.

Mathematica cannot make sense out of just any command; commands must be written using the correct *syntax*. Compare the interaction above with

```
In[2]:= 57,8

        Syntax::tsntxi : "57,8" is incomplete;
           more input is needed
```

(2.2)

Here, *Mathematica* is complaining that we have made a *syntax error*, which is comparable to a misspelled word or a grammatical mistake in English. Instead of answering our question (which makes no sense, since 57,8 isn't a valid numeral), *Mathematica* tells us that we've made a mistake and tries its best to tell us what the error is.

(All of the *Mathematica* interactions contained in this chapter, of which Examples 2.1 and 2.2 are the first two instances, are collected in a *Mathematica* notebook on the diskette included with this book. The same is true of each of the *Mathematica*-oriented chapters that follow. You should use these notebooks together with *Mathematica* to experiment interactively with the examples as we present them.)

2.4.2 Compound Arithmetic Expressions

We can also use *Mathematica* to perform an arithmetic operation on two numbers. For example, we can find the number of square miles of land there are for each person by dividing the earth's land area (57.8 million square miles) by its population (5.713 billion).

```
In[3]:= 57800000 / 5713000000
```

$$Out[3]= \frac{289}{28565} \text{ mi}^2$$

(2.3)

Notice that *Mathematica* expresses the quotient in Example 2.3 as a fraction in lowest terms. The two numbers in the command and the number in the result are examples of *Mathematica*'s *rational numbers*. Rational numbers are written *without* decimal points, and answers to rational number computations are always exact, which often means they must be given as fractions. We can verify that the quotient is exact by multiplying it by the original divisor

$$
\begin{array}{l}
\texttt{In[4]:= 289/28565 * 5713000000} \\[1em]
\texttt{Out[4]= 57800000 mi}^2
\end{array}
\qquad (2.4)
$$

and observing that the result is the original dividend.
    We could also have computed the quotient by entering

$$
\begin{array}{l}
\texttt{In[5]:= 57800000. / 5713000000.} \\[1em]
\texttt{Out[5]= 0.0101173 mi}^2
\end{array}
\qquad (2.5)
$$

Mathematica expresses this quotient not with a fraction but with a decimal number. The three numbers in Example 2.5 are examples of *Mathematica*'s *floating-point numbers*. Floating-point numbers are written *with* decimal points, and answers to floating-point computations are always rounded off to a fixed number of digits. The answer above is rounded to six digits, not counting the leading zeroes.
    It is common in science and engineering to write very large or very small numbers by using scientific notation, as in 57.8×10^6 square miles or 5.713×10^9 people. *Mathematica* floating-point numbers can also be written in scientific notation, though they appear different:

$$
\begin{array}{l}
\texttt{In[6]:= 57.8*\^{}6 / 5.713*\^{}9} \\[1em]
\texttt{Out[6]= .0101173 mi}^2
\end{array}
\qquad (2.6)
$$

Here, *^6 stands for "times ten to the sixth" and *^9 stands for "times ten to the ninth."
    Because floating-point numbers are restricted to a fixed number of digits, *Mathematica*'s floating-point computations are not necessarily exact. In fact, neither of the last two computations yielded an exact answer, as can be verified by multiplying the quotient by the divisor:

$$
\begin{array}{l}
\texttt{In[7]:= .0101173 * 5.713*\^{}9} \\[1em]
\texttt{Out[7]= 5.78001} \times 10^7 \texttt{ mi}^2
\end{array}
\qquad (2.7)
$$

Mathematica chose to write the answer using scientific notation. The important thing to notice is that the answer (57,800,100) differs from the original dividend (57,800,000) by 100.

2.4.3 *Mathematica* Capability: Units

Each of the results in Examples 2.3–2.7 is labeled as being in square miles. We have attached units to each answer to help you understand the meaning of the calculation that produced it, and we will continue to do so where appropriate throughout this book.

If you were to enter the calculations into *Mathematica* exactly as shown, however, *Mathematica* would produce only the numerical parts of the answers without attaching any units. This should not be particularly surprising. When you ask *Mathematica* to divide 5.78 million by 5.713 billion, it has no way of knowing whether the answer is in square miles, square meters, or square furlongs. *Mathematica* is a powerful system, but it is not clairvoyant!

If you attach units to the numbers that are involved in a calculation, however, *Mathematica can* determine the proper unit to attach to the result. For example, if we ask *Mathematica* to multiply 72 inches by 2.54 centimeters/inch,

```
In[8]:= 72 in * 2.54 cm/in

Out[8]= 182.88 cm
```
(2.8)

Mathematica will report that the answer is in centimeters.

Despite the fact that *Mathematica* has this capability, we will continue to write our calculations as we did in Examples 2.3–2.7. We will *not* attach units to the numbers involved in calculations, but we *will* attach units to the results. We are taking this approach because there are some subtle issues that arise in connection with attaching units to numbers. Dealing with these issues as they arise would detract from the central points about scientific programming that we will be making.

Mathematica's ability to deal with units is the first of several capabilities that we will note, briefly illustrate, and then decline to pursue in the main body of the text. In Appendix A, however, we explain how to exploit each of these capabilities. We are deferring a detailed discussion of these capabilities to Appendix A because, in our judgment, they are too useful to ignore but too awkward to pursue in sequence.

2.4.4 Composing Expressions

Besides multiplication ($*$) and division ($/$), *Mathematica* also supports addition ($+$), subtraction ($-$), and exponentiation ($\wedge$).

For example, to make progress toward solving the population density problem we need to compute the number of square feet in a square mile. We can square 5280 ft/mi to obtain the number of square feet per square mile.

```
In[9]:= 5280^2

Out[9]= 27878400 ft²/mi²
```
(2.9)

We now know both the number of square *miles* per earth inhabitant and the number of square feet per square mile. Putting this together, we can obtain the number of square *feet* per inhabitant using either floating-point arithmetic

```
In[10]:= .0101173 * 27878400

Out[10]= 282054. ft²
```

(2.10)

or rational number arithmetic.

```
In[11]:= 289/28565 * 27878400

Out[11]= 1611371520 ft²
         ─────────
           5713
```

(2.11)

Notice that we mixed floating-point and rational numbers in Example 2.10. The first number is a floating-point number, and the second number is a rational number. When floating-point and rational numbers are mixed in this way, *Mathematica* converts the rational number into floating-point form before carrying out the operation. This is why the final answer appears as a floating-point number.

It is both tiresome and error-prone to type back into *Mathematica* the numbers it has previously computed for us. Fortunately, it is possible to compose more than one operation into a single arithmetic expression. For example, we could have computed the floating-point result with

```
In[12]:= (57.8*^6 / 5.713*^9) * 27878400

Out[12]= 282053. ft²
```

(2.12)

or even more directly with

```
In[13]:= (57.8*^6 / 5.713*^9) * (5280^2)

Out[13]= 282053. ft²
```

(2.13)

Notice how we have used parentheses in the input to group the operations, which controls the order in which they are performed. We will discuss the order in which arithmetic operations are carried out in the *absence* of parentheses in Chapter 3.

Finally, notice that the answers that we obtained in Examples 2.10 and 2.12 differ by one in their last digits. It turns out that the two results are different for a simple but important reason. We will explain the reason for this discrepancy in Section 2.8.

2.4.5 Built-In Functions

To determine the size of the square of land area for each inhabitant (that is, the length of one side), we must take the square root of the area determined above. To do that, we can use *Mathematica*'s built-in square root function:

```
In[14]:= Sqrt[282053.]

Out[14]= 531.087 ft
```

(2.14)

This is the answer to our original problem: Every person would have a square of land approximately 531 feet on the side if the earth's land area were divided up evenly.

In the example above, the number 282053. is the *parameter* to the function Sqrt; we say that the function is *applied* to its parameter. In a function application, the function's name is followed by the parameter, which must be enclosed in square brackets. The parameter to Sqrt can be any expression whose value is a number:

```
In[15]:= Sqrt[(57.8*^6 / 5.713*^9) * (5280^2)]

Out[15]= 531.087 ft
```

(2.15)

Examples 2.14 and 2.15 both show the results of applying Sqrt to a floating-point parameter. The next example involves a rational number:

```
In[16]:= Sqrt[(57800000 / 5713000000) * (5280^2)]
```

$$Out[16]= 17952\sqrt{\frac{5}{5713}} \text{ ft}$$

(2.16)

With the exception of N (explained below), *Mathematica never* gives an approximate value for an expression composed entirely of rational numbers. That is why the answer above is expressed in terms of the square root of $\frac{5}{5713}$.

Another useful built-in function is N, which takes a rational number as its parameter and converts it into a floating-point number. For example

```
In[17]:= N[17952 * Sqrt[5/5713]]

Out[17]= 531.087 ft
```

(2.17)

produces the same answer as the one computed in Example 2.15.

Mathematica provides hundreds of built-in functions. Rather than attempt to list them all here, we will highlight selected functions as they become useful for solving problems throughout the book. In addition, Appendix B contains a summary of some of *Mathematica*'s more useful functions.

2.4.6 *Mathematica* **Capability: Typeset Mathematics**

In Example 2.14 we took the square root of a number by using the `Sqrt` built-in function. While it is not difficult to understand what `Sqrt[282053.]` means, it is not the way that a square root expression would be written in a math textbook, which would use the familiar radical symbol. It is not even the way that *Mathematica* writes square root expressions in its output, as you can see in Example 2.16.

If you are using the notebook interface to *Mathematica*, you can use the radical symbol interchangeably with the `Sqrt` function when you enter expressions. By using this capability, we can rewrite Example 2.14 as

$$\text{In[18]} := \sqrt{282053.}$$
$$\text{Out[18]} = 531.087 \text{ ft}$$

(2.18)

and Example 2.16 as

$$\text{In[19]} := \sqrt{\tfrac{57800000}{5713000000} \, 5280^2}$$
$$\text{Out[19]} = 17952\sqrt{\frac{5}{5713}} \text{ ft}$$

(2.19)

Mathematica will also allow you to typeset exponents and fractions (as we have in Example 2.19), as well as derivatives and integrals, to name just a few examples. As with *Mathematica*'s support for units, however, the issue of typesetting is tangential to our primary concerns. We will defer all further discussion of this capability to Appendix A.

2.5 Rational Numbers

In mathematics, the set of rational numbers consists of every fraction that can be formed from any two integers (excluding zero as a denominator). Since there are an infinite number of integers, there are an infinite number of rational numbers.

Computers are finite devices, of course, so there is no way that a programming language can provide an infinite number of rational numbers. Thus, when a programming language provides rational numbers, it typically provides a finite set of supported rational numbers consisting of every fraction that can be formed from any two integers that are between a minimum allowable negative integer (which we will call `minint`) and a maximum allowable positive integer (which we will call `maxint`). This means that all of the supported rational numbers lie between `minint` and `maxint`, and it means that every *integer* between `minint` and `maxint` is a supported rational number. This does *not* mean, however, that every *fraction* between `minint` and `maxint` is a supported rational number.

To see why this is true, let's imagine a programming language that provides a rational number system in which `minint` is -3 and `maxint` is 3. In this language, the supported set of rational numbers (Figure 2.1) would include the integers

$$\{-3, -2, -1, 0, 1, 2, 3\}$$

and the fractions

$$\left\{-\frac{3}{2}, -\frac{2}{3}, -\frac{1}{2}, -\frac{1}{3}, \frac{1}{3}, \frac{1}{2}, \frac{2}{3}, \frac{3}{2}\right\}.$$

This set excludes an infinite number of fractions, such as $\frac{1}{6}$, that lie between -3 and 3 but cannot be written as a ratio of integers between -3 and 3, as well as *all* integers and fractions that are either smaller than -3 or larger than 3.

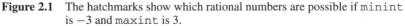

Figure 2.1   The hatchmarks show which rational numbers are possible if `minint` is -3 and `maxint` is 3.

Because a programming language can support only a finite number of rational numbers, an arithmetic expression involving valid rational numbers does not necessarily have a valid value itself. We can illustrate this fact by considering two examples that build on the artificially limited programming language considered above. If we were to add $\frac{1}{2}$ and $\frac{1}{2}$, we would get the answer 1, because this is a valid rational number. If we were to *multiply* the same two numbers, however, we would not be able to obtain a valid answer. This is because the exact answer, when expressed in lowest terms, is $\frac{1}{4}$, and 4 is not between `minint` and `maxint` in our hypothetical programming language's rational number system.

The error that occurs when the numerator or denominator of a rational number expression is too large is called *arithmetic overflow*. Evaluating a rational number expression in a programming language, then, yields either an exact answer or an overflow error.

Because of the possibility of overflow, many of the standard properties of arithmetic do not hold on computers. For example, the identity

$$a^{m-n} = \frac{a^m}{a^n} \tag{2.20}$$

does not hold in our hypothetical programming language. In this language, $3^{2-2} = 3^0 = 1$, since 3, 2, 1, and 0 (which are all of the numbers involved in evaluating the expression) are valid rational numbers. On the other hand, the expression $\frac{3^2}{3^2}$ would cause an overflow error, since 3^2 evaluates to 9, which is not a valid rational number.

2.6 Rational Numbers in *Mathematica*

Mathematica will try to calculate any rational number that you request. It will either succeed in calculating and displaying the number or it will run out of memory in the attempt. In either event, you may be tempted to abort the computation of a very large rational number before its very slow calculation finishes. In other words, the values for `minint` and `maxint` in *Mathematica* depend, in a practical sense, on both your computer's size and your patience.

   Mathematica can calculate rational numbers with hundreds or thousands of digits rather quickly; it is only when an answer requires tens of thousands of digits that the computation time becomes appreciable. For example, we can easily ask *Mathematica* to calculate a 600,000-digit number:

```
In[21]:= 10^600000

Out[21]= $Aborted
```
(2.21)

Unless we are willing to wait an extremely long time, however, we will have to abort the calculation before the answer is displayed.

   We observed earlier that Equation 2.20 did not hold in a language with a `minint` of −3 and a `maxint` of 3. It is easy to show that it does not hold in *Mathematica* either; it simply requires much larger numbers. Thus,

```
In[22]:= 10^(600000-600000)

Out[22]= 1
```
(2.22)

works perfectly, because all of the numbers involved in the calculation, and all of the intermediate results, are of a reasonable size. However, the calculation of the algebraically identical

```
In[23]:= 10^600000 / 10^600000

Out[23]= $Aborted
```
(2.23)

is likely to tax our patience.

2.7 Floating-Point Numbers

The set of real numbers consists of every number that lies on the real number line. Floating-point numbers are used on computers to approximate a subset of the real numbers. Just as there are many possible rational number systems, there are many possible floating-point number systems. All floating-point systems, however, consist of zero, a set of positive numbers, and the corresponding set of negative numbers.

The set of positive base ten floating-point numbers consists of every number that can be written in the form $m \times 10^e$, where

- m (the *mantissa*) is in the interval $[1 \dots 10)$ and contains p digits (p is called the *precision*), and
- e (the *exponent*) is an integer that lies between e_{min} (the minimum exponent) and e_{max} (the maximum exponent) inclusive.

The notation $[1 \dots 10)$ that we use above stands for all real numbers x such that $1 \le x < 10$. The [symbol means that the left endpoint 1 is included in the interval, whereas the) symbol means that the right endpoint 10 is excluded.

Our definition is for *base ten* floating-point numbers. We will use such numbers as examples because you are almost certainly most accustomed to base ten. Computers—and programming languages such as *Mathematica*—use *base two* floating-point numbers, which are numbers of the form $m \times 2^e$ in which m is written in base two. (When you enter a base ten floating-point number, *Mathematica* converts it into an internal base two representation. Similarly, before displaying a floating-point result that is represented internally in base two, *Mathematica* converts it into a base ten representation. All of this happens automatically.) All of the essential properties of floating-point numbers are, fortunately, independent of base.

No matter what values it uses for p, e_{min}, and e_{max}, a floating-point number system will provide only a finite set of numbers. When we want to represent a real number x with a floating-point number, we must first round x to the closest real number y that has a p-digit mantissa. There are then two possibilities:

1. If y is a floating-point number, we say that x is *represented* by y in the floating-point number system. The absolute difference $|x - y|$ is called the *roundoff error*; the relative difference $|\frac{x-y}{x}|$ is called the *relative error*.

2. If y is not a floating-point number, we say that x is *not representable* in the floating-point number system. This can happen if the absolute value of x is too large, which is called an overflow error, or if the absolute value of x is too small, which is called an underflow error.

Fortunately, the essential properties of a floating-point number system are independent of its precision, minimum exponent, and maximum exponent. We will take advantage of this by studying three extremely simple systems of base ten floating-point numbers before considering the specifics of *Mathematica*'s system.

2.7.1 A Simple Floating-Point Number System

We will first consider a floating-point number system in which p is 1, e_{max} is 1, and e_{min} is -1. Under these assumptions, the *positive* floating-point numbers (Figure 2.2a) are

{.1, .2, .3, .4, .5, .6, .7, .8, .9, 1, 2, 3, 4, 5, 6, 7, 8, 9, 10, 20, 30, 40, 50, 60, 70, 80, 90}.

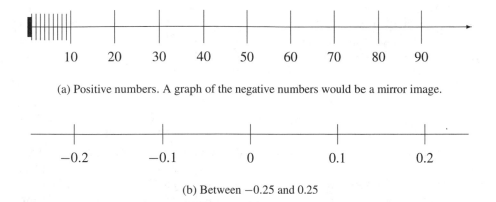

(a) Positive numbers. A graph of the negative numbers would be a mirror image.

(b) Between −0.25 and 0.25

Figure 2.2   The hatchmarks show the positions of floating-point numbers when the precision is 1, the minimum exponent is −1, and the maximum exponent is 1.

These are the *only* positive numbers that can be expressed with such short mantissas and small exponents. (The full set of floating-point numbers in this system would also contain the corresponding negative numbers and zero.) For example, 50 is a floating-point number because it is equal to 5×10^1 and 5 is a floating-point number because it is equal to 5×10^0. On the other hand, 55 (5.5×10^1) is not a floating-point number because it would require a two-digit mantissa, 500 (5×10^2) is not a floating-point number because it would require an exponent of 2, and 0.05 (5×10^{-2}) is not a floating-point number because it would require an exponent of −2.

There are many ways to write any given real number using scientific notation. $50 \times 10^0, 5 \times 10^1$, and 500×10^{-1} are just three of the ways that 50 can be written, for example. Whether or not a particular positive real number is also a floating-point number in our simple system is not a matter of checking whether it *is* written with a one-digit mantissa and an exponent between −1 and 1, but of checking whether it *can be* written with a one-digit mantissa and an exponent between −1 and 1.

By studying Figures 2.2a and 2.2b, you can appreciate how the positive floating-point numbers are distributed. There are nine numbers in the interval [0.1 . . . 1) (all with exponent −1), nine in the interval [1 . . . 10) (all with exponent 0), and nine in the interval [10 . . . 100) (all with exponent 1). The smallest positive floating-point number ($0.1 = 1 \times 10^{-1}$) is obtained by multiplying the smallest possible mantissa (1) by 10 raised to the smallest possible exponent (−1). Similarly, the largest positive floating-point number ($90 = 9 \times 10^1$) is obtained by multiplying the largest possible mantissa (9) by 10 raised to the largest possible exponent (1).

To determine whether a positive real number x can be represented by one of our floating-point numbers, we must find the closest real number y that can be written in scientific notation with a one-digit mantissa. If y is one of our 27 positive floating-point numbers, then x is represented by y. The same sort of rule applies to negative real numbers. Zero is always a floating-point number.

For example, 16 is represented by 20 (with roundoff error 4), 1.45 is represented by 1 (with roundoff error 0.45), and 0.56 is represented by 0.6 (with roundoff error 0.04). On the other hand, neither 96 (which rounds to 100) nor 0.094 (which rounds to 0.09) is represented. When a real number is halfway between two numbers, we will follow the convention of rounding it to the number with the *even* mantissa. Thus, both 3.5 and 4.5 are represented by 4.

Every real number in the interval $[0.095\ldots95)$ is represented by one of our floating-point numbers. For example, all of the numbers in the interval $[0.095\ldots 0.15)$ are represented by 0.1, and all of the numbers in the interval $[0.15\ldots 0.25]$ are represented by 0.2. On the other hand, the numbers in the interval $(0\ldots 0.095)$ are too *small*, and the numbers in the interval $[95\ldots\infty)$ are too *large*, to be represented.

In a floating-point number system, as you can see from Figure 2.2a, greater detail is concentrated in the intervals containing the smaller numbers. As a result, in our example system every real number in the interval $[0.095\ldots95)$ is represented by a floating-point number y such that the relative error, *i.e.* $|\frac{x-y}{x}|$, is no greater than $\frac{1}{3}$.

The real number 14.9999 rounds off to the floating-point number 10, for example, with a relative error of approximately 0.3333; the real number 94 rounds off to the floating-point number 90 with a relative error of approximately 0.04255. This property does not generally apply to the numbers that are not represented. The closest floating-point number to 150 is 90, which represents a relative error of 0.4. Similarly, the closest floating-point number to 0.05 is 0.1, with a relative error of 1.0.

2.7.2 Floating-Point Number Errors

We have already seen that rational number calculations are subject to arithmetic overflow. *Three* kinds of errors can occur when doing arithmetic with floating-point numbers: *overflow*, *underflow*, and *roundoff error*.

Overflow occurs when the absolute value of the result of an arithmetic operation is too *large* to be represented, and underflow occurs when the absolute value of the result of an arithmetic operation is too *small* to be represented. In our simplified system with one-digit mantissas and exponents between -1 and 1, for example, 50×5 would result in an overflow because the product, 250, would require an exponent of 2; similarly, 0.05×0.5 would result in an underflow because 0.025 would require an exponent of -2.

The result of any calculation that causes a floating-point overflow or underflow is unusable for further computation. Roundoff error, as we have already seen, occurs when a real number is rounded to the nearest floating-point number; it also occurs when the result of an arithmetic operation involving floating-point numbers must be rounded to the nearest floating-point number. For example, if we were to multiply 7 and 8 in our simplified floating-point system, the product, 56, would be rounded to 60, which is the closest floating-point number. Unlike the case with overflow and underflow, roundoff error yields results that are approximately correct and can be used in further calculations. As we have seen, even in our extremely simple floating-point number system the relative error in a single arithmetic calculation cannot exceed $\frac{1}{3}$, so long as it neither underflows nor overflows.

2.7.3 Reducing Roundoff Error

We can reduce the severity of roundoff error in our floating-point system by increasing its precision. If, for example, we allow two-digit mantissas while keeping e_{min} and e_{max} at -1 and 1 respectively, the positive floating-point numbers will be 0.10–0.99 (in increments of 0.01), 1.0–9.9 (in increments of 0.1), and 10–99 (in increments of 1). Thus, instead of having 27 numbers in the interval $[0.1 \ldots 100)$, we now have 270 numbers in that same interval. As a result, all relative errors for numbers in the interval $[0.0995 \ldots 99.5)$ will now be no greater than $\frac{1}{30}$, which is ten times smaller than our maximum possible relative error in the original system.

Figure 2.3a shows the positive floating-point numbers in this new system. Because the individual floating-point numbers are now ten times closer together, the magnitude of the roundoff errors will now tend to be smaller. For example, if we multiply 7 and 8 in this system, the answer will be 56 with no roundoff error. If we multiply 7.1 and 8.1, the product, 57.51, will be rounded to 58, with a relative error of approximately 0.008520.

Increasing the precision of a floating-point system, however, does nothing to cut down on the incidence of underflow and overflow. Figure 2.3a shows that just as was the case with our original system, there are no floating-point numbers greater than or equal to 100. Thus, 50×5 will still result in an overflow. The closeup in Figure 2.3b shows that there are still no positive floating-point numbers smaller than 0.1, which means that 0.05×0.5 will still result in an underflow.

One other characteristic of floating-point number systems is apparent from Figure 2.3b. Notice the "holes" on either side of zero. In any floating-point number system, there will always be a range of small numbers on either side of zero, and a range of large numbers on either extreme of the number line, that do not have floating-point representations.

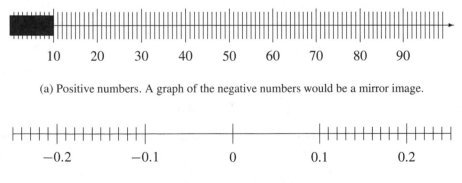

(a) Positive numbers. A graph of the negative numbers would be a mirror image.

(b) Between -0.25 and 0.25

Figure 2.3   The hatchmarks show the positions of floating-point numbers when the precision is 2, the minimum exponent is -1, and the maximum exponent is 1.

2.7.4 Reducing Underflow and Overflow

Just as an increase in precision (by allowing more digits in the mantissa) reduces the maximum possible relative error, allowing larger exponents reduces the incidence of underflow and overflow by enlarging the set of real numbers that have floating-point representations. Changing e_{min} to -2 and e_{max} to 2 while leaving the precision at 2 adds the numbers 0.010–0.099 (in increments of 0.001) and 100–990 (in increments of 10) to the set of positive floating-point numbers from Section 2.7.3. The quality of the approximation remains the same: all relative errors for numbers in the interval $[0.00995 \ldots 995)$ will still be no greater than $\frac{1}{30}$.

The effect of enlarging the range of the exponents is to reduce the incidence of underflow and overflow. In our new system, 50×5 will *not* cause overflow, but will result in 250, which is exactly represented with a floating-point number. Similarly, 0.05×0.5 will result in 0.025 instead of causing an underflow. Overflow and underflow are still possible, of course; it just takes bigger and smaller numbers to cause them.

Figure 2.4a shows the positive floating-point numbers in our enlarged system. (Keep in mind that the scale of this plot is ten times smaller than the scales in Figures 2.2a and 2.3a.) Figure 2.4b shows the floating-point numbers between -0.25 and 0.25. By reducing e_{min} to -2, we have added the numbers 0.010–0.099 and reduced the hole at zero. It still exists, however, as it will for any floating-point system.

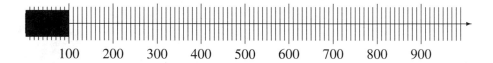

(a) Positive numbers. A graph of the negative numbers would be a mirror image. The scale has changed relative to Figures 2.2a and 2.3b.

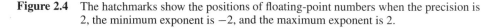

(b) Between -0.25 and 0.25

Figure 2.4   The hatchmarks show the positions of floating-point numbers when the precision is 2, the minimum exponent is -2, and the maximum exponent is 2.

2.7.5 *Mathematica* Capability: Floating-Point Simulation

We will consider the particulars of *Mathematica*'s floating-point number system in Section 2.8. As you might expect, *Mathematica* provides floating-point numbers with much longer mantissas and much larger exponents than the simple systems that we have been considering. We have studied these simple systems because they make it much easier to see how mantissa length and exponent size affect roundoff error, underflow, and overflow.

Mathematica provides a capability that you can use to create and experiment with your own simplified floating-point number systems. This capability is a good way for you to gain an even better understanding of the issues that we have raised in this section. Unfortunately, exploiting this capability involves aspects of *Mathematica* with which you are not yet familiar. For this reason, we have deferred a discussion of *Mathematica*'s floating-point number system simulation capability to Appendix A. You should experiment with it after you have gained a bit more grounding in *Mathematica*.

2.8 Floating-Point Numbers in *Mathematica*

If we continue to increase both the allowable mantissa length and exponent range, we will eventually arrive at a floating-point system such as *Mathematica*'s, which provides close approximations for large intervals of positive and negative real numbers. All of the aspects of floating-point number systems that we examined in Section 2.7—in particular, roundoff error, underflow, and overflow—manifest themselves in *Mathematica*'s floating-point number system.

2.8.1 Roundoff Error in *Mathematica*

Mathematica's floating-point numbers typically have a precision of 16 decimal digits, but this can vary from computer to computer. You can determine the number of digits of mantissa that *Mathematica* uses on *your* computer by entering the special symbol $MachinePrecision

```
In[24]:= $MachinePrecision

Out[24]= 16
```

(2.24)

With its 16-digit mantissas, all relative errors in *Mathematica*'s supported interval are no greater than one part in 3 quadrillion. This means that we must work a bit harder to observe roundoff error in *Mathematica*. The exact square of the number 1234567890

```
In[25]:= 1234567890^2

Out[25]= 1524157875019052100
```

(2.25)

is a 19-digit number. If we do this computation with floating-point numbers

```
In[26]:= 1234567890.^2.
```
$$\text{Out[26]= } 1.52416 \times 10^{18}$$

(2.26)

we see that the exact 19-digit answer has been rounded off to six digits.

Our claim that *Mathematica* uses 16-digit mantissas appears to be at odds with the fact that all of the floating-point answers in this chapter contain only six-digit mantissas. The truth is that *Mathematica* does all of its calculations using 16-digit mantissas, but rounds each result to six digits when displaying it. If we want to see more than the first six digits of an answer, we can use the built-in function `InputForm`.

When given a floating-point number as its parameter, `InputForm` displays the number complete with all of its digits. For example, we can see all 16 digits of the value of Example 2.26 by doing

```
In[27]:= InputForm[12345676890.^2.]
```
Out[27]= 1.524157875019051*^18

(2.27)

Notice that the sixteenth digit of this number is 1, whereas the sixteenth digit of the exact 19-digit result from Example 2.25 is 2. This is evidence of roundoff error in *Mathematica*.

We can now explain the answer to a question that we raised in Section 2.4.4. Recall that when we used a previously calculated floating-point number in an expression (Example 2.10), we obtained a slightly different result than when we used what we thought was an equivalent expression in place of the floating-point number (Example 2.12).

The underlying problem is that the number (0.0101173) that we used in Example 2.10 was only a six-digit approximation to the 16-digit value of the expression (57.8^6 / 5.713*^9) that we used in Example 2.12. If we use `InputForm` to obtain the full value of the expression

```
In[28]:= InputForm[57.8*^6 / 5.713*^9]
```
$$\text{Out[28]= } 0.01011727638718711 \text{ mi}^2$$

(2.28)

and then use this number instead of 0.0101173 in the calculation from Example 2.10

```
In[29]:= 0.01011727638718711 * 27878400
```
$$\text{Out[29]= } 282053. \text{ ft}^2$$

(2.29)

we obtain the same six-digit answer as in Example 2.12.

2.8.2 Overflow and Underflow in *Mathematica*

The exact values of the minimum and maximum exponents $e_{\min}$ and $e_{\max}$ vary among different versions of *Mathematica*, but are typically quite large. You can find out what their values are on your computer by evaluating the special symbols $MaxMachineNumber

```
In[30]:= $MaxMachineNumber

Out[30]= 1.79769 × 10^308
```

(2.30)

and $MinMachineNumber

```
In[31]:= $MinMachineNumber

Out[31]= 2.22507 × 10^-308
```

(2.31)

This reveals that in this case *Mathematica*'s $e_{\max}$ is 308 and its $e_{\min}$ is -308.

In *Mathematica*, overflow and underflow reveal themselves via error messages. Of course, it requires extremely large or small numbers to cause these errors. For example:

```
In[32]:= 10.^1000000000.

        General::ovfl : Overflow occurred
            in computation
```

(2.32)

causes an overflow and

```
In[33]:= 10.^-1000000000.

        General::unfl : Underflow occurred
            in computation
```

(2.33)

causes an underflow.

You may be puzzled by $MaxMachineNumber and $MinMachineNumber, which are, respectively, the largest and smallest positive floating-point numbers. If *Mathematica* is really using exponents ranging from -308 to 308, we would expect the largest positive floating-point number to be $9.99999999999999 \times 10^{308}$ and the smallest positive floating-point number to be 1×10^{-308}.

This apparent discrepancy exists because *Mathematica* uses *base two* floating-point numbers, not base ten numbers as we have used in our previous examples. Thus, $MaxMachineNumber and $MinMachineNumber are base ten approximations to the actual base two values. In reality, the version of *Mathematica* that we used to create the examples in this book has floating-point numbers whose base two mantissas contain 53 base two digits and whose exponents range from a minimum of -1022 to a maximum of 1023. Nevertheless, you should leave base two to the computers and continue thinking in base ten!

2.9 Assessment

We are now ready for the last step in the solution process, which is to assess our problem, model, method, and implementation. We will work backward from the implementation to the original problem.

2.9.1 Accumulation of Roundoff Error

We have obtained identical six-digit results (531.087 feet) for the length of the side of the square of land that every person would receive if the earth's surface were divided evenly. We obtained one result by working exclusively with floating-point numbers (Example 2.15), and the other by working with rational numbers and then converting to floating-point at the last step (Examples 2.16 and 2.17).

Although these numbers are identical to six digits, are they the same to 16 digits? Let's use InputForm to find out. We will first repeat the floating-point calculation from Example 2.15

```
In[34]:= InputForm[
            Sqrt[(57.8*^6 / 5.713*^9) * (5280.^2)]]

Out[34]= 531.0870719877836 ft
```
(2.34)

and then the rational calculation from Example 2.17

```
In[35]:= InputForm[N[17952 * Sqrt[5/5713]]]

Out[35]= 531.0870719877838 ft
```
(2.35)

Why do these two results differ in their last digits? Is either one of them correct?

Strictly speaking, no 16-digit result would be correct. We computed the exact answer using rational arithmetic in Example 2.16, and because it involves the irrational square root of $\frac{5}{5713}$, no number of finite length can possibly be exact. The question we are actually interested in is whether the exact answer, when rounded off to 16 digits, is equal to either of our results.

Mathematica's 16-digit floating-point arithmetic always correctly rounds the result of each arithmetic operation to 16 digits. Nevertheless, a sequence of correctly rounded arithmetic operations can easily lead to a result that is *not* correctly rounded. Although we could demonstrate this fact with 16-digit numbers, it is much easier to do so with smaller ones.

Let's consider what happens when we compute the product $(14 \times 11) \times 2$ in a floating-point system with two decimal digits of precision. The exact product is 308, so the correct answer rounded to two digits should be 310. But with two-digit mantissas, the intermediate product 14×11 is 154, which rounds to 150, so the overall product is 300.

If we group the product differently as $14 \times (11 \times 2)$, we actually get the correctly rounded answer. The intermediate product of 11 and 2 can be expressed exactly as the two-digit number 22, and as a result the overall product, which is 308, rounds to 310. This is an example of how easily the familiar properties of arithmetic can be violated on a computer.

Returning to our land area problem, it should now be evident why our two results differed. Both were subject to the accumulation of floating-point roundoff error. The pure floating-point calculation (Example 2.15) involved a division, an exponentiation, a multiplication, and a square root. Converting the exact rational result (Example 2.17) required converting a fraction, converting a radical, and multiplying the results.

Because roundoff error is inevitable in floating-point calculations, we must be careful to plan for it. You have probably noticed that roundoff error begins by accumulating in the low-order (rightmost) digits of the mantissa. Only after a long chain of calculations does it usually begin to affect the higher-order digits. The proper approach to dealing with floating-point roundoff error, then, is to use more digits in the mantissa than are desired in the final answer and then round the result to the appropriate number of digits. This is, in effect, what *Mathematica* is doing by using 16 digits to perform its calculations and then displaying only a six-digit result.

It turns out that the correct answer to the population density problem, when rounded to 16 digits, is 531.0870719877837 feet, which reveals that *neither* of our original answers was correct to 16 digits.

Our discussion of floating-point numbers in this chapter has left a number of issues unresolved. We identify those issues below and will address them in the next three chapters.

- When implementing a computational method in *Mathematica*, on what basis should we choose between using rational and floating-point numbers?
- What is the minimum acceptable mantissa length for performing a particular calculation?
- How quickly can roundoff error accumulate in a long sequence of computations?
- Under what circumstances can roundoff error be catastrophic, even in a simple computation?

2.9.2 Inexact Physical Measurements

Although our computational method was simple, it was nevertheless flawed in an important way. It did not take into account the uncertainty that is inherent in every physical measurement. While it is extremely unlikely that the land area of the earth is exactly 57.8 million square miles, it may well be safe to say that the actual area is somewhere between 57.75 and 57.85 million square miles. Similarly, although the population of the earth was exactly 5.761 billion at some point in time, it is almost certainly not exactly 5.761 billion right now. We will study how to compute with inexact physical measurements in Chapter 3.

2.9.3 Modeling Errors

We discussed some of the ways in which our idealized model is inaccurate when we first developed it in Section 2.1. A more subtle problem is that we are implicitly treating the earth's land mass as if it were a huge square that we can subdivide into 5.761 billion square tracts of equal size. In reality, of course, coastlines are irregular, mountains are jagged, and everything is laid out on the surface of an oblate spheroid.

The whole problem that we have been dealing with, of course, is entirely artificial, so the compromises that are built into our model are not unreasonable. Nevertheless, careful modeling is often the most important step in solving problems in computational science.

2.9.4 An Irrelevant Problem

We now consider whether the problem we have solved has anything to do with the question that we originally asked. In the beginning, we were wondering whether or not the world is overpopulated. Does the fact that everyone could be placed on a private plot of land approximately the size of a city block help us to decide whether or not the world is overpopulated?

The fact that we can calculate a result containing 16 digits doesn't make that result any more pertinent to the overpopulation question. The population question is complex, and calculations such as ours serve only to trivialize it.

2.10 Key Concepts

Arithmetic expressions. Arithmetic expressions in *Mathematica* are formed much as they are in algebra, with numbers, arithmetic operators (+, −, *, /, and ^), and function symbols. Parentheses are used for grouping, and square brackets are used to delimit function parameters.

Rational numbers. On a computer, a rational number system consists of every fraction that can be formed from any two integers that are between the smallest allowable negative integer (which we call `minint`) and the largest allowable positive integer (which we call `maxint`). *Mathematica* provides rational numbers. The exact values of `minint` and `maxint` vary among different versions of *Mathematica*, but in a practical sense they depend on both the size of your computer and on the degree of your patience.

Floating-point numbers.   The positive numbers of a base ten floating-point number system consist of every number that can be written in scientific notation using a p-digit (p is the precision) mantissa between 1 and 10 and an exponent between e_{min} (the smallest allowable negative exponent) and e_{max} (the largest allowable positive exponent). The full floating-point number system contains the corresponding negative numbers and zero. *Mathematica* provides floating-point numbers, but the exact values of p, e_{min}, and e_{max} vary from version to version.

Arithmetic error.   Rational numbers are susceptible to overflow error, which occurs when either the numerator or the denominator of a fractional result becomes too large. In floating-point arithmetic, overflow occurs when the exponent exceeds the largest allowable positive exponent; underflow occurs when the exponent exceeds the largest allowable negative exponent; and roundoff error occurs when a number cannot be exactly expressed with p digits of mantissa.

2.11 Exercises

2.1   What are some problems with the model that we used in this chapter, other than the ones we have already pointed out?

2.2   You will inevitably make syntax errors when using *Mathematica*. Make some deliberate errors with your version of *Mathematica* as a way of learning to understand the error messages.

2.3   Suppose that we had solved the problem in this chapter by converting the earth's area to square feet before dividing by the population and taking the square root. Would the answer obtained this way using rational numbers have been any different? Using floating-point numbers?

2.4   A city block in Salt Lake City is a square one-seventh of a mile on a side. What is the surface area of the earth when measured in Salt Lake City blocks?

2.5   Suppose that we rearranged the earth's land area into circular islands of identical size, with one island per person. What would be the radius of each island? (The value of π rounded to 16 digits is 3.141592653589793.)

2.6   Suppose we rearranged the earth's land area into identical islands shaped like equilateral triangles, one per person. What would be the length of each side of each island?

2.7   Unless its exponent is too big or too small, a *Mathematica* floating-point number is exactly equal to some *Mathematica* rational number. For example, 3.27 is exactly $\frac{327}{100}$. Use rational number arithmetic to trace through the floating-point solution to the land area problem and determine where the roundoff error was introduced.

2.8 Write the following numbers as *Mathematica* floating-point numbers with a mantissa between 1 and 10: 154, 0.0005, 10001, 52.

2.9 What are the smallest rational number systems that contain, respectively, the sets of numbers listed below? Assume that `minint` is the negative of `maxint`.

(a) $\{\frac{1}{4}, \frac{-3}{7}, \frac{2}{8}, \frac{5}{7}\}$

(b) $\{\frac{1}{2}, \frac{1}{3}, \frac{1}{4}, \frac{-3}{4}\}$

(c) $\{\frac{100}{121}, \frac{-100}{121}, 100, 121\}$

2.10 What are the smallest floating-point number systems that contain, respectively, the sets of numbers listed below?

(a) $\{2.2, 12.3, -8.96, -9150\}$

(b) $\{.0004, -600000, .001, -20\}$

(c) $\{.777, .77, .7, 7, 77, 777\}$

2.11 Suppose that a floating-point number system has two-digit mantissas and exponents ranging from -2 to 1. Classify the following expressions as to whether each would result in overflow, underflow, roundoff error, or an exact answer.

(a) `20. + 20.`

(b) `20. + .02`

(c) `.02 + .02`

(d) `20. * 20.`

(e) `20. * .02`

(f) `.02 * .02`

2.12 In many of the calculations in this text, we took the square root of an expression, one of whose factors was the square of 5280. It would have been computationally and mathematically simpler to remove that factor from under the radical and to multiply the square root by 5280 instead. Might doing this have produced a slightly different result?

2.13 Suppose that a floating-point number system allows three-digit mantissas and has exponents that range from -3 to 3. Create number lines showing the distribution of all floating-point numbers x such that $.01 \leq x < 1$, $1 \leq x < 10$, and $10 \leq x < 100$. What general property do you observe?

2.14 Suppose that a rational number system has a `minint` of -5 and a `maxint` of 5, and that a floating-point number system has one-digit mantissas and exponents ranging from -1 to 0. Which of the rational numbers are exactly representable in the floating-point system? Which of the floating-point numbers are exactly representable in the rational system?

2.15   Suppose that a floating-point number system allows ten-digit mantissas and exponents that range from −1000 to 1000. What would be the smallest rational number system for which all of the floating-point numbers would have exact rational equivalents?

2.16   Assuming that `minint` is negative and `maxint` is positive, how many rational number systems consist entirely of numbers that can be represented exactly by some base ten floating-point system?

2.17   Find some expressions that cause rational overflow on your version of *Mathematica*.

2.18   Find some expressions that cause floating-point overflow and underflow on your version of *Mathematica*.

2.19   We have already seen that neither rational nor floating-point multiplication is associative. Find *Mathematica* examples that violate each of the following standard arithmetic identities.

(a) Addition associates: $(x + y) + z = x + (y + z)$.
(b) Multiplication distributes over addition: $x(y + z) = xy + xz$.
(c) All numbers but zero have multiplicative inverses: $\frac{1}{x}x = 1$.

2.20   What would go into a realistic model of the earth and its population to make it possible to study seriously whether the population (or its current rate of growth) is too large?

3

Eratosthenes: Significant Digits and Interval Arithmetic

Eratosthenes of Cyrene was a Greek mathematician who lived during the latter part of the third century B.C. He is best known for discovering an algorithm, called the Sieve of Eratosthenes, for enumerating prime numbers. We, however, are interested in him for another reason. In about 225 B.C., more than 17 centuries before Columbus, he determined the circumference of the earth.

Eratosthenes was the head of the Library at Alexandria, on the southern coast of the Mediterranean Sea in Egypt. Using a sundial, he observed that at solar noon on the first day of summer in Alexandria the sun was 7.2 degrees south of being directly overhead. At solar noon on the same day in Syene, 5000 stadia (the plural of stade, an ancient Greek unit of length) south of Alexandria on the Nile River near the Tropic of Cancer, he knew that the sun was directly overhead. Based on these measurements, he calculated the earth's polar circumference in stadia. Although modern scholars are not certain exactly how long a stade is, their best estimate is 0.1575 kilometers.

In this chapter we will work from this description of Eratosthenes's experiment and determine the circumference of the earth in kilometers.

3.1 Model

When developing a model, it is often helpful to begin by drawing a diagram. This frequently leads directly to a model, since drawing a diagram usually involves simplifying and idealizing a problem statement. Our diagram is in Figure 3.1.

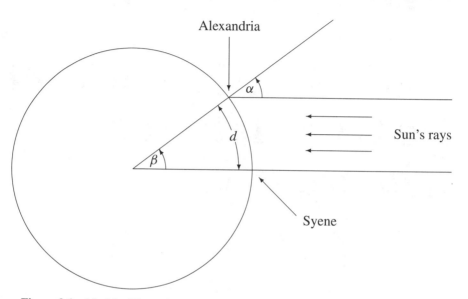

Figure 3.1   Model of Eratosthenes's experiment. The circumference of the circle is C.

By using this diagram and applying a bit of basic geometric reasoning, we can determine C, the circumference of the earth. The angle α is the one measured by Eratosthenes at Alexandria, and d is the distance between Alexandria and Syene. Because the lines representing rays of light are parallel, the angles α and β are identical. The ratio between 360 (the number of degrees in a circle) and β (or α) is the same as the ratio between C and d.

$$\frac{C}{d} = \frac{360}{\alpha} \tag{3.1}$$

With a bit of algebra we obtain

$$C = \frac{360}{\alpha} d \text{ stadia} \tag{3.2}$$

Of course, since Eratosthenes's value for d is given in stadia, we will need to convert to kilometers. Letting μ stand for the number of kilometers in a stade, our formula becomes

$$C = \frac{360}{\alpha} d \mu \text{ km} \tag{3.3}$$

3.2 Method

We now have a model for which Eratosthenes has provided values for α and d, and historians have provided a value for μ. The next step is to identify a method for computing the value of C, the circumference of the earth. This doesn't look particularly challenging: simply divide 360 degrees by 7.2 degrees and then multiply the resulting quotient by the product of 5000 stadia and 0.1575 km/stade. We will discover later in this chapter that this seemingly solid method can be improved, but for the time being let's stick with it and move on to the implementation.

3.3 Implementation

We will continue using *Mathematica* to do our computations. Because the calculation that we need to do is so simple, we will use this opportunity to address two additional issues. We will discuss some of the considerations that go into the choice of whether to use rational or floating-point numbers, and we will explore the use of *Mathematica*'s assignment expressions to save results.

3.3.1 Choosing Numbers

We saw in Chapter 2 that *Mathematica*'s rational and floating-point numbers have different properties. The rational numbers are restricted to a smaller range than are the floating-point numbers. In the absence of overflow and underflow errors, rational arithmetic is exact, whereas floating-point arithmetic is approximate.

Although there are rational numbers that cannot be represented as floating-point numbers and vice versa, in almost every problem that we will consider we could use either rationals or floating-points. We need some basis for deciding how to represent each number we encounter.

The usual practice in computing is to use rational numbers to represent exact values and floating-point numbers to represent approximate values. To understand the distinction, let's consider the four constants involved in Eratosthenes's problem.

The angle of the sun measured at Alexandria (7.2 degrees) and the distance between Alexandria and Syene (5000 stadia) are physical measurements. All such physical measurements are approximate because all measuring devices have limitations. We should really be saying that the measurements are *about* 7.2 degrees or *approximately* 5000 stadia. Although it is not a physical measurement, the conversion factor between stadia and kilometers (0.1575 km/stade) is also an approximation because it represents the best estimate that modern scholars have been able to determine. Later in this chapter we will see how to deal explicitly with the fact that these three values are approximations.

The number of degrees in a circle is, by definition, an exact value. There were exactly the same 360 degrees in a circle in Eratosthenes's time as there are today. While we could no doubt, with modern equipment, measure both α and d more precisely than Eratosthenes, the fact that there are 360 degrees in a circle is not subject to measurement. Another good example of exact values are counts (there are 50 states in the Union and two hydrogen atoms in a water molecule).

The practice of using rationals for exact values and floating-points for measurements is rooted in the fact that traditional programming languages such as Fortran and C provide integers but *not* fractions, which leaves no choice but to represent most physical measurements using floating-point numbers. Nevertheless, it is a good idea to follow the convention because, as we will see in Chapter 4, a long sequence of rational number computations can take *Mathematica* much longer than an equivalent sequence of floating-point computations. *Mathematica*'s exact rational arithmetic is computationally expensive, and there is no reason to pay the price if exact results are neither required nor meaningful.

Having said all this, we will express our calculation of the circumference of the earth

```
In[4]:= 360 / 7.2 * 5000. * 0.1575

Out[4]= 39375. km
```

(3.4)

using a mixture of rational and floating-point numbers. Using 360 instead of 360. makes no practical difference here, of course, since the 360 is converted into floating-point form before the division is carried out. In Chapter 4 we will investigate this issue in more depth.

Example 3.4 gives us 39,375 kilometers as our first approximation to the circumference of the earth.

3.3.2 Variables and Assignment

When *Mathematica* computes a value, it displays the value and goes on to the next question. If we want to use that value in a subsequent computation, we must either memorize it, search back through the *Mathematica* notebook to find it, or recompute it. If we were to do a large number of computations, any of these would quickly get tiresome.

For this reason, *Mathematica* provides *variables* to save values and the *assignment expression* to associate values with them. For example, let's repeat the computation that we just did, but this time let's save the result:

```
In[5]:= circumference = 360 / 7.2 * 5000. * 0.1575

Out[5]= 39375. km
```

(3.5)

This command is an assignment expression, and we have used it to associate the value of the expression on the right-hand side with the variable `circumference` on the left-hand side. Notice how the symbol =, which is called the *assignment operator*, is used to punctuate the assignment expression.

Mathematica maintains an internal list of variables and their values. When *Mathematica* evaluates an assignment expression, it updates the list to record the mapping between the variable on the left-hand side and the value of the right-hand side. We will call this list the *global variable list*, because the variables in it can be referenced from anywhere in a *Mathematica* notebook. When *Mathematica* evaluates Example 3.5, it records in the global variable list the fact that the value of `circumference` is 39375.

A variable name in *Mathematica* must be composed of a letter followed by any number of letters or digits (or both). Although we are free to use any legal variable name, it is always a good idea to choose a descriptive name to make it easier to remember what it stands for. It is also a good idea to begin a variable with a lower-case letter, so that they can be easily distinguished from built-in *Mathematica* variables, which all begin with upper-case letters.

Since our variable stands for a number, we can use it anywhere a number is needed. For example, we can easily determine its value

```
In[6]:= circumference

Out[6]= 39375. km
```

(3.6)

or use it in a formula to compute the diameter of the earth.

```
In[7]:= diameter = circumference / 3.14159

Out[7]= 12533.5 km
```

(3.7)

In both examples, *Mathematica* looks up the value of `circumference` in the global variable list. In the second example, *Mathematica* adds `diameter` and its new value to that list.

The variables `circumference` and `diameter` will continue to have 39375. and 12533.5, respectively, as their values until we change them with subsequent assignment expressions.

3.3.3 Pure and Imperative Expressions

Every expression in *Mathematica* has a value; we enter an expression into *Mathematica* in order to obtain that value. For example, when we evaluate the expression 2+2, *Mathematica* tells us that its value is 4. An assignment expression is not an exception. For example, when we evaluate the expression x = 2+2, *Mathematica* tells us that its value is 4, and when we evaluate the assignment in Example 3.5, *Mathematica* reports that its value its 39375. The value of an assignment expression, then, is the value of its right-hand side.

Because expressions have values, they can be put together to produce more complicated expressions by nesting them within one another. This works because each subexpression can contribute its value to the containing expression. We can, for example, combine expressions with arithmetic operators or by making them parameters to a function.

Besides producing values, some expressions also have *side effects*. A side effect is something that happens during the evaluation of an expression that has the potential to affect the evaluation of subsequent expressions. The evaluation of the assignment expression x = 2+2, for example, has the side effect of changing the value of the global variable x. This will affect the subsequent evaluation of any expression that depends on the value of x. The evaluation of the arithmetic expression 2+2, on the other hand, has no side effects. The fact that we once added 2 and 2 will in no way affect any future calculations.

We will refer to expressions with side effects as *imperative* expressions, and to expressions without side effects as *pure* expressions. To this point, assignments are the only kinds of imperative expressions that we have seen. We will soon encounter more, and as we do the distinction between pure and imperative expressions will become ever more crucial.

With imperative expressions, we are almost always more interested in the side effect caused by the expression than in the value it produces. The fact that an assignment expression, for example, produces a value is essentially irrelevant. All that we will ever care about in this text is that an assignment changes the value of a variable.

As we begin to program with imperative expressions, we will see that using them involves arranging for them to be evaluated in a particular order. This is an issue that does not arise at all with pure expressions. We could, for example, have evaluated the expressions from Chapter 2 in any order whatsoever and obtained the same results, because none of the expressions caused side effects. This is not true of the expressions in this chapter, however, because some of the expressions are assignments.

3.4 Implementation Assessment

Having computed the circumference of the earth as 39,375 kilometers, we are now prepared to enter the assessment phase. Notice that we have an advantage that Eratosthenes did not: We can assess our result by comparing it to the modern figure for the polar circumference of the earth. In order to keep ourselves honest, however, we will refrain from doing this until the end of the chapter.

The assessment problem that confronted Eratosthenes is entirely typical of scientific problem-solving in general. There is often no easy way to check whether or not an answer is correct. Instead, we must carefully analyze the solution process and convince ourselves (and others) that an answer is correct. This is exactly the purpose of the assessment phase.

We will begin with our implementation assessment by making sure that we didn't make any mistakes in the implementation. We will then take up the question of what our answer of 39,375 kilometers actually means.

3.4.1 Operator Precedence

With such a simple implementation, it is hard to believe that we could have made a mistake; but look carefully at the expression we gave to *Mathematica* in Example 3.5. We intended for *Mathematica* to do the division of 360 and 7.2 degrees; to multiply the measurement of 5000 stadia and the conversion factor of 0.1575 km/stade; and to multiply the resulting quotient and product. Since we included no parentheses in the expression, how do we know in what order *Mathematica* carried out the operations?

When giving arithmetic expressions to *Mathematica*, we always have the option of using parentheses to indicate the order in which operators are to be evaluated. Let's try again:

```
In[8]:= circumference = (360 / 7.2) * (5000. * 0.1575)

Out[8]= 39375. km
```
(3.8)

Notice that we used parentheses to indicate the order in which the three operations were to be done, and that we obtained the same answer as before. Even though we didn't uncover a mistake here, it is important that we made sure. The wrong answers that *Mathematica* produces look just as authoritative as the correct ones.

This gives rise to a general question about *Mathematica*: In the absence of parentheses, in what order does *Mathematica* carry out the arithmetic operations of addition, multiplication, subtraction, division, and exponentiation?

Mathematica, like all programming languages, is governed by the concept of *operator precedence*. *Mathematica* specifies a level of precedence for each operator. In addition, for each level of precedence *Mathematica* specifies whether the operators at that level associate to the left or right. From highest to lowest, the precedence of *Mathematica*'s five arithmetic operators is given in Table 3.1.

Table 3.1 Arithmetic operator precedence in *Mathematica*

^	right associative
* and /	left associative
+ and –	left associative

In the absence of parentheses, this means that exponentiations are done first, followed by multiplications and divisions, and finally additions and subtractions. Furthermore, all of the operators except exponentiation are performed from left to right. Exponentiations, in contrast, are all performed from right to left. Table 3.2 shows several example expressions and their fully parenthesized equivalents.

It turns out that our circumference computation didn't unfold exactly as we had intended. The division was indeed done first, but the two multiplications were then carried out from left to right. The order of the multiplications didn't matter in

Table 3.2   Arithmetic expressions and their fully parenthesized equivalents

a+b*c	a+(b*c)
a*b+c	(a*b)+c
a-b-c	(a-b)-c
a+b^c*d	a+((b^c)*d)
a^b^c	a^(b^c)

this case, but it certainly could have. As we saw in Chapter 2, multiplication on a computer is not always associative.

Mathematica's actual precedence table has dozens of levels because there are many more than five operators. The assignment operator, for example, appears in the table with a precedence lower than any of the arithmetic operators. It is not important that you memorize the table, only that you be aware of its existence and that you use parentheses whenever you are in doubt. When you learn other languages, you should keep in mind that all have their own precedence rules that are not necessarily identical to *Mathematica*'s.

3.4.2 Significant Digits

The result that we have obtained for the circumference of the earth, 39,375 kilometers, is quite appealing. It doesn't have any fractional component and it doesn't exhibit any floating-point roundoff error, as you can easily verify by repeating the calculation using rational numbers. Before we proclaim the circumference of the earth to be 39,375 kilometers, however, we must carefully consider the reality that our final result can be no more precise than our original measurements.

Any physical measurement is inexact because all measuring devices have limited precision. For this reason, physical measurements are usually described by giving the range in which the actual value is believed to lie. We say that an approximate measurement is *accurate* if the true value lies within the specified range; we say that it is *precise* if the range is small.

For example, if you weigh 160 pounds on an accurate digital bathroom scale that rounds to the nearest pound, then you actually weigh somewhere between 159.5 and 160.5 pounds. (Anything outside of that range would not round to 160.) If the scale rounds to the nearest half *pound*, however, then your actual weight must be between 159.75 and 160.25 pounds. If the scale is accurate to within plus or minus 1 percent, then your actual weight must be between 158.4 and 161.6 pounds. Although all three versions of the scale may well be accurate, it is the second version that is the most precise.

It is standard practice in science and engineering to describe the range in which a measurement lies by giving only the *significant* digits of a number that is, or is derived from, a physical measurement. All of the digits of a measurement are significant if we are certain that the true value, when rounded to the same number of digits as the measurement, is equal to the measurement. For example, when we give the value of α as 7.2 degrees, we are in effect asserting that the true value is somewhere between 7.15 and 7.25, since only a number in that range would round to 7.2. If we were to give the value as 7.20 degrees, we would be saying that the true value is between 7.195 and 7.205. Thus, 7.20 degrees is a more precise measurement than 7.2 degrees.

We have been giving the value of d as 5000 stadia, which implies that the actual value is between 4999.5 and 5000.5 stadia. Historians believe that Eratosthenes obtained this distance by employing surveyors to pace it off. Consequently, it is perhaps more realistic to treat the actual value as lying between 4950 and 5050 stadia. If that is our assumption, we should be writing the distance using only a two-digit mantissa, as 50×10^2 or 5.0×10^3 or $.50 \times 10^4$ stadia.

When we write μ as 0.1575 km/stade, we are asserting that it is accurate to four significant digits. The digit count always begins with the first *non-zero* digit. Thus, 0.1575 and $.01575 \times 10^1$ are equivalent ways of writing the same number, precise to four significant digits.

The concept of significant digits applies only to numbers derived from physical measurements. It does not apply, for example, to the number 360 when we are talking about the number of degrees in a circle.

The number of significant digits in the measurements used in a calculation must be considered when interpreting the result of the calculation. A useful (though only approximate) rule of thumb is that there will be no more significant digits in the result than there were in the least precise input value. For example, suppose that we multiply two numbers, one of which has seven significant digits and the other two. The product will contain two or fewer significant digits.

Because of this rule of thumb, it is usually misleading to report more significant digits in the result than there were in the least precise input. Since the three measurements on which we based our result consisted of two, two, and four significant digits, we are obliged to round our final result to two digits when reporting our result. Thus, the best that we can do is to report that the circumference of the earth is approximately 3.9×10^4 kilometers.

Even if we assume that the three measurements are accurate, we cannot *guarantee* that both of the digits in our result are significant; *i.e.*, that the true circumference rounds off to 39,000 kilometers. To establish bounds in which we can have more confidence, we must do interval arithmetic, which will be explained in Section 3.5.1.

3.4.3 Appropriate Mantissa Lengths

It should now be apparent that there are *two* sources of error when a computation is based on physical measurements: the errors inherent in the measurements and the roundoff error inherent in floating-point arithmetic. Although both are unavoidable, we can minimize or eliminate the effect of roundoff error by being aware of the amount of measurement error.

The crucial observation is that roundoff error, except in extremely long sequences of computations (Chapter 4) or in unstable calculations (Chapter 5), affects only the one or two lowest-order digits in the mantissa. By making sure that the mantissa length is two or three digits longer than the number of significant digits that we expect in the result, the roundoff error becomes insignificant relative to the unavoidable measurement errors.

For the circumference problem, the default 16 digits of mantissa is more than sufficient. In fact, we could get by with as few as four. To illustrate this, let's do the calculation from Example 3.8 by hand, rounding each intermediate result to four digits.

1. $360/7.2$ yields 50.
2. 5000×0.1575 yields 787.5.
3. 50×787.5 yields 3938×10^1.

Although our result (39,380 kilometers) differs by 5 kilometers from the original result (39,375 kilometers), it still rounds to 3.9×10^4 kilometers, which is the honest way to present the result.

We earlier obtained 12,533.5 kilometers as the diameter of the earth using 16-digit mantissas and an approximation to π of 3.14159. Because the value for the circumference of the earth contained only two significant digits, we should have reported the diameter as 1.3×10^4 kilometers. We could have obtained the same two-digit result even with two-digit mantissas, as this *Mathematica* calculation suggests:

```
In[9]:= 3.9*^4 / 3.1

Out[9]= 12580.6 km
```
(3.9)

These examples do not mean that you should deliberately use short mantissas and crude approximations. To the contrary, they mean that you should be careful to use adequate mantissas and approximations. By the same token, you should not put misplaced faith in every single digit that results from a computation. When you report a result, you should give only the significant digits.

3.5 Method Assessment

By paying careful attention to the number of significant digits in the input measurements, we were able to put crude but not necessarily accurate bounds on the circumference of the earth. By employing an improved computational method—interval arithmetic—we will be able to put accurate bounds on the circumference, assuming of course that the measurements and model were accurate to begin with. We will begin by exploring the idea behind interval arithmetic, and will then investigate its relationship to the idea of significant digits.

3.5.1 Interval Arithmetic

We have been working under the assumption that we have the following accurate bounds on the values of our three input values:

- $7.15 \le \alpha \le 7.25$ degrees,
- $4950 \le d \le 5050$ stadia, and
- $0.15745 \le \mu \le 0.15755$ km/stade.

We can obtain an upper bound on the circumference by using the largest possible values for d and μ (since both are in the numerator of Equation 3.3) and the smallest possible value for α (since it is in the denominator). Any other in-range values would necessarily lead to a *smaller* result.

```
In[10]:= (360 / 7.15) * (5050. * 0.15755)

Out[10]= 40059.6 km
```
(3.10)

Similarly, we can obtain a lower bound by using the *smallest* possible values for d and μ and the *largest* possible value for α. (Any other in-range values will lead to a *larger* result.)

```
In[11]:= (360 / 7.25) * (4950. * 0.15745)

Out[11]= 38700.1 km
```
(3.11)

Both of these answers are certainly correct to the six digits that *Mathematica* displays, since any roundoff error will be contained in the undisplayed digits that lie far to the right of the decimal point. We can make our answers a bit more palatable by rounding the lower bound *down* and the upper bound *up*. We can now assert that the polar circumference of the earth is somewhere between 38,700 and 40,060 kilometers, assuming that our measurements and model are accurate. We cannot give a more precise estimate of the circumference unless we are first given more precise physical measurements to work with.

In Section 3.4.2 we estimated the circumference of the earth, to two significant digits, as 39,000 kilometers. This was based on the assumption that α, d, and μ were accurate to, respectively, two, two, and four significant digits. Based on these same assumptions, we have just used interval arithmetic to show that the circumference must lie between 38,700 and 40,060 kilometers, which means that it can round to either 39,000 or 40,000 kilometers. Our significant digit analysis thus gave us only approximate bounds on the true answer.

Performing interval arithmetic depends on knowing the range of possible values for the input measurements as well as understanding how different values from those ranges can affect the magnitude of the result. Doing interval arithmetic is not always as straightforward as this example might lead you to believe. As the number of variables and the complexity of the calculation increase, ever more ingenuity is required to carry out interval arithmetic. This is why the less accurate method of estimating the number of significant digits in the result is often used instead.

3.5.2 Interval Arithmetic in *Mathematica*

Another example will help to further illuminate the idea behind interval arithmetic. Suppose that a father carries his infant son onto a bathroom scale, and that their combined weight is 200 pounds. On the same scale, the father alone weighs 186 pounds. If the scale is accurate to the nearest pound, how much does the infant weigh?

Since the scale is accurate to the nearest pound, the combined weight must lie between 199.5 and 200.5 pounds. Similarly, the father's weight must lie between 185.5 and 186.5 pounds. We can determine the *most* the infant can weight by subtracting the *largest* possible combined weight and the *smallest* possible father weight.

```
In[12]:= 200.5 - 185.5

Out[12]= 15. lb
```

(3.12)

Similarly, we can determine the *least* the infant can weigh by subtracting the *smallest* possible combined weight and the *largest* possible father weight.

```
In[13]:= 199.5 - 186.5

Out[13]= 13. lb
```

(3.13)

Thus, the infant weighs somewhere between 13 and 15 pounds.

Doing interval arithmetic requires a bit of careful thought. In both of the examples that we have examined, we have had to figure out what combination of numbers would lead to the largest possible result, and what combination of numbers would lead to the smallest possible result. It was fairly straightforward to do this for Eratosthenes's circumference problem and for the infant weight problem, but this is not always the case.

Fortunately, *Mathematica* provides built-in support for doing interval arithmetic. For example, we can ask *Mathematica* to do the interval arithmetic required to solve the infant weight problem by entering

```
In[14]:= Interval[{199.5, 200.5}] -
         Interval[{185.5, 186.5}]

Out[14]= Interval[{13., 15.}] lb
```

(3.14)

We are asking *Mathematica* to subtract a number that is known to lie in the interval 185.5 . . . 186.5 from a number that is known to lie in the interval 199.5 . . . 200.5 and to tell us the interval in which the answer must lie. In response, *Mathematica* tells us that the answer must lie in the interval 13 . . . 15. Notice that we specify an interval to *Mathematica* by using the built-in function `Interval`. This function takes as its parameter the two endpoints of an interval, where the endpoints are enclosed in braces and separated with a comma. *Mathematica* specifies its result intervals in the same way.

We can easily repeat our interval arithmetic solution to Eratosthenes's circumference problem from Examples 3.10 and 3.11, this time using *Mathematica*'s built-in interval arithmetic.

```
In[15]:= (360 / Interval[{7.15,7.25}]) *
             (Interval[{4950,5050}] *
                 Interval[{0.15745,0.15755}])

Out[15]= Interval[{38700.1, 40059.6}] km
```
$$(3.15)$$

We end up with the same result that we earlier calculated "by hand."

3.6 Model Assessment

Our conclusion that the polar circumference of the earth is between 38,700 and 40,060 kilometers applies only to the idealized model that is based on our diagram. Before we can assert that we have found an answer, we must carefully assess our model for possible problems. In fact, we made a number of assumptions on our way from the problem to the model.

- We assumed that the earth is a perfect sphere so that a cross section through the center would be a circle as diagrammed in Figure 3.1. But the earth is clearly not a perfect sphere, since much of its land area is irregular.
- Even if we ignore the landforms and treat the surface as being entirely uniform, it turns out that a polar cross section is actually oval, since the earth is flattened slightly at the poles.
- We assumed that the sun's rays over Alexandria are parallel to the sun's rays over Syene. Strictly speaking this is not true, but the immense distance from the sun to the earth relative to the distance between Alexandria and Syene makes the difference extremely small.
- We assumed that Syene is due south of Alexandria, but Syene is actually 3 degrees east of Alexandria. This affects the distance measurement of 5000 stadia, as we are actually interested in how far south Syene is, not how far away.

You can probably uncover other problems with our model. For each problem we must either improve our model to account for the problem, justify why the problem can be safely ignored, or admit that our answer applies only when the problem is ignored. Let's consider each of the four problems in turn.

- Irregular surface. From the data given it is impossible to calculate the linear distance, measured along the ground, of a path from Alexandria across the two poles and back. It is not even clear that such a calculation would be interesting. We have no choice but to do the computation for an idealized world with a uniform surface.
- Not a perfect sphere. If we knew the eccentricity of the earth's polar circumference as well as the latitudes of Alexandria and Syene, we could complicate our model to account for the earth's oval polar cross section. We will not do that here, though.
- Suns rays not parallel. If the sun's rays at Alexandria and Syene are not parallel, then our measurement of α is wrong. Given that our measurement of α is only good to two significant digits, however, the error introduced by this simplification is almost certainly insignificant.
- Not due south. This is already accounted for in the precision of the measurement of the distance between Alexandria and Syene, which is only good within ± 50 stadia.

3.7 Problem Assessment

Our best answer to the original problem is that the circumference of the earth is between 38,700 and 40,060 kilometers, assuming that

- Eratosthenes's measurements of α and d are good to two significant digits;
- the conversion factor is good to four significant digits; and
- the earth is modeled as a perfect sphere.

The modern estimate for the earth's polar circumference is 40,009 kilometers, which indeed falls into the range that we calculated.

Eratosthenes, of course, did not have access to this modern estimate. His result implied a surface area so much greater than the known area of the earth that his value of 250,000 stadia for its circumference was not widely believed. Approximately 150 years after Eratosthenes, Poseidonios of Apameia calculated an "improved" circumference of 180,000 stadia. This mistake, combined with his overestimate of the size of Asia, led him to conclude that it was only 70,000 stadia west across the Atlantic to India. His conclusion was repeated by other authors down through the ages, and ultimately helped to convince Christopher Columbus to attempt a westward voyage to India in 1492.

3.8 Key Concepts

Rational versus floating-point numbers. In general, floating-point numbers should be used to represent approximate physical measurements and rational numbers should be used to represent exact constants and counts.

Assignment. *Mathematica* provides assignment expressions as a means of associating a value with a variable. The variable can then stand in place of the value in subsequent expressions.

Pure versus imperative expressions.   All expressions have values, but some (such as assignments) cause side effects. Expressions with side effects are called imperative expressions; expressions without side effects are called pure expressions. The order in which imperative expressions are evaluated can be critical; the order in which pure expressions are evaluated is irrelevant.

Operator precedence.   *Mathematica*, like all languages, has an operator precedence table that it uses to determine the order in which arithmetic operations are carried out in the absence of parentheses.

Significant digits.   Physical measurements are generally written so that they contain only significant digits. A useful, though only approximate, rule of thumb is that the result of a computation will contain no more significant digits than the least precise of the measurements involved.

Roundoff error.   Roundoff error is inevitable with floating-point numbers. To cope with this, the floating-point numbers used in a computation should have at least a few more digits in their mantissas than there will be significant digits in the answer. In this way, the roundoff error can be confined to the insignificant digits of the result.

Interval arithmetic.   Interval arithmetic provides a way to put accurate bounds on the results of computations involving approximate physical measurements. *Mathematica* provides support for interval arithmetic via its `Interval` function.

3.9 Exercises

3.1   Which of the following values should be represented with rational numbers? Which with floating-point numbers?

(a) The speed of light

(b) The number of protons in an atom

(c) The distance from the earth to the moon

(d) Acceleration due to gravity on Jupiter

(e) The number of megabytes of memory in a computer

(f) The balance in a bank account

3.2   Use the precedence rules of *Mathematica* to predict the values of each of the following expressions.

(a) 2 * 3 + 4 / 2

(b) 5 - 6 + 7 ^ 2

(c) 1 - 2 ^ 3 / 4

(d) 8 / 4 / 2

3.3 How many different values could each of the expressions in the preceding exercise take on if parentheses were added?

3.4 Add the minimum number of parentheses needed so that the four expressions in Exercise 3.2 have values of 7, 64, −1, and 4, respectively.

3.5 Redo the calculation in Example 3.4 using purely rational arithmetic.

3.6 There are 5280 feet in a mile. Explain why it would be wrong to assert that this conversion factor contains four significant digits.

3.7 The ages of adults are typically measured in years. Why would such measurements be more precise if adults considered themselves one year older six months before their birthdays?

3.8 Use interval arithmetic to find bounds on the average age of four people who are 44, 46, 20, and 15 years old. Solve this problem both "by hand" and by using *Mathematica*'s `Interval` function.

3.9 What happens if you use something other than a variable, such as a number or a compound expression, on the left-hand side of an assignment expression?

3.10 Write the following numbers in *Mathematica*'s scientific notation so that they reflect the given number of significant digits.

(a) 25.3, four significant digits
(b) .00005, two significant digits
(c) 1.351, two significant digits
(d) 84000, three significant digits

3.11 In Chapter 2 our calculations were based on a surface area of 57.8 million square miles and a population of 5.713 billion people. Assuming that the area measurement contains three significant digits and the population measurement four, how many digits, at most, in the final answer were significant?

3.12 Experiment by hand to see how small a mantissa length would suffice to calculate the significant digits of the solution to the population density problem.

3.13 Use interval arithmetic together with the assumptions of Exercise 3.11 to establish bounds on the side of the square plot of land available for each person on earth.

3.14 Repeat Exercises 2.5 and 2.6 using interval arithmetic to establish bounds.

3.15 Suppose that we take the value of α to be 7.20 and of d to be 5.00×10^3, both good to three significant digits. Assume that the conversion from stadia to kilometers is exact. Using the method of interval arithmetic, in what range must the circumference of the earth fall?

3.16 Assuming that Eratosthenes's measurements were accurate and that the model we used is perfect, what does the result of the previous exercise tell you about the precision of Eratosthenes's measurements?

3.17 Suppose that we take the value of α to be 7.20, good to three significant digits, and the conversion from stadia to kilometers to be exact. Use interval arithmetic to determine, at most, how many of the digits of the value of d can be significant given what we know about the true circumference of the earth.

3.18 Two rays of light leave the center of the sun, whose average distance from the earth is 1.496×10^8 kilometers. One travels to Alexandria, the other to Syene. What is the angle between them in degrees? How many of the digits in your result are significant?

3.19 How precisely would Eratosthenes have had to measure α before the error introduced by the fact that the sun's rays over Alexandria and Syene are not parallel would have become significant?

3.20 Had Eratosthenes believed the earth to be cylindrical, with the axis of the cylinder aligned from north to south, he could have used his measurements to determine the distance from the earth to the sun. What would have been his result in stadia?

4

Stairway to Heaven: Accumulation of Roundoff Error

Martin Gardner's popular "Mathematical Games" column appeared monthly in *Scientific American* from January 1957 through December 1981. During those 25 years he entertained his readers with discussions of such topics as how to fold adding machine tape into hexaflexagons, John Conway's Game of Life, and the discoveries of the mysterious numerologist Dr. Matrix. He inspired a generation of budding engineers, scientists, and mathematicians.

One of Gardner's columns was on infinite series. In it, Gardner discussed the idea behind the following problem, which he attributed to the *Pi Mu Epsilon Journal* for April 1954. Imagine that we have a supply of identical blocks of uniform density, each 1 foot long. We also have a strong table that is anchored securely to the floor. Our job is to arrange a stack of blocks on the table so that the stack extends out beyond the right edge of the table. When building the stack, the following rules apply:

• The bottommost block must be placed on the table so that it extends beyond the right edge.
• Every remaining block must be placed so that it extends further right than the block below it.

How far out beyond the edge of the table can the right edge of the topmost block extend?

Let's consider two simple examples to be sure the problem is clear. If we have only one block, it's not hard to see that the best we can do is to place it so that it extends halfway off the table, as in Figure 4.1. It is perfectly balanced on the table, and any attempt to slide it even a tiny bit to the right will make it fall off.

Figure 4.1   One balanced block

If we have two blocks, though, we'd better not try positioning them as in Figure 4.2. The weight of the top block will overbalance the bottom block, which was just barely balanced to begin with, and the entire stack will collapse. Try this with a couple of identical books if you need help developing the proper physical intuition.

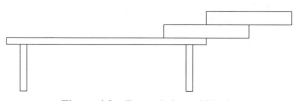

Figure 4.2   Two unbalanced blocks

How can we best arrange two blocks? How about three or more? How can we efficiently compute the maximum amount of extension that we can achieve, even for large numbers of blocks? We'll be able to answer all of these questions by the end of the chapter.

4.1 An Inductive Model

The most challenging part of arriving at our model will be discovering the best way to arrange the blocks. We will lead you carefully through the process because it requires a key insight that is not immediately obvious. Along the way we will discuss the concept of center of gravity, which is central to the physics of balancing.

4.1.1 One Block

Every rigid object has a *center of gravity*, which has the following key property: When an object A is supported by an object B, A will be balanced so long as its center of gravity is somewhere above B. If A's center of gravity is beyond the edge of B, though, A will fall.

In an object of uniform density, such as one of our blocks, the center of gravity coincides with the geometric center. In Figure 4.1, then, the block is balanced because its center is (just barely) above the table.

In other kinds of objects, such as human bodies, the center of gravity is harder to calculate but no less important. Your personal center of gravity is probably somewhere in the middle of your lower torso, but it shifts whenever you move an arm or leg. If you are standing and your center of gravity strays from over your feet, you will lose your balance and begin to fall. Athletes must be carefully attuned to their centers of gravity. A diver spins around his center of gravity; a gymnast works to keep her center of gravity above the balance beam.

As we explore the block-stacking problem, we will be concerned with calculating centers of gravity and verifying that they lie above a supporting surface. Fortunately, we will be dealing with identical, uniform blocks instead of with human bodies, which greatly simplifies the mathematics.

4.1.2 Two Blocks

Before finding the best way to stack two blocks, let's understand why the blocks in Figure 4.2 are not balanced. We can do this by seeing what goes wrong when we try to show that they *are* balanced. If the blocks are balanced, two things must be true:

1. The top block must be balanced atop the bottom block. In Figure 4.2, this is in fact true for the same reason that the block is balanced atop the table in Figure 4.1: its center of gravity is supported.

2. The two blocks, taken as a unit, must be balanced atop the table. In Figure 4.2 this is *not* true, because, as we argue below, the overall center of gravity of the two blocks lies to the right of the table's edge.

To find the overall center of gravity of two blocks, we need only average their individual centers of gravity. We can simplify our task by considering only the horizontal coordinate of each center of gravity, since horizontal positioning is all that matters in this problem. The bottom block's center of gravity is 0 feet from the table's edge, and the top block's is $\frac{1}{2}$ foot away, so the center of gravity of the two blocks taken together is $\frac{1}{4}$ foot to the right of the table's edge. Since this point is not above the table, the two blocks are not balanced.

Suppose instead that we add a second block to Figure 4.1 by inserting it beneath the existing block so that its right edge coincides exactly with the table's edge as in Figure 4.3a. If we now slide the stacked blocks to the right until the pair is exactly balanced, as in Figure 4.3b, we will have done as well as is possible with two blocks. The balance of the top block is unaffected by this process, as it is now supported by the bottom block exactly as it was originally supported by the table.

In this arrangement, one-quarter of the bottom block extends beyond the table. The center of gravity of the top block, which was exactly at the table's edge, is now $\frac{1}{4}$ foot to the *right*. The center of gravity of the bottom block, on the other hand, is $\frac{1}{4}$ foot to the *left* of the table's edge. Averaging, we find that the overall

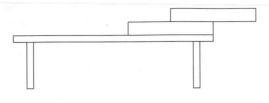

(a) First step in arranging two blocks

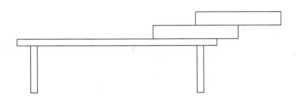

(b) Second step in arranging two blocks

Figure 4.3   Two balanced blocks

center of gravity is exactly at the table's edge, which means that the blocks are now exactly balanced.

Notice that the right edge of the top block is now $\frac{3}{4}$ foot ($\frac{1}{2} + \frac{1}{4}$) from the table's edge, compared with the $\frac{1}{2}$ foot that we attained with one block in Figure 4.1. We are making progress with extending our stack into space.

4.1.3 Three Blocks

Continuing in the same vein, we can extend the configuration in Figure 4.3b by inserting a third block (as in Figure 4.4a) and then sliding the three-block unit to the right until it is exactly balanced (as in Figure 4.4b). The balance of the upper two blocks, as we have previously argued, is unaffected by this procedure.

How far can we slide the bottom block in moving from Figure 4.4a to Figure 4.4b? In Figure 4.4a, the right edge of the bottom block is exactly at the center of gravity of the two-block unit. This means that the bottom block's center of gravity is $\frac{3}{6}$ foot to the left of two-block unit's center of gravity.

The overall center of gravity of the new three-block unit must lie between the centers of gravity of the bottom block and the two-block unit. Since the two-block unit is twice as massive as the bottom block, the overall center of gravity will be proportionately closer to the two-block center of gravity. As a result, the overall center of gravity will lie $\frac{2}{6}$ foot to the *right* of the bottom block's center of gravity and $\frac{1}{6}$ foot to the *left* of the two-block unit's center of gravity. This point is $\frac{1}{6}$ foot to the left of the bottom block's right edge. Thus, the bottom block can be moved $\frac{1}{6}$ foot beyond the edge of the table.

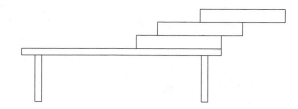

(a) First step in arranging three blocks

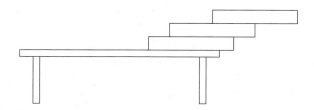

(b) Second step in arranging three blocks

Figure 4.4   Three balanced blocks

The extension beyond the table's edge is now $\frac{1}{2} + \frac{1}{4} + \frac{1}{6}$, for a total of $\frac{11}{12}$ foot. We are continuing to make progress.

4.1.4 More Blocks

By now you can probably see the pattern that is developing:

- With one block, the block extends $\frac{1}{2}$ foot beyond the table.
- With two blocks, the bottom block extends $\frac{1}{4}$ foot beyond the table and the top block extends $\frac{1}{2}$ foot further.
- With three blocks, the bottom block extends $\frac{1}{6}$ foot beyond the table, the middle block extends $\frac{1}{4}$ foot further, and the top block extends $\frac{1}{2}$ foot more.

If we were to add a fourth block, it would extend $\frac{1}{8}$ foot beyond the edge of the table. This would give a total extension of $\frac{1}{2} + \frac{1}{4} + \frac{1}{6} + \frac{1}{8}$, or $\frac{25}{24}$ of a foot. With n blocks, the total extension would be

$$\frac{1}{2} + \frac{1}{4} + \frac{1}{6} + \cdots + \frac{1}{2n} \tag{4.1}$$

Series 4.1, then, is our mathematical model of the best possible extension obtainable with n blocks.

Note that we have proved that this stacking strategy is balanced only for three blocks or fewer. It is possible to prove that it works for any number of blocks using the technique of mathematical induction, but that is beyond the scope of this chapter. Notice also that we have not proved that this stacking strategy is the best one possible.

4.2 Summing the Harmonic Series

We now have a tidy model of the block-stacking problem. We have developed a way of stacking n blocks that can be modeled with a series of increasingly smaller fractions. To determine the distance that a stack of n blocks arranged according to our design can extend beyond the edge of the table, we need to find the sum of the terms of Series 4.1.

Perhaps you recognize that Series 4.1 is composed of the first n even terms of the *harmonic series* $(\frac{1}{1} + \frac{1}{2} + \frac{1}{3} + \cdots)$. The harmonic series diverges, which means that we can make the sum of Series 4.1 as big as we choose by making n sufficiently large. In theory, at least, if we have enough blocks we can stack them to extend any distance that we choose beyond the edge of the table! Don't become so involved with the mathematics that you lose sight of this remarkable fact.

There is a powerful and efficient computational method for summing Series 4.1 for different values of n, but it is based on specialized knowledge about the harmonic series. Although we will touch on this method later in the chapter, it will not be our main focus. Instead, we will explore the method of simply adding up the n terms of Series 4.1.

4.3 Accumulation of Roundoff Error

We will experiment with using both rational and floating-point numbers to sum Series 4.1. This will give us the opportunity to explore further the tradeoff between *Mathematica*'s slow, exact rational numbers and its faster, inexact floating-point numbers. It will also illustrate how roundoff error can accumulate during a long sequence of arithmetic operations.

4.3.1 Roundoff Error and Mantissa Length

It is straightforward to add up any number of terms in Series 4.1; the only problem is typing them all in. For example, the maximum extension possible with ten blocks is

```
In[2]:= rat10 = 1/2 + 1/4 + 1/6 + 1/8 + 1/10 + 1/12 +
              1/14 + 1/16 + 1/18 + 1/20

Out[2]=  7381
         ----  ft
         5040
```
(4.2)

Converting the result to floating point, we obtain

```
In[3]:= N[rat10]

Out[3]= 1.46448 ft
```
(4.3)

We can also compute the sum of the first ten terms in the series using *Mathematica*'s floating-point arithmetic. Writing all of the denominators with decimal points will force floating-point arithmetic throughout.

```
In[4]:= float10 = 1/2. + 1/4. + 1/6. + 1/8. + 1/10. +
                  1/12. + 1/14. + 1/16. +
                  1/18. + 1/20.

Out[4]= 1.46448 ft
```
(4.4)

Note that `float10` and `N[rat10]` are identical. Their shared value is not exact, of course, since it is rounded off, but it is as close as we can get to the exact answer with six digits.

It would be tempting to conclude at this point that there is no difference between doing a computation using floating-point numbers versus doing the same computation using rational numbers and then converting to floating-point at the end. We learned differently, of course, in Chapter 2.

When *Mathematica* sums the ten terms of our series using floating-point arithmetic, it is doing ten floating-point divisions and nine floating-point additions. Every one of these operations is subject to roundoff error. Over the course of a sufficiently long summation, these errors can accumulate to the point where the final answer is meaningless. We can demonstrate this problem either by summing extremely long sequences using 16-digit mantissas or by summing smaller sequences using two-digit mantissas. We will experiment with both approaches, but will begin with the latter.

Let's imagine what would happen if we were using a programming language that provided floating-point numbers with only two digits of mantissa. Table 4.1 shows the two-digit versions of the ten quotients that are involved in Example 4.4; the two-digit sum of these ten quotients, computed by adding the quotients from top to bottom, is 1.6.

Since *Mathematica*'s result was 1.46448 feet, our answer is off by more than 0.13 feet. Even with only ten terms in the sum, roundoff error has accumulated to the point where the result is incorrect whether it is rounded to one digit or two.

Things get worse very quickly. Suppose that we add the eleventh term, `1/22.`, into our series. The two-digit version of this quotient is 0.045. When we add this to 1.6, we obtain 1.645, which when rounded to two digits is 1.6. This is exactly the same sum that we obtained for the ten-term series. Since all of the subsequent

Table 4.1   Two-digit quotients and running sum of terms from Example 4.4

Term	Quotient	Running Sum
1/2.	.50	0.50
1/4.	.25	0.75
1/6.	.17	0.92
1/8.	.12	1.0
1/10.	.10	1.1
1/12.	.083	1.2
1/14.	.071	1.3
1/16.	.062	1.4
1/18.	.056	1.5
1/20.	.050	1.6

terms in the series are even smaller than 0.045 feet, even if we were to add up the first million terms of the series, we would still obtain 1.6 feet as our answer.

Clearly, a two-digit mantissa is not large enough for this (or almost any other) computation. But roundoff error will accumulate even with a 16-digit mantissa; it will simply require more terms before the error accumulates to any significant degree.

4.3.2 Libraries

It is instructive to observe what happens when hundreds, thousands, or even millions of terms are summed. Unfortunately, that is not feasible if we must type all of them in by hand. It would be handy if *Mathematica* provided a built-in function called `BlockSeries` that would add up the first *n* terms of Series 4.1. We could then evaluate `BlockSeries[10]`, for example, and get back the sum of the first ten terms in the series as the result.

Mathematica cannot possibly provide every built-in function that any user might ever need. Instead, it provides a set of generally useful functions and the means for individual users to define others as necessary. This is the strategy adopted by most programming languages. In fact, a large part of becoming a programmer is learning how to define your own functions. We will begin treating this aspect of *Mathematica* in Chapter 5, but for now let's focus on the problem of using functions that *others* have defined.

Like most programming languages, *Mathematica* makes it easy to exploit functions developed by its worldwide community of users. Functions created within *Mathematica* can be saved to files that can be kept on the machine where they were developed or moved to any other machine. (The recent explosive growth in computer networks makes sharing material like this easier than ever.) When the functions are needed for a computation, they can be loaded into *Mathematica* and then used as if they had been built into *Mathematica* in the first place.

A collection of predefined functions is called a *library*. *Mathematica* comes with an extensive library of functions, some of which were contributed by users. Many of the functions in the *Mathematica* library are loaded automatically when they are first used.

We have created a library of functions to accompany this text. It is contained on the enclosed diskette. Before using our library you (or someone else) will have to install it in a place where *Mathematica* can find it. Because the procedure for doing this varies among computer systems, we have included details on the diskette.

Even after the library is properly installed on your machine, the functions in our library will not be loaded automatically. Instead, you must ask *Mathematica* to load them prior to use. Our library is divided into collections of related functions called *packages*. Among other things, our library defines a package of four functions related to the block-stacking problem: `BlockFloat`, `BlockRat`, `BlockFast`, and `BlockPrecision`. To load this package of functions from our library, you must use the `Needs` function:

```
In[5]:= Needs["ISP`Blocks`"]
```
(4.5)

The name of our library is `ISP` (short for *Introduction to Scientific Programming*) and the name of our block-stacking function package is `Blocks`; the parameter to `Needs` contains these two pieces of information. The package name must be surrounded by backquotes (`` ` ``), and the entire parameter must be surrounded by double quotes (`"`). We will tell you the names of the other packages in our library as they become relevant in succeeding chapters.

Once you have loaded a package, you can determine the names of the functions that it defines by using the `Names` function:

```
In[6]:= Names["ISP`Blocks`*"]

Out[6]= {BlockFast, BlockFloat, BlockPrecision,
         BlockRat}
```
(4.6)

Now that we have loaded the functions from the `Blocks` package, we can use them as if they were built in. For the moment we are interested only in `BlockFloat` and `BlockRat`. Each of these functions expect a single number as its parameter, much like the built-in *Mathematica* function `Sqrt` does. The expression `BlockFloat[n]` (where n is a positive integer) computes the sum of the first n terms in Series 4.1 using floating-point arithmetic, while `BlockRat[n]` computes the same sum using rational number arithmetic.

For example, we can compare the values computed by the two new functions with those we obtained earlier via laborious typing with

```
In[7]:= BlockRat[10]

Out[7]= 7381/5040 ft
```
(4.7)

and

```
In[8]:= BlockFloat[10]

Out[8]= 1.46448 ft
```
(4.8)

These are identical to the values that we obtained in Examples 4.2 and 4.4.

With BlockFloat and BlockRat in hand, we can easily compute the results of summing different numbers of terms using both floating-point and rational numbers. Table 4.2 summarizes the results obtained for eight different series lengths. The first column gives the number of terms being summed, the second column gives the results of using BlockRat and then converting the final answer into one of *Mathematica*'s 16-digit floating-point numbers, and the third column gives the results of using BlockFloat with *Mathematica*'s 16-digit floating-point numbers. (We will explain the fourth and fifth columns in the next section.)

Table 4.2   Summing series of differing lengths. The results of using four different ways to compute each sum are shown: *Mathematica*'s rational arithmetic with the final result converted to a 16-digit floating-point number, *Mathematica*'s floating-point arithmetic with 16-digit mantissas, simulated floating-point arithmetic with 10-digit mantissas, simulated floating-point arithmetic with 5-digit mantissas.

| | | | Floating Point | |
Terms	Rational	16 digits	10 digits	5 digits
10	1.464484126984126	1.464484126984126	1.464485127	1.4645
50	2.249602669164712	2.249602669164711	2.249602669	2.2495
100	2.593688758819810	2.593688758819810	2.593688758	2.5937
500	3.396411714995262	3.396411714995259	3.396411707	3.3960
1000	3.742735430275172	3.742735430275171	3.742735433	3.7418
5000	4.547254426492218	4.547254426492202	4.547254405	4.5289
10000	4.893803018022190	4.893803018022174	4.893802997	5.0288

4.3.3 Managing Roundoff Error

Our function BlockPrecision takes *two* parameters: the number of terms of Series 4.1 to sum and the precision that is to be used when summing them. For example, we can sum the first ten terms of the series using two-digit mantissas by evaluating the function call

```
In[9]:= BlockPrecision[10, 2]

Out[9]= 1.6000000000000000 ft
```

(4.9)

(`BlockPrecision` works by exploiting *Mathematica*'s capability for simulating floating-point number systems, which we mentioned in Chapter 2 and discuss in more detail in Appendix A.)

The fourth column of Table 4.2 gives the results of using `BlockPrecision` with a precision of ten to sum the terms of the series, while the fifth column gives the results of using `BlockPrecision` with a precision of five.

By taking the converted rational sums in the second column of Table 4.2 to be exact, we can determine the amount of roundoff error in the floating-point sums. Notice that, as the number of terms becomes larger, ever greater error accumulates in the floating-point computations. For most of the floating-point sums, the error is confined to the final two or three digits. But consider the 10,000-term sum. The five-digit sum is correct only when it is rounded to one digit. By comparison, the ten-digit sum is correct when rounded to seven digits and the 16-digit sum is correct when rounded to 14 digits. If we were to consider enough terms, however, even the 16-digit mantissa sum would eventually degrade completely.

It is possible to quantify the maximum possible error that can result from a sequence of floating-point computations, but that is beyond the scope of this text. There are two rules of thumb to understand in relation to our discussion of floating-point numbers over the last three chapters.

1. Roundoff error accumulates first in the low-order digits of a computation. When doing floating-point calculations, you must be sure that there are more digits in the mantissa than there are significant digits in the result. In other words, the minimum acceptable mantissa length increases with the number of reliable digits desired in the answer.

2. The longer the sequence of floating-point calculations that is required to compute a result, the more digits of the mantissa will be compromised by roundoff error. As a result, the minimum acceptable mantissa length also increases with the number of floating-point operations required to calculate the result.

As a programmer, you will not typically be able to pick the exact precision of the floating-point numbers used in your programs. Instead, you will have to accept what the programming language gives you. We have seen that the version of *Mathematica* that we are using to produce this book, for example, provides 16-digit floating-point numbers.

Most computers have special-purpose hardware that does floating-point arithmetic. Such hardware will only deal with floating-point numbers that have a particular precision, minimum exponent, and maximum exponent. If a programming language like *Mathematica* wishes to harness this hardware to perform its floating-point calculations, it must use exactly the same kind of floating-point numbers. The specifics of *Mathematica*'s floating-point number system vary from machine to machine simply because the specifics of the floating-point hardware varies from machine to machine. Thus, the version of *Mathematica* that we are using to produce this book provides 16-digit floating-point numbers because the *computer* that we are using to produce this book has 16-digit floating-point hardware.

There is a powerful incentive for the designers of a programming language to exploit a computer's floating-point hardware: speed. The simulated floating-point number systems that we have used in this chapter to experiment with the effects of small mantissa sizes do not use the computer's floating-point hardware. As a result, as we will see in Section 4.3.5, such calculations are extremely slow when compared with normal floating-point arithmetic. (*Mathematica*'s simulated floating-point capability exists only so that students can do small-scale experiments with the characteristics of floating-point number systems, so this is not a severe drawback.)

4.3.4 *Mathematica* Capability: Arbitrary Precision

Although 16-digit floating-point numbers will be more than sufficient for all of the calculations that we will do in this book, situations can arise in practice when they are *not* sufficient. For this reason *Mathematica* provides *arbitrary-precision numbers*, which are an alternative to floating-point numbers for approximating real numbers. In an arbitrary-precision calculation, the programmer can specify a desired precision p. *Mathematica* will then try its best to produce a p-digit result in which every digit is correct.

Mathematica does this by using more than p digits in the calculation and by keeping careful track of which digits have been contaminated by roundoff error. By doing this, *Mathematica* can guarantee that the result—which will contain p or sometimes fewer digits—is correct to the last digit. Not surprisingly, though, arbitrary-precision calculations are quite a bit slower than floating-point calculations.

Arbitrary-precision numbers are a unique characteristic of *Mathematica*; unlike floating-point numbers they are not provided by other programming languages. Because they are particular to *Mathematica*—and because floating-point numbers are often entirely adequate—we will defer all further discussion of arbitrary-precision numbers to Appendix A.

4.3.5 Speed of Arithmetic

You might reasonably ask at this point why we can't always use rational numbers for all of our computations, thus entirely avoiding the issue of roundoff error. You will appreciate the reason if you experimentally determine the *time* required for *Mathematica* to do long summations with such numbers.

Mathematica provides a built-in function `Timing` that takes an expression as its parameter and returns two pieces of information: the value of the expression and the number of seconds of computer time that *Mathematica* required to compute it. For example,

```
In[10]:= Timing[BlockFloat[5000]]

Out[10]= {0.891 Second, 4.54725}
```
(4.10)

tells us that *Mathematica* used the computer's central processing unit for approximately 0.891 seconds while calculating `BlockFloat[5000]`.

Table 4.3 gives the time in seconds required for *Mathematica* to sum increasingly longer portions of Series 4.1 when using rational numbers (`BlockRat`), floating-point numbers (`BlockFloat`), and five-digit simulated floating-point numbers (`BlockPrecision`). If you repeat our experiments, your absolute results will undoubtedly differ, but the trend that you discover should be similar to ours.

Notice the difference between the times required to compute the floating-point sum, which is calculated by `BlockFloat`, and the simulated five-digit floating-point sum, which is calculated by `BlockPrecision`. `BlockFloat`, which exploits the computer's floating-point hardware and delivers 16-digit answers, is over 175 times faster than `BlockPrecision`, which does not exploit the hardware and delivers only five-digit answers. This clearly illustrates why performance considerations mandate that a programming language make use of the floating-point hardware provided by the computer.

Now let's compare the time required to compute the rational sum and the floating-point sum. For small numbers of terms, both methods work in fractions of a second. In fact, the rational number method is no more than twice as slow as the floating-point method for 1000 terms and fewer.

As the number of terms increases, both methods slow down. This is not particularly surprising, of course, because more operations are required for the longer series. But notice carefully *how* the methods slow down. The floating-point method

Table 4.3 Time required (in seconds) to sum series of differing lengths using `BlockRat` for the rational computation, `BlockFloat` for the floating-point computation, and `BlockPrecision` for the simulated five-digit floating-point computation. (A 200 MHz Pentium Pro processor was used to obtain these timing measurements.)

Terms	Rational	Floating Point	Simulated Floating Point
50	0.012	0.007	1.61
100	0.027	0.015	3.19
500	0.141	0.094	15.90
1000	0.297	0.171	31.80
5000	2.391	0.891	158.80
10000	7.094	1.781	321.20
50000	22590.000	8.953	1597.77

slows down almost exactly in proportion to the number of terms being summed: When we double the number of terms, the time required doubles; when we quintuple the number of terms, the time required quintuples. We can extrapolate from this behavior that one million terms would require about 178 seconds.

The rational number method doesn't behave this way. The time required to sum a series grows much more rapidly than the number of terms. If we had cut off the table at 10,000 terms, you might have been tempted to predict that it would require perhaps 60 seconds to sum 50,000 terms. In reality, it requires over six hours!

The reason for this behavior is not hard to understand if you reflect on what *you* would have to do to sum up the series with pencil and paper. If it takes you ten minutes to add up a column of 50 16-digit numbers, it will take you around twenty minutes to add up a column of 100 16-digit numbers. Exactly the same kind of reasoning—with much smaller time scales—applies to a computer's floating-point hardware.

Rational numbers, however, are completely different. Each term in Series 4.1 has a larger denominator than the preceding term. If you were to add the first 50 fractions in the series, you would spend a lot of time finding least common denominators, converting fractions, adding integers, and reducing to lowest terms. It would likely take much more time to add in the next 50 fractions in the series, because the numerators and denominators would be much larger than before. Exactly the same kind of reasoning—though again with much smaller time scales—applies to a computer's ability to add rational numbers.

The ability to predict the amount of time that a computer program requires to run as a function of its inputs is a critical skill. Tested only on sequences of length 50, both the rational and floating-point implementations are acceptably efficient, but for moderate problem sizes the rational arithmetic approach rapidly becomes intolerably slow. The area of computer science concerned with making such predictions is called *complexity analysis*.

4.4 Assessment

The problem that we set out to solve was idealized, and the model that we developed has little practical importance. In reality, it would be impossible to manufacture a large number of blocks with exactly identical physical characteristics. Even a small amount of variance in the length or weight of the blocks could frustrate our stacking plan, which is based on positioning the center of gravity of each block exactly at the edge of the object supporting it.

Even if the blocks *were* identical, it would be impossible to arrange the blocks to the exact tolerances required. Some amount of error in measurement when placing the blocks is inevitable, and any deviation from the exact plan compromises our solution. A more realistic model would have to take manufacturing and placement tolerances into account.

Given that our problem was a theoretical exercise, however, our model is perfect. Our computational *method*, on the other hand, leaves much to be desired. Our method of directly summing the requisite number of terms posed extreme implementation difficulties.

As we experimented with different ways of implementing our method in *Mathematica*, we came up against an engineering tradeoff. We were forced to trade off the accuracy that can be obtained with rational numbers against the relative speed that can be obtained with floating-point numbers. Faced with such a tradeoff, the reasonable course of action is to choose the fastest method that produces an answer with an acceptable amount of roundoff error.

It is possible to make such a tradeoff when the problem is small, but for large problems it can be impossible to find an acceptable implementation. Suppose, for example, that we want to sum the first billion terms in Series 4.1. Using `BlockRat` is out of the question, and `BlockFloat` will take upwards of 50 hours on our computer. Even if we were willing to wait over two days for our answer, roundoff error would likely overwhelm each of the 16 digits in the result.

In such a situation, we have two choices. The first is to improve the implementation. We could, for example, move to a faster computer with larger floating-point numbers and reimplement our approach in a faster programming language such as C or Fortran. We could certainly speed up our implementation, and we could probably make it fast enough to deal with a billion terms in a tolerable length of time. But what if we're interested in a billion times a billion terms? It would not be hard to stagger even the fastest program running on the most expensive supercomputer.

The second choice is to admit that our *method* is faulty and then look for a better one. This is where a little bit of knowledge about the behavior of floating-point numbers or of the mathematics that underlies the harmonic series can pay large dividends. Accordingly, we will look at two improved methods for summing Series 4.1.

4.4.1 Summing in Reverse

Table 4.1 from Section 4.3.1 shows the result of adding the first ten terms of Series 4.1 using two-digit floating-point arithmetic. Table 4.4 shows the same calculation with one important difference: the terms are summed beginning with the *smallest* ($\frac{1}{20}$) and ending with the *largest* ($\frac{1}{2}$).

The sum (1.5) that we obtain by doing the calculation in this fashion is different from the sum (1.6) that we obtained originally, which demonstrates conclusively that floating-point addition is not associative. More importantly, however, it suggests a better way to sum the series.

When we summed Series 4.1 using two-digit mantissas in Section 4.3.1, we could sum no more than the first ten terms. The problem, we found, is that the sum of the first ten terms (1.6) is so much larger than the eleventh term (0.045) that it was lost in the roundoff error. Thus, the sum of the first ten terms

Table 4.4 Two-digit quotients and running sum of terms from Example 4.4 when summed from right to left

Term	Quotient	Running Sum
1/20.	.050	0.050
1/18.	.056	0.11
1/16.	.062	0.17
1/14.	.071	0.24
1/12.	.083	0.32
1/10.	.10	0.42
1/8.	.12	0.54
1/6.	.17	0.71
1/4.	.25	0.96
1/2.	.50	1.5

```
In[11]:= BlockPrecision[10, 2]

Out[11]= 1.600000000000000 ft
```
(4.11)

and the sum of the first 20 terms

```
In[12]:= BlockPrecision[20, 2]

Out[12]= 1.600000000000000 ft
```
(4.12)

each come to 1.6 feet when working with two-digit mantissas.

If the first parameter to BlockPrecision is the negative number $-n$, it adds the first n terms of Series 4.1 beginning with the smallest term. (The same applies to both BlockRat and BlockFloat.) Using BlockPrecision in this way, we discover that the sum of the first ten terms

```
In[13]:= BlockPrecision[-10, 2]

Out[13]= 1.500000000000000 ft
```
(4.13)

and the sum of the first 20 terms

```
In[14]:= BlockPrecision[-20, 2]

Out[14]= 1.700000000000000 ft
```
(4.14)

are different.

By summing the terms from smallest to largest, we ensure that the small terms are added into the running sum while it is still small, thus minimizing roundoff error. This is a marked improvement over the behavior that we observed when adding the terms from largest to smallest.

4.4.2 A Radically Better Method

Although adding the terms from smallest to largest is a better method, it is not the best solution. Even when we add just 20 terms from smallest to largest, the answer (1.7 feet) differs from the correctly rounded two-digit answer (1.8 feet). As with our original method, if we add enough terms we can overwhelm the result with roundoff error. Furthermore, the new method is no faster than the original.

We need to look deeper for a better method. It turns out that the sum of the first n terms of the harmonic series can be expressed in terms of a function called Ψ and a constant called γ:

$$\sum_{i=1}^{n} \frac{1}{i} = \Psi(n+1) + \gamma \tag{4.15}$$

If we factor the constant 2 out of the denominator of every term in Series 4.1, we see that it can be converted to

$$\frac{1}{2} \sum_{i=1}^{n} \frac{1}{i} \tag{4.16}$$

Thus, the sum of the first n terms in Series 4.1 is

$$\frac{1}{2} \left(\Psi(n+1) + \gamma \right) \tag{4.17}$$

Conveniently, Ψ and γ are built into *Mathematica* as a function `PolyGamma` and a constant `EulerGamma`, so we can easily compute the maximum possible extension for practically any number of blocks. For example, the extension for one billion blocks is

```
In[18]:= 0.5 * (PolyGamma[1.*^9 + 1] + EulerGamma)

Out[18]= 10.6502 ft
```
(4.18)

If you try this out, you will discover that *Mathematica* computes this answer almost instantaneously, which is a big improvement over the 50 hours it would require `BlockFloat`!

This example illustrates the danger of trying to apply computers to scientific problem solving without knowing something about the application area and its mathematical underpinnings. The best programmer in the world, working with the most up-to-date programming languages and the fastest computers, could not come close to implementing our original naive model in a way that would run anywhere close to as fast as the simple solution that we developed in this section.

4.5 Key Concepts

Roundoff error accumulation.   Although roundoff error begins by accumulating in the rightmost digits of the mantissa, a sufficiently long series of floating-point calculations can eventually compromise *every* digit in the result. Accordingly, the minimum acceptable length of the mantissa is governed by both the number of meaningful digits required in the final result *and* by the length of the computation.

Function libraries.   No useful programming language is self-contained. All provide a means for the programmer to define new functions and share them with others. Collections of programmer-defined functions are called libraries. Many of the functions in the *Mathematica* library are organized into packages. To access the functions in a package, you must use the `Needs` command, which takes as its parameter the name of the package.

Relative costs of numbers.   In *Mathematica*, arithmetic with large rational numbers is generally far slower than arithmetic with large floating-point numbers. This often poses an implementation tradeoff since rational number arithmetic is exact whereas floating-point arithmetic is approximate. The time differential does not become crucial except for long sequences of calculations.

Importance of appropriate method.   When solving a problem, the most important decisions are the ones that you make earlier in the problem-solving process. In this chapter we saw how a bad method could not be implemented efficiently, whereas a less obvious but much better method allowed a simple and blazingly fast implementation.

4.6 Exercises

4.1   Repeat our timing measurements on your own machine. Because the `Timing` function is only approximate, you will notice that the time required to run a command can be different each time you run it. You will obtain more accurate answers by averaging the results of multiple timings.

4.2   If you were using a programming language whose floating-point numbers had only a one-digit mantissa, how many terms of Series 4.1 could be summed before the sum stopped changing?

4.3   Extend Table 4.2 so that it contains columns for one- and two-digit mantissas.

4.4   The `BlockFast` function that is included with our `Blocks` package takes as its parameter the number of blocks being stacked and computes the maximum possible extension using the method identified in Example 4.18. Use `BlockFast` to find out how far the stack will extend if 10^{100} blocks are used.

4.5 To the nearest 100,000 blocks, how many blocks does it take to build a stack that extends 10 feet beyond the edge of the table? If each block weighs 200 grams, how much will the stack weigh?

4.6 Suppose that our blocks are 1 inch high. Determine the ratio between the height of the stack and the length of its extension beyond the table edge for 1, 10, 100, 1000, 10,000, 100,000, and 1,000,000 blocks.

4.7 Create a table similar to Table 4.2 in which the floating-point sums are obtained by adding the terms from smallest to largest. How do the accumulated roundoff errors compare?

4.8 One way to add the series $\frac{1}{2} + \frac{1}{4} + \frac{1}{6} + \cdots + \frac{1}{2n}$ is to add the series $\frac{1}{1} + \frac{1}{2} + \frac{1}{3} + \cdots + \frac{1}{n}$ and then divide the sum by two. If the arithmetic were done using floating-point numbers, would this make any difference from the standpoint of roundoff error?

4.9 When `BlockPrecision` is used to add up the terms of Series 4.1 beginning with the smallest term, why does this not make any difference in its running time?

4.10 Does the order in which Series 4.1 is summed make any difference for rational numbers? Try it out and explain any difference.

4.11 Suppose that we make a mistake when stacking blocks and place each block 0.2 inch further to the left than would be ideal. How far out will it be possible to extend the stack?

4.12 Suppose that when the blocks are stacked, the placement of every block may be in error by 0.1 inch in either direction. Develop a stacking strategy that takes account of this error. Assuming the worst, what will the extension be as a function of the number of blocks?

4.13 A rabbit begins 1 meter from a wall and hops once every second. Each time it hops, it covers half of the remaining distance to the wall. How far is it from the wall after 20 seconds? Don't try to add up a lot of fractions; identify a clever method instead.

4.14 A super rabbit hops once every second. The first hop is 1 meter, the second hop is 2 meters, the third hop is 4 meters, and so on. How far has the rabbit hopped in 20 seconds? Again, identify a clever method.

4.15 Suppose that you have 1031 floating-point numbers to average. You can do this by adding up all the numbers and dividing by 1031, or by dividing each number by 1031 and adding up the results. Which approach is likely to produce more roundoff error?

4.16   What would be the best way to calculate $ab + cb + b^2 + db + ad + cd$ to minimize roundoff error? Why?

4.17   The factorial of n, where n is a positive integer, is the product of all of the numbers from 1 up to n, inclusive. Thus, the factorial of 4 (written 4!) is $1 \cdot 2 \cdot 3 \cdot 4 = 24$, and $5! = 120$. The factorial operator is built into *Mathematica*. For example,

$$
\begin{array}{l}
\texttt{In[19]:= 5!} \\[1em]
\texttt{Out[19]= 120}
\end{array}
\tag{4.19}
$$

When computing a factorial, does *Mathematica* multiply all of the terms in the series or does it use a more clever method? Do some timing experiments to decide which.

4.18   Using 16-digit mantissas, each time you add two floating-point numbers the result may be off by one digit in the last place of the mantissa. How many additions would it take before the error could reach the first digit in the mantissa?

4.19   Using the results of timing experiments, can you determine a formula that predicts the amount of time required to sum n blocks using floating-point arithmetic? Using rational number arithmetic?

4.20   For values of x less than 1, the value of arctan(x) is given by the sum of the infinite series

$$\arctan(x) = x - \frac{1}{3}x^3 + \frac{1}{5}x^5 - \frac{1}{7}x^7 + \frac{1}{9}x^9 + \cdots \tag{4.20}$$

We can approximate arctan(x) by taking the first few terms of this series. If we are using 16-digit mantissas and $x < 0.1$, how many terms of the series should we use for maximum accuracy? (The `arctan` function is built into *Mathematica*.)

5

Kitty Hawk: Programmer-Defined Functions

The Outer Banks is a sand bar extending for hundreds of miles along the east coast of the United States, mostly along the coast of North Carolina. The sand bar is broken by narrow inlets that connect the Atlantic Ocean on the east to the shallow salt water sounds on the west. The result is a chain of long, thin islands, in many places only a few hundred yards wide and a few feet above sea level.

These islands were the site of Sir Walter Raleigh's Lost Colony in 1585 and the Wright Brothers' first powered flight in 1903. They are also known for surf fishing, hazardous navigation, and impressive riptides. The Wright Brothers' flight was made from Kill Devil Hill near Kitty Hawk, North Carolina. Kill Devil Hill is a broad sand dune that peaks at 66 feet above Albemarle Sound, which lies due west.

Imagine that you are on top of Kill Devil Hill on a calm day with no waves. Through a telescope, you are watching a friend swim across the sound toward the mainland. Because of the curvature of the earth, she will eventually disappear across the horizon. How far away (in feet) can your friend get before she disappears from view?

5.1 Model

Like Eratosthenes's problem in Chapter 3, this is a geometric problem that all but demands a diagram. Figure 5.1 reveals that we can model our problem as one of finding the length of a leg of a right triangle. The hypotenuse of the triangle is the distance from the center of the earth to the top of Kill Devil Hill, one leg is the radius of the earth, and the other leg is the distance from the top of the hill to the point where the swimmer will disappear across the horizon.

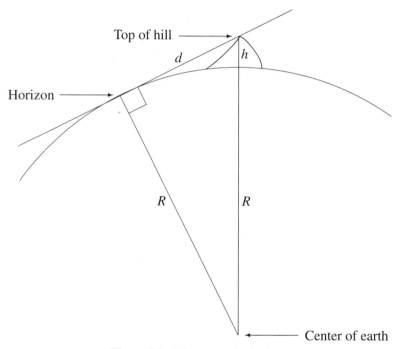

Figure 5.1   Distance to the horizon

Before proceeding, let's consider ways in which we have idealized the problem and assess their possible effects on our eventual solution.

• We are assuming that the surface of the earth—at least the part where the swimmer is located—is perfectly uniform. This is a reasonable assumption, since the problem statement specifies that it is a calm day with no waves.

• We are ignoring the effects of tides, which would raise or lower the surface of the sound relative to the center of the earth and the summit of the hill. Since both the earth's radius and the elevations of its surface features are typically measured relative to mean sea level, we will assume that this problem is also to be solved relative to mean sea level.

- The effective height of the hill in this problem will be the 66 feet of Kill Devil Hill *plus* the height of the telescope. Since the height of the telescope is not given, we will treat it as zero.

When we assess our final solution, we will try to quantify the effects that tide and telescope height would have on our final solution.

5.2 Method

Having modeled our problem with a right triangle, the Pythagorean Theorem immediately suggests a computational method. If R is the radius of the earth, d the line-of-sight distance to the horizon, and h the height of the hill, we know that

$$R^2 + d^2 = (R + h)^2 \qquad (5.1)$$

so we can determine d by evaluating the expression

$$\sqrt{(R + h)^2 - R^2} \qquad (5.2)$$

Notice that our model would have been useless if we had not known the crucial property relating the sides of right triangles. In fact, it probably would not have even occurred to us to devise such a model in the first place. This is another example that illustrates the fact that computational problem solving requires much more than knowing how to use a computer.

All we need to know in order to proceed is the height of the hill, the radius of the earth, and the significance of the two measurements. We already know that the height of Kill Devil Hill is 66 feet, and we will assume that this figure is good to two significant digits. We will also assume that the radius of the earth is 2.09×10^7 feet, significant to three digits.

5.3 Implementation

As we develop our implementation, we will see how a catastrophic loss of significance can occur when we subtract floating-point numbers, explore how algebraic transformations can be used to avoid this loss of significance, and learn how to create programmer-defined functions that extend the reach of *Mathematica*.

5.3.1 Loss of Significance

Formula 5.2 involves R, which is known to three significant digits, and h, which is known to two. Consequently, we can expect no more than two significant digits in the final result. It would appear, then, that any language with floating-point numbers with four or more digits of mantissa would be more than sufficient to compute a solution. *Mathematica* certainly provides that many digits, so it would appear that the answer to our problem is

```
In[3]:= Sqrt[(2.09*^7 + 66)^2 - (2.09*^7)^2]

Out[3]= 52524.3 ft
```
(5.3)

approximately 5.3×10^4 feet.

   If *Mathematica* had only four digits of mantissa, would we will able to get the same first two digits in our answer? To answer this question, let's simulate what the calculation of Example 5.3 would be like with only four digits of mantissa. We will do this in *Mathematica* by carrying out the steps of Example 5.3 one at a time, rounding each answer to four digits before proceeding.

   The first step is to do the addition:

```
In[4]:= 2.09*^7 + 66.

Out[4]= 2.09001 × 10⁷ ft
```
(5.4)

The first four digits of this result's mantissa are 2.090, which are what we carry forward and square:

```
In[5]:= (2.090*^7)^2

Out[5]= 4.3681 × 10¹⁴ ft²
```
(5.5)

The third step is to square the radius:

```
In[6]:= (2.09*^7)^2

Out[6]= 4.3681 × 10¹⁴ ft²
```
(5.6)

The fourth step is to take the difference of the results of Examples 5.5 and 5.6:

```
In[7]:= 4.386*^14 - 4.386*^14

Out[7]= 0 ft²
```
(5.7)

And the final step is to take the square root of the result of Example 5.7.

```
In[8]:= Sqrt[0]

Out[8]= 0 ft
```
(5.8)

If *Mathematica* had only four-digit mantissas, then, we would get zero feet as the distance to the horizon. If we look back on the step-by-step simulation, the problem is readily apparent. When we added the earth's radius to the hill's height in Example 5.4, we got back the earth's radius as our four-digit answer. Because the radius is so large, the hill's height was lost in the roundoff error. As a result, the subsequent subtraction and square root operations gave us zero as answers.

If we were using a language with four-digit mantissas, what could we do to get around this difficulty? One approach would be to find a language with longer mantissas. Table 5.1 summarizes the results that would be obtained by using Formula 5.2 for a selection of mantissa sizes between 4 and 16. (We show how to obtain these results using *Mathematica*'s floating-point simulation capability in Appendix A.) It takes a mantissa with six digits to obtain an answer different from zero, and this answer is quite a bit different from the answer obtained with seven digits. The answer obtained with seven digits differs significantly from the answer obtained with eight digits, at which point the solution appears to settle down, at least in its first few digits.

Table 5.1   Horizon distance calculated using Formula 5.2 assuming different mantissa lengths, as in Examples 5.4–5.8. Both the calculated value and the result of rounding it to two digits are shown.

Digits	Formula 5.2	Rounded
4	0	0
5	0	0
6	63245.6	63000
7	53851.65	54000
8	52535.702	53000
9	52526.1839	53000
10	52524.28010	53000
16	52524.32156629916	53000

Although it appears to have worked in this case, using languages with ever-longer mantissa sizes until reasonable-looking answers appear is not generally a good idea. What does a reasonable answer look like? Although it is clear that zero is the wrong answer to our problem, how are we to know that the value 63245.6 obtained with a six-digit mantissa is also unreasonable? For that matter, how much confidence do you have at this point in the answer obtained with *Mathematica*'s 16-digit mantissas? It is disturbing to see the computed answer change so radically as we add digits to the mantissa. We usually expect that increasing the mantissa size will simply add more digits to the result, even though those digits may not be significant.

The proper approach to dealing with our difficulty is to begin by understanding the underlying problem. The addition of floating-point numbers of radically different orders of magnitude, and the subtraction of floating-point numbers that differ only in their nonsignificant digits, must be avoided if possible and treated carefully if not.

It is pointless to add the hill's height to the earth's radius, no matter how many digits of mantissa we are using. Since only the first three digits of the number 20,900,000 are significant, the last five digits could be *anything*. Even if the hill we are talking about is Mount Everest, we will be adding the height to the insignificant (and therefore unknown) digits of the radius.

When we do the subtraction of $(R + h)^2$ and R^2, then, we are subtracting quantities that differ—if at all—only in their nonsignificant digits. Since the significant digits cancel out, *none* of the digits of the result, even assuming that the result is nonzero, will be significant. We saw in Chapter 4 how the number of significant digits in a computation can gradually degrade over a long series of operations. Here we see how a single subtraction can lead to a complete loss of significance in one step.

5.3.2 An Algebraic Transformation

To get around the loss of significance, we need to improve our computational method by rearranging the original formula to eliminate the problematic subtraction. A bit of simple algebra will do the trick in this case, although it is not always so easy. As a first step, let's expand the $(R + h)^2$ term of Formula 5.2 to obtain

$$\sqrt{R^2 + 2Rh + h^2 - R^2} \tag{5.9}$$

This expansion affords us the opportunity to do the subtraction symbolically, which cancels out the first and last terms, yielding

$$\sqrt{2Rh + h^2} \tag{5.10}$$

If we use *this* formula as the basis for our computations, we observe much improved behavior. For example, let's repeat the four-digit simulation of Examples 5.4–5.8, this time using Formula 5.10 as the basis of our calculations.

This time, the first step is to double the radius:

```
In[11]:= 2 * 2.09*^7

Out[11]= 4.18 × 10^7 ft
```
(5.11)

Next, we multiply this result by the hill's height:

```
In[12]:= 4.18*^7 * 66.

Out[12]= 2.7588 × 10^9 ft^2
```
(5.12)

The third step is to square the hill's height:

```
In[13]:= 66.^2

Out[13]= 4356. ft²
```

(5.13)

The fourth step is to add the four-digit versions of the results from Examples 5.12 and 5.13:

```
In[14]:= 2.759*^9 + 4356.

Out[14]= 2.759 × 10⁹ ft²
```

(5.14)

And the final step is to take the square root of this sum:

```
In[15]:= Sqrt[2.759*^9]

Out[15]= 52526.2 ft
```

(5.15)

To four digits, then, our answer is 5.253×10^4 feet, an answer that required *nine* digits of mantissa to obtain with Formula 5.2.

Table 5.2 compares the results of using implementations based on Formulas 5.2 and 5.10 to compute the solution to our problem for different mantissa lengths. Notice that *all* of the values calculated using Formula 5.10 round to the same two significant digits, regardless of mantissa length.

Table 5.2   Horizon distance calculated using Formulas 5.2 and 5.10 for different mantissa lengths

Digits	Formula 5.2	Formula 5.10
4	0	52530.
5	0	52524.
6	63245.6	52524.3
7	53851.65	52524.32
8	52535.702	52524.322
9	52526.1839	52524.3216
10	52524.28010	52524.32157
16	52524.32156629916	52524.32156629916

5.3.3 Programmer-Defined Functions

Formula 5.10 defines the line-of-sight distance to the horizon of a planet from the top of a hill in terms of the planet's radius and the hill's height. Although we derived the formula to solve a problem involving Kill Devil Hill, it could equally well be applied to a volcano on Mars or a satellite orbiting the moon.

During the remainder of this chapter we will apply Formula 5.10 many more times. We will use interval arithmetic to obtain accurate bounds on the distance to the horizon, and we will use the formula to study the effects of tide and telescope height on our answers.

The problem with using Formula 5.10 in *Mathematica* is that we must keep remembering exactly how it is written. For example, if we want to compute the distance to the earth's horizon from the top of a 100-foot hill, we must know to enter

```
In[16]:= Sqrt[2 * 2.09*^7 * 100 + 100^2]

Out[16]= 64653. ft
```
(5.16)

It would be much simpler if our formula were available as a built-in *Mathematica* function called `horizon` that took a planet radius and a hill height as its two parameters. If this were the case, we would be able to perform the computation above simply by supplying the name of the function and its two parameters, as in `horizon[2.09*^7, 100]`, without having to remember the exact form of the formula. Unfortunately, `horizon` is *not* built in, so this won't work just yet.

Like all programming languages, *Mathematica* provides a way for programmers to define their own functions. This is a powerful feature because it allows programmers, in effect, to extend *Mathematica*'s capabilities. You should already appreciate the utility of programmer-defined functions from *using* the programmer-defined `BlockFloat` and `BlockRat` in Chapter 4. We now turn our attention to *creating* programmer-defined functions.

Defining a function that computes Formula 5.10 is not much harder than simply writing the formula down.

```
In[17]:= horizon[R_, h_] := Sqrt[2*R*h + h^2]
```
(5.17)

Having defined `horizon` in this way, we can use it just as if it were built in. We can repeat our earlier horizon computation, this time obtaining an answer.

```
In[18]:= horizon[2.09*^7, 100]

Out[18]= 64653. ft
```
(5.18)

If the definition of the `horizon` function from Example 5.17 were appearing in a math book, it would probably be written as

$$horizon(R, h) = \sqrt{2Rh + h^2}$$
(5.19)

which isn't very different from the way it is written in *Mathematica*. Let's dissect the *Mathematica* definition to get a better feel for how function definitions are created.

Both the *Mathematica* and the mathematical definitions consist of three pieces of information:

- the *name* of the function (horizon);
- a list of the *formal parameters* to the function (R, h); and
- the *body* of the function (Sqrt[2*R*h + h^2]).

The name of the function must be a valid *Mathematica* variable, as must each of the formal parameters. The body must be an expression written in terms of the formal parameters.

The difference between the two definitions is in the punctuation. In the *Mathematica* definition, the : = operator separates the name and the formal parameter list from the body, while in the mathematical version the equal sign plays this role. The formal parameter list in the *Mathematica* version is enclosed in square brackets and each parameter name is followed by an underscore, while in the mathematical version the list is enclosed in parentheses and there are no underscores. Finally, the body of the function in the *Mathematica* version is written using *Mathematica* syntax, while the body of the function in the mathematical version is written using mathematical syntax.

The operator that is used in the function definition (: =) differs from the familiar assignment operator (=) in one important respect. When the = operator is used, the right-hand side of the assignment is evaluated; when the : = operator is used, the right-hand side is *not* evaluated. As a result, *Mathematica* does not display a value in response to a function definition that uses the : = operator. This is exactly what we want in the case of a function definition: the body of the function should not be evaluated until the function is *called*.

A function application, as you already know from your use of built-in and library functions, consists of a function name followed by a list of *actual parameters* enclosed in square brackets. There must be the same number of actual parameters in the function application as there are formal parameters in the function definition. When you apply a programmer-defined function, as illustrated in Example 5.18, *Mathematica* evaluates it with the following steps.

1. *Mathematica* evaluates the actual parameters as if they had been typed in directly. In our example, the parameters evaluate to 2.09×10^7 and 100, respectively. An actual parameter, as you already know, can be an arbitrarily complicated expression, which is why it is necessary to evaluate it.

2. *Mathematica* makes a copy of the function's body and modifies it by replacing each occurrence of a formal parameter with the value of the corresponding actual parameter. In our example, the actual parameter 2.09×10^7 corresponds to R and the actual parameter 100 corresponds to h. The function body, Sqrt[2*R*h + h^2], is transformed into Sqrt[2*2.09*^7*100 + 100^2].

3. *Mathematica* evaluates the modified copy of the function's body as if it had been typed directly into *Mathematica*. The modified function body evaluates to 64,653 feet, which is what is displayed as the result in Example 5.18.

We will look at the process of *creating* programmer-defined functions in more detail in Chapter 6. For now, let's get some more experience with *using* them.

5.3.4 Bounds on the Solution

The last step in solving the horizon problem is to use interval arithmetic to put bounds on the distance to the horizon from the top of Kill Devil Hill. Recall that the radius of the earth, 2.09×10^7 feet, is known to three significant digits, while the height of the hill, 66 feet, is known to two significant digits. This means that the radius is known to $\pm 0.005 \times 10^7$ feet and that the hill height is known to ± 0.5 feet.

If we think about the geometry of the problem a bit, it is clear that the distance to the horizon will be greater if either the radius or the hill height is larger. Thus, the greatest possible distance to the horizon is

```
In[20]:= horizon[2.095*^7, 66.5]

Out[20]= 52785.9 ft
```
(5.20)

while the smallest possible distance is given by

```
In[21]:= horizon[2.085*^7, 65.5]

Out[21]= 52262.4 ft
```
(5.21)

Rounding the upper bound up and the lower bound down, as usual, we can say that the distance to the horizon from the top of Kill Devil Hill is between 52,262 and 52,786 feet, assuming that the measurements for the radius of the earth and the height of the hill are accurate to the specified number of digits.

5.4 Assessment

We will assess the implementation, method, and model separately. Note, however, that we did the most important part of the assessment when we found a shortcoming in our method and converted the computationally flawed Formula 5.2 into the computationally sound Formula 5.10.

5.4.1 Significant Digits and Interval Arithmetic

When we used Formula 5.10 to solve the horizon distance problem, we found the answer, when rounded to two significant digits, was 53,000 feet. The lower and upper bounds that we obtained via interval arithmetic round to 52,000 and 53,000 feet, respectively. This is another illustration that the method of significant digits, which would say that the true answer must round to 53,000 feet, is only an approximation.

5.4.2 Simplifying the Method

This chapter has given a clear example of the difference between model and method, which was not always so apparent in earlier chapters. We modeled the horizon problem as the problem of finding the length of a leg of a right triangle. We then explored two different methods—Formulas 5.2 and 5.10—for computing a result based on the model. Although the two methods were algebraically equivalent, Formula 5.2 led to a catastrophic loss of significance that Formula 5.10 avoided.

Although Formula 5.10 is computationally sound, it still involves the addition of a small number containing two significant digits (h^2) to a much larger number also containing two significant digits ($2Rh$). So long as $2Rh$ is more than two orders of magnitude larger than h^2, this addition is pointless. Since in this circumstance the addition doesn't contribute anything to the significant digits of the answer, we might as well eliminate it to obtain

$$\sqrt{2Rh} \qquad (5.22)$$

This simplification is not *algebraically* valid, but it is *computationally* valid so long as we know something about the possible values of R and h and the number of significant digits to expect in the answer. The simplification would not be computationally valid if h could take on values closer to R, which would be the case if you were watching your friend from orbit instead of from a sand dune.

We can easily create a *Mathematica* function `horizonFast` to compute Formula 5.22.

```
In[23]:= horizonFast[R_, h_] := Sqrt[2*R*h]
```
(5.23)

This implementation is faster because it eliminates a multiplication and an addition, but the price we pay for the additional speed is that the formula is valid only for values of R that are large compared to h. The speedup is insignificant if we are only doing an occasional calculation with `horizonFast`, but if we are doing *millions* of calculations in the context of a larger program, the payoff could be tremendous.

Table 5.3 shows the results of using Formulas 5.10 and 5.22 with different mantissa lengths. The calculated values differ, if it all, only in their insignificant digits. Notice that all of the calculated values round to the same two significant digits.

5.4.3 Tide and Telescope Height

When we devised our model we raised several questions. Specifically, we were concerned about the extent to which tide and telescope height could affect our answers. In this section we will assess those concerns further.

Table 5.3   Horizon distance calculated using Formulas 5.10
and 5.22 for different mantissa lengths

Digits	Formula 5.10	Formula 5.22
4	52530.	52530.
5	52524.	52524.
6	52524.3	52524.3
7	52524.32	52524.28
8	52524.322	52524.280
9	52524.3216	52524.281
10	52524.32157	52524.28010
16	52524.32156629916	52524.2800997786

We have done all of our computations in this chapter relative to average sea level. A higher tide would serve to increase the radius of the earth while decreasing the height of the hill; a lower tide would do just the opposite. A five-foot increase in the tide would result in

```
In[24]:= horizon[2.09*^7 + 5, 66 - 5]

Out[24]= 50495.6 ft
```
(5.24)

a horizon distance of less than 51,000 feet, whereas a five-foot decrease in the tide would result in

```
In[25]:= horizon[2.09*^7 - 5, 66 + 5]

Out[25]= 54477.6 ft
```
(5.25)

a horizon distance of more than 54,000 feet. The solution to this problem is thus *extremely* sensitive to tidal variations.

Adding or subtracting five feet from the radius of the earth is computationally irrelevant, as it doesn't come close to affecting the three significant digits. The radical differences observed above must thus be resulting from the influence of the tide on the effective height of Kill Devil Hill. This should lead us to expect that adding the height of the telescope to the height of hill would have a similar effect. For example, if our telescope is three feet above the top of the hill, the distance to the visible horizon is

```
In[26]:= horizon[2.09*^7, 69]

Out[26]= 53704.8 ft
```
(5.26)

an increase of more than 1000 feet over the mean-tide result.

5.5 Key Concepts

Addition of floating-point numbers. There is no point in adding two floating-point numbers if one is so much larger than the other that their significant digits do not overlap, because the contribution of the smaller will be lost in the insignificant digits of the larger. If the exact result of the addition plays a critical role in subsequent calculations, as in Formula 5.2, this can pose a serious problem. Otherwise, this may provide an opportunity to streamline a computation, as in Formula 5.22.

Subtraction of floating-point numbers. If a subtraction problem involves two floating-point numbers that differ only in their nonsignificant digits, the result will contain *no* significant digits, which can be catastrophic if the result is used in subsequent calculations. Algebraic transformations can often be used to eliminate the subtraction and avoid the problem, which is what we did with Formula 5.10.

Programmer-defined functions. Programmer-defined functions are as easy to create as the definitions of mathematical functions are to write down. They allow the programmer to extend the reach of *Mathematica*.

5.6 Exercises

5.1 What will affect the solution to the horizon problem more: an error of 1 foot in measuring the height of the hill or an error of 100 miles in measuring the radius of the earth?

5.2 Suppose that a ship whose mast is 53.5 feet above the water is sailing on the sound, and that this measurement is accurate to three significant digits. Use interval arithmetic to determine bounds on how far from Kill Devil Hill the top of the mast will be visible.

5.3 At low tide, Kill Devil Hill is effectively five feet higher than normal; at high tide, it is effectively five feet lower. Assuming that these two measurements are good to one significant digit and the precisions of the other measurements are as given in the text, use interval arithmetic to determine bounds on the distance out to sea one can see from the top of Kill Devil Hill at any time.

5.4 Implement a function called `blocksPerPerson` that takes as parameters an area in square miles, a population, and the length of one side of a city block in feet. It should return the number of city blocks that each person would receive if the area were divided evenly among the population.

5.5 Implement a function called `circumference` that takes as parameters the angle `alpha` and distance `d` from Eratosthenes's experiment described in Chapter 3, and returns the circumference of the earth in kilometers.

5.6   Implement a function called `blockRatio` that takes a positive integer n as its parameter and returns the ratio between the extension and the height of a stack of n blocks arranged as in Chapter 4. You may assume that the blocks are 1 inch thick and 12 inches long.

5.7   If R is known to three significant digits and h to four, what must be the ratio R/h to ensure that Formula 5.22 is valid?

5.8   Implement a function that will determine the radius of the earth when given the height of a hill and the distance to the horizon from the top of that hill.

5.9   Suppose that a series of 1000-foot towers are built at equal intervals along the equator. How many towers are needed if each tower is to be visible from its closest neighbors on the east and west?

5.10   Repeat Exercise 5.9 using interval arithmetic. Assume that each tower is between 995 and 1005 feet in height and that the earth's radius is known to three significant digits.

5.11   Consider the expression

$$1 - \frac{1 - x}{1 - x^2} \qquad (5.27)$$

Implement a function that calculates this expression. Use your function to create a table showing how your function behaves for small values of x using various mantissa sizes.

5.12   Algebraically transform Expression 5.27 to render it more computationally sound. Implement a function to calculate your new expression, and create a table showing how your function behaves for small values of x using various mantissa sizes. Compare your table with the one you created in Exercise 5.11 and explain any differences.

5.13   Suppose that we would like to calculate the value of $x - \arctan(x)$ for $0 < x < 0.1$. For values of x in that range, is the expression different from zero? (The `arctan` function is built into *Mathematica*.)

5.14   Implement a function that accurately calculates $x - \arctan(x)$ for $0 < x < 0.1$. (Hint: Use the result of Exercise 4.20.)

5.15   Implement a function that takes as parameters the height of a hill and the distance to the horizon, and returns the extra distance it would be possible to see if the hill were 1 foot higher.

5.16   Generalize the function from Exercise 5.15 so that it takes as a third parameter the number of feet added to the height of the hill.

5.17   If a ball is thrown straight up in the air, its position as a function of time is given by

$$vt - \frac{1}{2}gt^2 \tag{5.28}$$

where v is the initial velocity in m/sec, g is 9.8 m/sec^2, and t is measured in seconds. Implement a function that takes v and t as parameters and reports the position of the ball. Assess your function. What is its most obvious shortcoming?

5.18   Implement a function that takes an angle in degrees as its parameter and converts it to radians.

5.19   Stirling's approximation is

$$n! \approx \frac{n^n}{e^n}\sqrt{2\pi n} \tag{5.29}$$

Implement a function that takes a positive integer n as its parameter and computes Stirling's approximation to $n!$.

5.20   Implement a function that takes a positive integer n as its parameter and computes the ratio between $n!$ and Stirling's approximation to $n!$.

6

Baby Boom: Symbolic Computation

The Bureau of the Census estimates that the resident population of the United States on January 1, 1994, was 259,167,000 and that it had increased to 261,638,000 by January 1, 1995. Although these are only estimates that are retroactively revised from time to time, we will entertain ourselves in this chapter by pretending that they are exact. More precisely, we will pretend that the first figure was the exact population at midnight when 1994 began, and that the latter figure was the exact population at midnight when 1994 ended. Our goal is to develop a way to determine the population of the United States at any instant of time during 1994.

The key to solving this problem is to figure out how to model the way that the population grew during the year. We will do this by analogy with the way that the balance of an interest-bearing bank account grows over time. We will consider three successively more refined interest models: simple interest, compound interest, and continuously compounded interest. As we develop each model, we will first implement it and then assess how well it can be used to model the problem of human population growth.

6.1 Simple Interest

Suppose that you deposit $1000 in a bank account and leave it there without making any subsequent deposits or withdrawals. The bank credits your account with 6% interest at the end of each year. What will your account balance be after six years?

In this circumstance, we say that the bank is paying simple interest at the rate of 6% per year. In this section we will develop a model of simple interest. We will implement the model by creating a function named `simple` that takes three parameters—initial balance, interest rate, and number of years—and returns the balance after the specified number of years. Developing the model will require some mathematical manipulations, and we will use *Mathematica* to assist us.

6.1.1 Numerical Computation

In the beginning your account will contain the $1000 that you initially deposited. Let's use an assignment expression to keep track of this fact in *Mathematica*.

```
In[1]:= year0 = 1000.

Out[1]= 1000.
```
(6.1)

At the end of the first year, your account will contain your beginning balance plus interest. The interest is calculated by multiplying the beginning balance by the interest rate of .06. Thus, after one year your account will contain

```
In[2]:= year1 = year0 + year0 * .06

Out[2]= 1060.
```
(6.2)

The calculations for subsequent years are similar to this one. At the end of each year, your account will contain the balance from the previous year plus interest calculated by multiplying the previous year's balance by the interest rate. Thus, after two years your account will contain

```
In[3]:= year2 = year1 + year1 * .06

Out[3]= 1123.6
```
(6.3)

and after three years

```
In[4]:= year3 = year2 + year2 * .06

Out[4]= 1191.02
```
(6.4)

A definite pattern has appeared in this sequence of computations, and we can now easily push the sequence as far into the future as we please. Table 6.1 shows the interest earned and the balance outstanding at the end of each of the first six years.

Notice how the use of assignment helped us here by eliminating the need to type in increasingly longer floating-point numbers. Thus, in Example 6.4 we were able to enter the variable `year2` instead of the number 1123.6. In fact, we do not even

Table 6.1 Accumulation of simple 6% annual interest

Year	Interest	Balance
0		$1000.00
1	$60.00	1060.00
2	63.60	1123.60
3	67.42	1191.02
4	71.46	1262.48
5	75.75	1338.23
6	80.29	1418.52

have to look at the result of the second year calculation before going forward with the third year calculation. The use of sequentially numbered variables representing numbers, instead of the numbers themselves, also made the similarities among the four calculations much more apparent.

6.1.2 Symbolic Computation

Given a starting balance and an interest rate, we can now extend our simple interest calculation out to any point in the future. Unfortunately, the amount of work that we must do is proportional to the number of years that we wish to cover. Calculating a balance after three years have elapsed is one thing, but calculating a balance 500 years into the future would be quite time consuming, requiring a sequence of 500 calculations.

Let's see if we can develop more insight into the accumulation of simple interest by repeating the calculations from the previous section using *symbolic* instead of *numerical* constants. We will use p to stand for the beginning balance and r to stand for the annual interest rate.

In the beginning our account will contain p dollars.

```
In[5]:= year0 = p

Out[5]= p
```
(6.5)

At the end of one year, our balance will consist of the previous year's balance plus interest.

```
In[6]:= year1 = year0 + year0 * r

Out[6]= p + pr
```
(6.6)

The calculations for two years

```
In[7]:= year2 = year1 + year1 * r

Out[7]= p + pr + r(p + pr)
```

(6.7)

and three years

```
In[8]:= year3 = year2 + year2 * r

Out[8]= p + pr + r(p + pr) + r(p + pr + r(p + pr))
```

(6.8)

continue to follow the pattern that we discovered with our earlier numerical calculations.

Until now, every calculation in this book has resulted in *numerical* answers. The four calculations in this section, however, have all resulted in *symbolic* answers. The variables year0, year1, year2, and year3 have all been made to stand for something via assignment. The variables p and r, in contrast, have never been assigned values, so they stand for themselves as symbols or, if you prefer, as mathematical unknowns.

In addition to manipulating rational and floating-point numbers, *Mathematica* can manipulate symbols. This fact greatly extends the scope of *Mathematica* and makes it much more powerful. With numbers, *Mathematica* can do arithmetic; with symbols, it can do algebra and calculus. With numbers, *Mathematica* can add and subtract; with symbols, it can solve equations and differentiate functions.

Unfortunately, the complicated symbolic value of year3 does not afford us much insight into the problem of finding a better way to calculate simple interest. We can make progress, however, once we realize that we can do more than simply stare at the value—we can manipulate it algebraically.

For example, *Mathematica* has a built-in function Expand that takes an expression as its parameter and expands it by distributing products over sums. Let's try it out on the value of year3.

```
In[9]:= Expand[year3]

Out[9]= p + 3pr + 3pr² + pr³
```

(6.9)

This expanded version of year3 is simpler looking, but it doesn't provide any immediate insight. Let's try factoring year3 instead.

```
In[10]:= Factor[year3]

Out[10]= p(1 + r)³
```

(6.10)

Now it looks like we're getting somewhere. Next let's factor `year2`.

In[11]:= Factor[year2]

Out[11]= $p(1+r)^2$

(6.11)

It's beginning to look as if we can calculate the balance after n years by multiplying *p* by the *n*th power of $(r+1)$. Let's confirm that the pattern continues through year four.

In[12]:= Factor[year3 + year3 * r]

Out[12]= $p(1+r)^4$

(6.12)

This is compelling evidence—though certainly not a proof—that the formula

$$p(r+1)^n$$ (6.13)

gives the amount in an account with initial balance *p* and annual interest rate *r* after *n* years. Once you realize that the formula asserts that the balance increases every year by a factor of $r+1$, however, it should not be hard to convince yourself that the formula is indeed correct and can be used as the basis for simple interest calculations in *Mathematica*.

6.1.3 A Simple Interest Function

The next step in our treatment of simple interest is to implement Formula 6.13 as a *Mathematica* function that will do our simple interest computations for us. This is easily done.

In[14]:= simple[p_, r_, n_] := p * (r+1)^n   (6.14)

We can now, for example, calculate the account balance in 500 years given an initial balance of $1000 and an interest rate of 6%.

In[15]:= simple[1000, .06, 500]

Out[15]= 4.4971×10^{15}

(6.15)

This would pay off the national debt of the United States 1000 times over! We can also calculate the factor by which an investment of size *B* would increase if invested at 10% interest for eight years.

```
In[16]:= simple[B, .10, 8]

Out[16]= 2.14359 B
```
(6.16)

The initial investment would more than double. Notice that this expression involves a mixture of numerical and symbolic calculation, since the parameters to `simple` include both numbers and symbols.

6.1.4 Simple Interest and Population Growth

We can use *Mathematica* to determine that the population at the end of 1994 represents an approximate 0.95% increase over the population at the beginning of the year.

```
In[17]:= simpleGrowth = (jan95 - jan94) / jan94

Out[17]= 0.00953439
```
(6.17)

(Before doing this computation, we assigned the 1994 population to `jan94` and the 1995 population to `jan95`.)

We can also use the simple interest formula to extrapolate this growth rate out into the future. For example, assuming that the population continues to grow at an annual rate of approximately 0.95%, on January 1, 2000, the United States should have more than 274 million people.

```
In[18]:= simple[jan94, simpleGrowth, 6]

Out[18]= 2.74351 × 10⁸
```
(6.18)

While this is an interesting application of our simple interest formula, it is not the problem that we set out to solve. We want to be able to find the U.S. population at any point *during* 1994. The simple interest model assumes that all growth occurs only at the end of each year. Although this is accurate for certain kinds of bank accounts, it is clearly unrealistic for human populations.

6.2 Compound Interest

Most banks credit interest to their accounts more often than once a year. A checking account earning 6% annual interest, for example, might earn interest every day. This means that every day, interest is added into the account at the rate of $\frac{1}{365}$ of 6% of the outstanding balance. In this case, we say that the bank is paying interest at the rate of 6% per year, compounded daily.

In this section we will develop a model of compound interest. We will implement the model by creating a function named `compound` that takes four parameters—initial balance, interest rate, number of years, and number of compounding intervals per year—and returns the balance after the specified number of years. This will yield a better, although still not ideal, model of population growth.

6.2.1   Compounding Intervals

We can use our function `simple` to derive a formula for annual interest that is compounded daily. We derived the `simple` function to deal with *annual* interest, paid *annually*, over a period of *years*. Nevertheless, it is not this inflexible. It can also be used, for example, to deal with *daily* interest, paid *daily*, over a period of *days*.

An account earning 6% annual interest, compounded daily, is the same as an account earning simple $\frac{6}{365}$% daily interest, paid daily. If we invest $1000 in such an account, after one year we will have

```
In[19]:= simple[1000., .06/365, 365]

Out[19]= 1061.83
```

(6.19)

The third parameter in the expression above is 365 days, instead of one year, because the second parameter is the amount of interest paid per day. The `simple` function requires the time units used for the second (percent/day) and third (days) parameters to correspond.

Notice that getting 6% interest a little bit at a time yields us a total of $61.83 in interest over the course of a year, as opposed to the $60.00 that accrues if the interest is paid in a lump sum at the end of the year. This is because the interest that accumulates one day begins earning interest itself the next.

Let's use symbols instead of numbers to generalize this example. An account earning r% annual interest, compounded daily, is the same as an account earning simple $\frac{r}{365}$% daily interest, paid daily. If we invest p in such an account, after one year we will have

```
In[20]:= simple[p, r/365, 365]
```

$$\text{Out[20]= } p\left(1 + \frac{r}{365}\right)^{365}$$

(6.20)

and after n years, we will have

```
In[21]:= simple[p, r/365, 365*n]
```

$$\text{Out[21]= } p\left(1 + \frac{r}{365}\right)^{365n}$$

(6.21)

While this gives us a way to compute interest that is compounded daily, there are other possible compounding intervals. Interest is commonly compounded monthly or quarterly, for example. Generalizing one step further, let's assume that interest is compounded m times per year. Using m in place of 365 in Example 6.21, we find that after n years of receiving $r\%$ annual interest compounded m times per year, we will have a balance of

```
In[22]:= simple[p, r/m, m*n]
```

$$Out[22]= p\left(1 + \frac{r}{m}\right)^{mn}$$

(6.22)

We will use the result of Example 6.22 as the starting point for our *Mathematica* implementation of compound interest.

6.2.2 A Compound Interest Function

We can easily convert Example 6.22 into a four-parameter function called compound

```
In[23]:= compound[p_,r_,n_,m_] := p * (1 + r/m)^(m*n)
```

(6.23)

where p is the beginning balance, r is the annual interest rate, n is the length of the investment in years, and m is the number of times per year that the interest is compounded.

We have used this function to compute the return on $1000, invested at 6% annual interest compounded daily, over a period of six years. The results are summarized in Table 6.2; you should compare them with Table 6.1.

Table 6.2   Accumulation of 6% annual interest, compounded daily

Year	Interest	Balance
0		$1000.00
1	$61.83	1061.83
2	65.66	1127.49
3	69.71	1197.20
4	74.02	1271.22
5	78.61	1349.83
6	83.46	1433.29

The yearly figures differ only by a few dollars, but the difference can magnify over time. For example, if we repeat our calculation from Example 6.15 by investing $1000 at 6% interest, compounded daily, for 500 years

```
In[24]:= compound[1000, .06, 500, 365]

Out[24]= 1.06602 × 10^16
```

(6.24)

we end up with an extra $6000 trillion or so.

6.2.3 Effective Yield

What is the quantitative difference between simple 6% annual interest and 6% annual interest, compounded daily? One way to get at the answer to this question is to ask a different one. What simple interest rate, paid annually, is equivalent to a 6% annual interest rate compounded daily? This rate is called the *effective* annual yield of the compound rate. It is often published by banks to give customers a means of comparing the effects of different interest rates and compounding intervals.

We can answer this question by solving for r in the equation

$$\text{simple}(p, r, 1) = \text{compound}(p, .06, 1, 365) \qquad (6.25)$$

We can easily do this in *Mathematica* with the built-in function `Solve`.

```
In[26]:= Solve[simple[p, r, 1] ==
              compound[p, .06, 1, 365], r]

Out[26]= {{r → 0.0618313}}
```

(6.26)

The `Solve` function expects as its two parameters an equation and the name of the unknown to solve for. Equations are written in *Mathematica* by using ==, which is the equality operator, instead of =, which is the assignment operator. `Solve` returns a list of solutions. Each solution in the list gives a value for the unknown. In the example above, the list produced by *Mathematica* contains only one solution. Notice that this solution (approximately 6.18%), which is the effective annual yield of 6% interest compounded daily, is independent of the initial balance p.

6.2.4 Compound Interest and Population Growth

If we are willing to pretend that each day's population growth occurs all at once at midnight, we can use the compound interest formula to model the country's population growth in 1994. To do that we must first find an interest rate that, when compounded daily, will produce the observed population growth between January 1994 and January 1995. To find this interest rate, we must solve an equation.

```
In[27]:= Solve[jan95 == compound[jan94, r, 1, 365], r]

Out[27]= $Aborted
```

(6.27)

Unfortunately, when we do this *Mathematica* works for an extremely long time without making progress, which leads us to abort the computation. The equation, which reduces to

$$\left(1 + \frac{r}{365}\right)^{365} = \frac{\text{jan95}}{\text{jan94}} \tag{6.28}$$

is rather formidable with its exponent of 365 on the left-hand side; even *Mathematica* has its limits. We will revisit this equation when we consider equation solving in more detail.

The problem with Equation 6.28 is that it is a polynomial of degree 365, which means that it has 365 complex roots. We're interested in a real root, and we can find it if we give *Mathematica* a hand. If we raise both sides of Equation 6.28 to the $\frac{1}{365}$th power, we obtain

$$1 + \frac{r}{365} = \left(\frac{\text{jan95}}{\text{jan94}}\right)^{\frac{1}{365}} \tag{6.29}$$

which *Mathematica* can solve easily. There is a compelling lesson here. It is important to understand what *Mathematica* is doing when it gets stuck; a little bit of human ingenuity can often help.

```
In[30]:= Solve[1 + r/365 == (jan95/jan94)^(1/365), r]

Out[30]= {{r → 0.00948935}}
```
(6.30)

While we could use this compound interest rate as a basis of a model for population growth, it is perhaps best that we do not. Such a model would assume that the country's population ratchets upward like a staircase once a day. We will continue to push forward, looking for a model that better matches reality.

6.3 Continuous Interest

Back in the early 1970s, the interest rates that banks in the United States could pay on regular savings accounts were strictly regulated. Typically, a bank could pay no more than 5% on such accounts. After a time, due to banking deregulation and double-digit inflation, interest rates rose higher, at least until the 1990s.

Because the amount of annual interest that banks could pay depositors was regulated by the government, one way they competed in the 1970s was by offering favorable compounding intervals. If one bank compounded monthly while another compounded daily, the second bank would enjoy a competitive advantage even though both were paying 5% annual interest.

Of course, there was no reason to stop at daily compounding. A bank could, if it chose, compound every hour or every minute or every second. In fact, many banks took it to the limit and began compounding *continuously*. Continuous compounding is what we get when the number of compounding intervals approaches infinity. We receive, in theory, an infinite number of infinitely small interest payments.

6.3.1 A Continuous Interest Function

Beginning with the function compound, we can come up with a formula that computes the results of growth under continuous compounding by taking the *limit* of the compound function as the number of compounding intervals approaches infinity.

$$\lim_{m \to \infty} \text{compound}(p, r, n, m) \tag{6.31}$$

Fortunately, *Mathematica* has a built-in function that finds limits. Limit takes two parameters: the expression whose limit we wish to find and a parameter that identifies the limit we are interested in.

> ```
> In[32]:= Limit[compound[p, r, n, m], m->Infinity]
>
> Out[32]= E^{nr}p
> ```
(6.32)

This solution shows that the growth of a continuously compounded investment is governed by a power of e, the base of the natural logarithm.

Based on this result, we can now define a function called continuous that computes exponential growth under continuous compounding.

> ```
> In[33]:= continuous[p_,r_,n_] := p*Exp[n*r]
> ```
(6.33)

Mathematica's Exp function returns the result of raising e to the power specified by its parameter.

We have used this function to compute the return on $1000, invested at 6% annual interest compounded continuously, over a period of six years. The results are summarized in Table 6.3; you should compare it with Tables 6.1 and 6.2.

Table 6.3 Accumulation of 6% annual interest, compounded continuously

Year	Interest	Balance
0		$1000.00
1	$61.84	1061.84
2	65.66	1127.50
3	69.72	1197.22
4	74.03	1271.25
5	78.61	1349.86
6	83.47	1433.33

6.3.2 Continuous Compounding and Population Growth

In order to use either simple interest or compound interest as models for population growth, we were forced to make the invalid assumption that population increases occur only at regularly spaced intervals. Continuous compounding gives us a much more natural model for population growth, because the growth of a large population tends to happen more or less continuously.

To use continuous compounding as the basis for a model, we have to find an interest rate that, when compounded continuously, will produce the observed population growth between January 1994 and January 1995.

```
In[34]:= Solve[jan95 == continuous[jan94,r,1], r]

Out[34]= {{r → 0.00948923}}
```
(6.34)

Although we tried and failed to solve the corresponding equation for compound interest, this time we were successful. The difference lies in the fact that the formula for continuous interest is significantly simpler than the formula for compound interest.

We can now create a variable that contains the continuous growth rate of the United States for which we solved in Example 6.34.

```
In[35]:= continuousGrowth = 0.00948923

Out[35]= 0.00948923
```
(6.35)

We now have all the ingredients we need to create a function to estimate the population of the United States at any instant in 1994. This function, `dailyPop`,

```
In[36]:= dailyPop[days_] :=
           continuous[jan94,
                continuousGrowth, days/365.]
```
(6.36)

takes as its parameter the number of the day in 1994 for which we want a population estimate. For example, the population at midnight on February 1 was

```
In[37]:= dailyPop[31]

Out[37]= 2.59376 × 10^8
```
(6.37)

and in the next six hours

```
In[38]:= dailyPop[31.25]

Out[38]= 2.59378 × 10^8
```
(6.38)

the population increased by approximately 2000 people.

There is a small flaw in our `dailyPop` function. The value that we are using for `continuousGrowth`, which we obtained by solving the equation in Example 6.34, is not as precise as it could be. After *Mathematica* solved the equation in Example 6.34, it printed out a six-digit approximation to the 16-digit solution that it actually found. We then typed this six-digit approximation back in as the value of `continuousGrowth`.

There are two ways to remedy this problem. One is to use `InputForm` to display all 16 digits of the equation's solution

```
In[39]:= InputForm[
            Solve[jan95 == continuous[jan94,r,1], r]]

Out[39]= {{r → 0.009489227414592271}}
```

(6.39)

and to type *this* back in as the value of `continuousGrowth`. A better approach would be to take advantage of *Mathematica*'s capabilities to extract the value of r directly from the list of solutions produced in Example 6.34. We will explore this idea in Chapter 8.

6.4 Assessment

We have already assessed each of our three models of population growth. In this section we will explore three issues related to the development and implementation of those models.

6.4.1 Discrete versus Continuous Growth

We put a lot of effort into coming up with an appropriate model for population growth. We ultimately rejected both the simple interest and compound interest models because they are discrete. That is, they both are based on the assumption that growth occurs only at regularly spaced intervals. We settled on the continuous model because it avoids this flaw, modeling population growth with a continuous curve.

All very nice, but populations don't really grow continuously. People are born and people die one at a time. The resident population also depends on migration. People enter and leave the United States by the carload, planeload, and boatload at every hour of every day of the year. If we were to graph the exact population of the United States as a function of time, we wouldn't see a smooth curve. Instead, we'd see a line like a staircase, sometimes ratcheting upward and sometimes downward.

The problem with the compound interest model isn't that it is discrete; the problem is that it assumes that the population changes only at regular intervals. An exact model of population growth would be like the graph that we imagined above. It would be discrete, but it would also be rather chaotic. If we were to step back far enough from our imaginary graph, though, we would see a continuous curve smoothing out its steps. This curve would be the graph of our continuous model.

The continuous model is appropriate not just because it does the best job of approximating reality, but also because it is simple. Recall that the daily interest model gave rise to an equation that even *Mathematica* could not solve. This was because it had the constant 365 in its exponent. This same constant would also make calculations done with daily interest computationally more expensive than calculations done with continuous interest.

There is another, more serious, problem with our model of population growth. We have made the unjustified assumption that the population grows at the same rate throughout the year. This is not true for a variety of reasons. More people die in the winter than in the summer, for example. More people travel abroad in the summer than in the winter, for another.

We could improve our model for population growth if we had access to the population at the end of each month of 1994. We could then determine a different continuous growth rate for each month, and in effect model each month separately.

6.4.2 The Power of Programmer-Defined Functions

If it were not possible for programmers to define their own functions, it would be practically impossible to write large programs. No programming language can possibly provide every function that every programmer might someday need. A language that allows users to create new functions is an *extensible* language—the entire community of users of that language can create and share new functions, allowing them to build on one another's work.

Developing a function can require quite a bit of effort. For example, we devoted this entire chapter to creating an implementation for the function `dailyPop`. You may not fully understand how we derived or implemented the function—or you may forget by tomorrow—but it will never be very difficult to *use* the function. It is always easy to use a function no matter how complicated the implementation is.

If you want to know the population of the United States two days into 1994, you need only evaluate

```
In[40]:= dailyPop[2]

Out[40]= 2.5918 × 10^8
```

(6.40)

But think about what would be involved if you had access to the analysis that we did in this chapter but *not* to the programmer-defined functions that we created. In that case, you'd have to evaluate

```
In[41]:= 259167000 * Exp[2/365 * .00948923]

Out[41]= 2.5918 × 10^8
```

(6.41)

instead. It is much easier to remember how to use `dailyPop` than it is to remember and continually retype the expression in Example 6.41.

If you want to share your functions with others—or if you want to keep track of what your own functions do—it is important to document them. To document a function, write down everything it is necessary to know in order to use it: the name of the function, the number of parameters expected by the function, and what the function computes with those parameters.

For example, the documentation for `dailyPop` might look like

- `dailyPop[d]` returns the estimated population of the United States *d* days into 1994. For example, `dailyPop[5]` returns the population on midnight after January 5, and `dailyPop[5.5]` returns the population on noon of January 6.

Similarly, the documentation for `continuous` might be

- `continuous[p,r,n]` returns the value after *n* years of *p* dollars invested at *r*% annual interest, compounded continuously. For example, `continuous[100, .06, 1.5]` returns the value after 1.5 years of $100 invested at 6% annual interest, compounded continuously.

Documentation such as this is crucial. Without it, most of the benefits of programmer-defined functions are lost because the potential user of a function is forced to read and understand the implementation to find out what the function does. Well-written documentation clearly describes *what* the function does, but needs not bother with *how* the function does it.

6.4.3 Clearing Variables

Mathematica treats a variable symbolically if it has never been given a value via assignment. Once a variable has been assigned, it no longer stands for itself in symbolic calculations.

For example, the variable `simpleGrowth` was assigned a value in Example 6.17. If we now attempt to use it in a symbolic calculation, *Mathematica* complains.

```
In[42]:= Solve[simpleGrowth^2 == 5, simpleGrowth]

        General::ivar : 0.00953439 is not a
            valid variable
```

(6.42)

We can ask *Mathematica* to forget we ever assigned a value to `simpleGrowth` and to resume treating it symbolically by using the built-in function `Clear`.

```
In[43]:= Clear[simpleGrowth]
```

(6.43)

This process is called *clearing* a variable. We can now easily verify that *Mathematica* has resumed treating simpleGrowth symbolically.

```
In[44]:= Solve[simpleGrowth^2 == 5, simpleGrowth]

Out[44]= {{simpleGrowth → −√5}, {simpleGrowth → √5}}
```

(6.44)

An unwary attempt to use, as a symbol, a variable that has been given a numerical value can lead to confusing error messages. Although clearing a variable can solve such problems, it is best to avoid the problems altogether. You should get into the habit of using descriptive names such as simpleGrowth for variables to which you wish to assign values, and shorter names such as x for variables that you wish to use as mathematical unknowns.

6.5 Key Concepts

Symbolic data.   If a variable has not been given a value via assignment, *Mathematica* treats it as a mathematical unknown. Because symbols can be used in most of the same contexts as numbers, this increases the variety of expressions that can be manipulated with *Mathematica*. Once a variable has been given a value, it can no longer be used symbolically until it is cleared.

Built-in functions for symbolic expressions.   *Mathematica* provides a large variety of functions for manipulating symbolic expressions. We looked at four in this chapter—Expand, Factor, Solve, and Limit—but there are many more.

The power of programmer-defined functions.   By knowing the name of the function, the number of parameters that it expects, and the value that it computes, you can use a function without needing to know how it is implemented. A description of what value a function computes is often much simpler than a description of how the function's implementation works.

6.6 Exercises

6.1   Suppose you have $1000 to invest at 10%. Is it better to get simple interest for 25 years, interest compounded monthly for 24 years, or continuous interest for 23 years?

6.2   If you have $1000 to invest for five years, at most how much of it would you pay up front as a fee to get 6% continuous interest instead of 6% simple interest?

6.3 If you invest some money at 8% continuous interest for six years and end up with $793.44, how much did you initially invest?

6.4 If the population of the United States continues to grow exactly as it did in 1994, on what day and at what time (to the nearest month) will the population reach 400 million people? 800 million people?

6.5 Repeat Exercise 6.4 assuming that the continuous growth rate of the United States population is twice its actual value.

6.6 Implement a function `difference[p, r, n]` that takes as parameters an initial balance p, an interest rate r, and a number of years n and returns the difference between investing p at continuous interest compared with simple interest.

6.7 Implement a function `daysForIncrease[rate, pop]` that takes as parameters a continuous growth `rate` and returns the number of days after January 1, 1994, that the population will reach `pop`. (*Growth rate* is the term used for *interest rate* when talking about natural growth; *i.e.*, it is the *r* in Example 6.31.)

6.8 Implement a function `whatYear` that takes a population p as its parameter and returns the year during which the population of the United States will first exceed p. You may assume that p is at least as large as the population on January 1, 1994.

6.9 Implement a function `whatRate` that takes as parameters `pop` and `year`, where `pop` is the population of the United States on January 1 of `year`. You may assume that `year` is an integer greater than 1994. Your function should return the necessary continuous growth rate between January 1, 1994, and January 1 of the specified year in order to produce the specified population.

6.10 The sum of a series is usually written in mathematics by using summation notation. *Mathematica* provides a function `Sum` for doing this. For example, the sum of the squares of the first 100 positive integers is written

$$\sum_{n=1}^{100} n^2 \tag{6.45}$$

To compute this value in *Mathematica*, we can do

```
In[46]:= Sum[n^2, {n,1,100}]

Out[46]= 338350
```

(6.46)

The first parameter must be a symbolic expression containing the symbol whose range is specified in the second expression. Does the Sum function simply add up a lot of terms, or does it do something more sophisticated? Do some timing experiments to find out.

6.11   The Sum function can also be used to do symbolic summations by specifying bounds that are not numbers. Use this capability to find symbolic expressions for the sum of the first *n* integers, the sum of the first *n* squares of integers, and the sum of the first *n* powers of 2.

6.12   The Product function is analogous to the Sum function, but it finds the product of a series of terms. Find a symbolic expression for the product of the reciprocals of the first *n* powers of 2.

6.13   Create a function that uses Sum to solve the block-stacking problem of Chapter 4.

6.14   The Sum function will even deal with limits at infinity. For example,

```
In[47]:= Sum[1/2^n, {n,1,Infinity}]

Out[47]= 1
```
(6.47)

computes the sum of the reciprocals of the powers of 2 from 1 to infinity. Use this capability to find the sums of the reciprocals of the squares of *n* and of the cubes of *n*.

6.15   The well-known quadratic formula gives the general form of the two solutions to the equation

$$ax^2 + bx + c = 0 \qquad (6.48)$$

Use the Solve function to find the general form of the three solutions to the equation

$$ax^3 + bx^2 + cx + d = 0 \qquad (6.49)$$

6.16   It is possible to take limits at infinity by specifying the bound of the limit as Infinity. Show that infinite extension is possible in the block-stacking problem of Chapter 4 by taking the limit of the sum of Series 4.1 as the number of terms approaches infinity.

6.17   Suppose we stacked the blocks in the block-stacking problem of Chapter 4 by using a series of 1/200, 1/400, 1/600, and so on. In other words, the extension at each step is only one hundredth as great as in Series 4.1. Create a function that computes the extension that can be attained with *n* blocks.

6.18   Show that even with the stacking approach of Exercise 6.17, an infinite amount of extension is possible.

6.19   Implement a function that computes the ratio between the extensions of the stacks in Chapter 4 and Exercise 6.17.

6.20   Implement a function that takes as its parameter the desired extension and computes the number of blocks required to achieve it, using the stacking approach of Chapter 4.

7

Ballistic Trajectories: Scientific Visualization

A baseball batted from home plate, a golf ball struck from the first tee, and a cannonball fired from a pirate ship all follow ballistic trajectories as they fly through the air. In each case, a projectile is given an initial velocity and is thereafter affected only by the downward force of gravity and by the retarding force of air resistance.

The shape of a ballistic trajectory is determined by the angle and magnitude of the initial velocity. If we assume that the projectile is fired exactly from ground level in the absence of wind, a projectile fired straight up will fall straight down while a projectile fired straight out will hit the ground immediately. In between lie a variety of different behaviors.

During World War II, the Ballistics Research Laboratory (BRL) of the United States Army was responsible for producing the firing tables used to aim artillery pieces and bombs. A gunner in the field, for example, would estimate the distance to the target, the wind speed and direction, and the air density (which in turn depended on such factors as temperature, humidity, and altitude). He would then consult the firing table appropriate to the combination of gun and ammunition he was using to determine the direction and angle at which to aim his gun.

To determine each entry in such a table, the ballistic trajectory corresponding to a particular combination of gun, ammunition, distance, wind condition, and air density had to be found. Finding a single trajectory required, among other calculations, on the order of 750 multiplications of ten-digit numbers. A skilled person—called, in those days, a computer—working with desk calculators required almost 20 hours to do these calculations. There were approximately 3000 trajectories in a typical firing table, which meant that it would have required almost seven years of continuous effort for one person to produce a single firing table.

In addition to nearly 200 human computers, almost all of them women recruited from college campuses and the Women's Army Corp, the BRL also had access to two differential analyzers. Differential analyzers were mechanical computers designed for solving the types of equations that arise in calculating ballistic trajectories. The larger of the two analyzers weighed almost 100 tons and could find a trajectory in about 20 minutes.

Every time a new type of gun, bomb, or shell was designed, the BRL had to calculate a firing table for it. During the second week of August 1944, for example, the BRL completed 14 firing tables, had 66 in progress, and had an even larger backlog on which work had not yet begun. Six requests for new tables were being received each day.

Desperate for a way to speed up the production of firing tables, in June 1943 the BRL contracted with the Moore School of Electrical Engineering at the University of Pennsylvania to design and build what became ENIAC, the first fully electronic digital computer. When it was completed, ENIAC could compute a single trajectory in a few seconds. Ironically, ENIAC was not completed until November 1945, several months after the war ended.

Our goal in this chapter is to find ways to visualize how a projectile's trajectory depends on the angle and magnitude of its initial velocity. To keep matters simple, we will ignore the effects of air resistance. Besides greatly simplifying the mathematics, the absence of a possible crosswind confines the motion of the projectile to a vertical plane.

7.1 Ballistic Motion

The key step in developing a mathematical model of ballistic motion is to realize that the two-dimensional trajectory of a projectile can be divided into its horizontal and vertical components. We can analyze and model these one-dimensional components in isolation, and then recombine them in the end to obtain a model for the overall trajectory. The advantage of this approach is that it is easier for us to work in one dimension than in two.

The initial velocity of the projectile can be broken down in the same way. Suppose that the projectile starts out at a speed of v m/sec at an angle of θ radians from the horizontal, as diagrammed in Figure 7.1. This velocity can be divided into a horizontal velocity of v_h m/sec and a vertical velocity of v_v m/sec. Notice that v_h and v_v are the legs and v is the hypotenuse of a right triangle with an interior angle of θ.

From basic trigonometry we know that

$$\frac{v_h}{v} = \cos \theta \tag{7.1}$$

$$\frac{v_v}{v} = \sin \theta \tag{7.2}$$

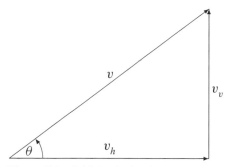

Figure 7.1   Initial velocity (v) broken into horizontal (v_h) and vertical (v_v) components

and by simplifying we find that

$$v_h = v \cos \theta \tag{7.3}$$

$$v_v = v \sin \theta \tag{7.4}$$

Given the horizontal and vertical components of the initial velocity, we can now determine the horizontal and vertical positions of the projectile as functions of time.

The horizontal motion is the easiest to deal with because there is no external force acting to retard the projectile's motion in that direction. Gravity only operates vertically, and we are ignoring the effects of air resistance. This means that the projectile will move at a constant speed of v_h in the horizontal direction, and its position as a function of time is thus $v_h t$ or

$$vt \cos \theta \tag{7.5}$$

To model vertical motion gravity must be taken into account. If the projectile were not retarded by gravity, its position at time t would be $v_v t$. Conversely, if the projectile had no initial vertical velocity (and the ground didn't get in the way) its position at time t would be $-\frac{1}{2}gt^2$. (The constant g stands for acceleration due to gravity.) Putting these two expressions together, the vertical position of the projectile as a function of time is $v_v t - \frac{1}{2}gt^2$ or

$$vt \sin \theta - \frac{1}{2}gt^2 \tag{7.6}$$

We can now compute the x- and y-coordinates of the projectile's position at any point of its flight by plugging the appropriate values into Formulas 7.5 and 7.6. Because we are assuming that the projectile stops as soon as it hits the ground, we must be careful not to apply either formula after that point in time.

The projectile will hit the ground when its vertical position is zero, which *Mathematica* can tell us occurs twice.

```
In[7]:= Solve[v*t*Sin[theta] - (1/2)*g*t^2 == 0, t]

Out[7]= {{t → 0}, {t → 2vSin[theta]/g }}
```
(7.7)

The second of the two solutions tells us that the projectile hits the ground at time

$$\frac{2v \sin \theta}{g}$$
(7.8)

after which it stops moving.

7.2 Scientific Visualization

The problem we are studying in this chapter is qualitatively different from those that we have encountered up to now. All of our previous problems have required us to come up with a number as a result. Although we have used a variety of different models, methods, and implementation techniques, every solution has come down in the end to evaluating algebraic expressions to obtain numbers. We have done everything in *Mathematica*, but with a few exceptions we could have equally well used pen and paper, a pocket calculator, or a conventional programming language such as C or Fortran. In other words, the computational method was essentially independent of the implementation strategy.

Although the formulas that we have just derived could certainly be used to calculate numbers, that is not our primary interest here. Our goal is to construct visualizations to help us understand how the initial velocity of a projectile affects its ballistic trajectory. It will not work to decide what sorts of visualizations we would like and to then go off and try to implement them in an arbitrary language. Unlike support for evaluating arithmetic expressions, which is essentially universal across languages, the support that languages provide for visualization varies widely.

We must pick a language with good support for visualization and then temper our ambitions by what is possible in that language. As we will shortly see, *Mathematica*'s simple but powerful two-dimensional plotting and animation capabilities are nicely matched to the problem at hand. What we are about to do would be extremely tedious or impossible with pen, paper, and calculator; it would require specialized knowledge and hours or days of effort with Fortran or C; it will take only minutes in *Mathematica*.

The computational method that we will employ, then, is highly *Mathematica*-dependent. We will implement as *Mathematica* functions the three formulas that we derived in Section 7.1, and we will use them together with *Mathematica*'s visualization capabilities to produce a variety of two-dimensional plots and animations. Along the way we hope to shed as much light on *Mathematica* as we do on ballistic trajectories.

7.3 Motion Functions

In this section we will lay the groundwork for our visualization efforts by creating implementations of *Mathematica* functions named `horizontal`, `vertical`, and `duration`. They will all assume that a projectile has been fired at time zero, from coordinate $(0, 0)$, at an angle of between 0 and 90 degrees ($\frac{\pi}{2}$ radians) from the horizontal.

   • `horizontal[v, θ, t]` returns the horizontal coordinate at time t seconds, assuming the projectile was fired at angle θ radians with initial speed v m/sec. Its results are valid only for values of t between 0 and the time at which the projectile strikes the ground.

   • `vertical[v, θ, t]` returns the vertical coordinate at time t seconds, assuming the projectile was fired at angle θ radians with initial speed v m/sec. Its results are valid only for the same time values as `horizontal`.

   • `duration[v, θ]` returns the time at which a projectile fired at angle θ radians with initial speed v m/sec will hit the ground.

As usual, implementing and testing these three functions will provide us with a learning opportunity. We will see how *Mathematica* treats the constant π, and we will learn what it means to incorporate a free variable into the body of a function.

7.3.1 Three Functions

The implementation of each of the three functions follows directly from the corresponding formula in Section 7.1. Formula 7.5 inspires the implementation of `horizontal`,

```
In[9]:= horizontal[v_, theta_, t_] :=
              v * t * Cos[theta]
```
(7.9)

Formula 7.6 inspires `vertical`,

```
In[10]:= vertical[v_, theta_, t_] :=
              v * t * Sin[theta] - 1/2 * g * t^2
```
(7.10)

and Formula 7.8 inspires `duration`.

```
In[11]:= duration[v_, theta_] :=
              2 * v * Sin[theta] / g
```
(7.11)

Notice once again how easy it is to move from the mathematical definition of a function to the *Mathematica* version.

7.3.2 The Value of π

Before we try to create visualizations based on these functions, it is a good idea to test them out to see if they work sensibly. For example, if we fire our projectile straight up (at 90 degrees), its horizontal coordinate should be zero at all times.

```
In[12]:= horizontal[100, Pi/2, 10]

Out[12]= 0 m
```
(7.12)

Mathematica expects the angle parameters to such trigonometric functions as Sin and Cos to be measured in radians, which is why we have given Pi/2 instead of 90 as the second parameter to horizontal. To make it easier to express angles in radians, *Mathematica* allows the programmer to specify π by entering Pi. As we will shortly see, Pi is not a variable. *Mathematica* usually treats Pi symbolically, but at other times replaces it with a floating-point approximation to π.

Let's see what happens when we fire our projectile at 45 degrees ($\frac{\pi}{4}$ radians).

```
In[13]:= horizontal[100, Pi/4, 10]

Out[13]= 500√2 m
```
(7.13)

Notice that the radical $\sqrt{2}$ appears in the answer. In this case, *Mathematica* is treating Pi symbolically. To evaluate Example 7.13, *Mathematica* must find the cosine of $\frac{\pi}{4}$,

```
In[14]:= Cos[Pi/4]

Out[14]=  1
         ───
         √2
```
(7.14)

for which *Mathematica* gives back an exact answer. If we fire our projectile at 22.5 degrees ($\frac{\pi}{8}$ radians), however,

```
In[15]:= horizontal[100, Pi/8, 10]

Out[15]= 1000 Cos[π/8] m
```
(7.15)

Mathematica declines to simplify the cosine expression and leaves it unreduced (and thus exact) instead of giving a floating-point approximation. As always, we can obtain a floating-point approximation for any of these expressions by using N.

```
In[16]:= N[horizontal[100, Pi/4, 10]]

Out[16]= 707.107 m
```

(7.16)

This tells us that a projectile fired at 45 degrees at 100 m/sec will be approximately 707 meters downrange after ten seconds, assuming that it hasn't hit the ground yet.

7.3.3 Free Variables

Let's continue by experimenting with `vertical`.

```
In[17]:= N[vertical[100, Pi/4, 10]]

Out[17]= 707.107 - 50g m
```

(7.17)

Even though we have used N to eliminate the radical that would otherwise appear in the result, the symbol g appears. Recall that *g* stands for acceleration due to gravity, which is approximately 9.8 m/sec² on earth. It appears in the answer because it appears in the definition of `vertical`, whose body is

```
v * t * Sin[theta] - 1/2 * g * t^2
```

The body of `vertical` contains four variables: v, theta, t, and g. When we created `vertical`, we chose to make v, theta, and t its formal parameters. Let's trace through what happens when `vertical` is called according to the steps that we detailed in Section 5.3.3.

When `vertical` is called, as in

```
In[18]:= vertical[100, Pi/4, 10]

Out[18]= 500√2 - 50g m
```

(7.18)

the values of the actual parameters (100, $\frac{\pi}{4}$, and 10) are used to replace the corresponding formal parameters (v, theta, and t) in the body of the function. This yields the expression

```
In[19]:= 100 * 10 * Sin[Pi/4] - 1/2 * g * 10^2

Out[19]= 500√2 - 50g m
```

(7.19)

which *Mathematica* evaluates as if it had been typed in. Because g has never been given a value, *Mathematica* treats it symbolically, which is why g appears symbolically in the result of Example 7.18.

The variable g within the body of vertical is called a *free variable* because it is not one of the formal parameters of vertical. When a function call is evaluated, free variables in the function's body are looked up in the global variable list. If we do not intend for a free variable to be treated symbolically, we must assign it a value *before* calling the function.

To make vertical work properly, then, we must assign a value to g,

```
In[20]:= g = 9.8

Out[20]= 9.8 m/sec²
```

(7.20)

which adds g to the global variable list. We can now call vertical and get a numerical answer.

```
In[21]:= N[vertical[100, Pi/4, 10]]

Out[21]= 217.107 m
```

(7.21)

After ten seconds, then, the projectile is approximately 217 meters high.

The function duration also contains g as a free variable, and by assigning a value to g we ensure that it will work properly as well.

```
In[22]:= N[duration[100, Pi/4]]

Out[22]= 14.4308 sec
```

(7.22)

The flight of the projectile will last a little more than 14 seconds.

7.3.4 Free Variables Versus Parameters

There are four ways that we could have treated g when we created vertical (and duration). First, we could have incorporated the constant value directly into the function definition by defining vertical as

```
In[23]:= vertical[v_, theta_, t_] :=
                v * t * Sin[theta] - 1/2 * 9.8 * t^2
```

(7.23)

Defined this way, vertical is a three-parameter function that is completely independent of the value of g on the global variable list. It can be invoked exactly as in Example 7.21, and it always uses 9.8 m/sec² as its gravitational constant.

Although this would have been the simplest approach, it would have meant that vertical would work only for problems posed on the surface of the earth. If we were interested in the trajectory of a golf ball hit on the surface of the moon, where the gravitational constant is different, we would have to create a new function containing a different constant.

Second, we could have made g the fourth parameter to `vertical`, as in

```
In[24]:= vertical[v_, theta_, t_, g_] :=
               v * t * Sin[theta] - 1/2 * g * t^2
```
(7.24)

Defined this way, `vertical` is a four-parameter function to which the value of g must be supplied in each function call. For example, it can be invoked as

```
In[25]:= N[vertical[100, Pi/4, 10, 4.9]]

Out[25]= 462.107 m
```
(7.25)

to determine what would happen if earth's gravitational constant were half as large as it is.

Although this makes `vertical` more flexible, it also makes it less convenient to use. If we expect that almost all of our computations with `vertical` will use the same value for g, as will be the case in this chapter, it would rapidly become tiresome to keep passing 9.8 as a parameter over and over.

The third option involves the use of *default values*. In *Mathematica*, a programmer can give a default value to a formal parameter of a function. In the implementation of `vertical` below, the parameter g is given a default value of 9.8.

```
In[26]:= vertical[v_, theta_, t_, g_:9.8] :=
               v * t * Sin[theta] - 1/2 * g * t^2
```
(7.26)

Defined this way, `vertical` can be invoked with either three or four parameters. If it is given three parameters, as in Example 7.21, the default value of 9.8 is used for g. If it is given four parameters, as in Example 7.25, the value of g specified in the function call (4.9 in this case) is used.

This third option is an improvement on both the first and second options. As in the previous two options, no free variable is involved. If we wish to do our calculations relative to earth's gravitational constant, we need only specify three parameters. Only if we wish to do our calculations relative to another gravitational constant do we need to specify the value of g as a parameter.

The fourth option, which is the version we will use for the remainder of this chapter, is to make g a free variable as we did originally in Example 7.10. Leaving g as a free variable in both `vertical` and `duration` has an important benefit that is missing from the previous three options: we can guarantee that both `vertical` and `horizontal` always use the same value for g. Since both functions use the value of g from the global variable list, changing that value will cause both functions to use the changed value.

7.3.5 Specifying Functions with Free Variables

The body of a function will always contain some number of variables. The choice of which to make parameters and which to leave free is only partly a matter of taste. Functions containing free variables are harder to document and understand than are functions without them. This is because the behavior of a function with free variables depends on more than just the values of its parameters. The specifications of vertical and duration that appear in Section 7.3.1 are, in fact, incomplete. They do not explain how both functions rely on the value of the variable g. The specification of duration, for example, should read

• duration[v, θ] returns the time in seconds at which a projectile fired at angle θ radians with initial speed v m/sec will hit the ground. The free variable g specifies the acceleration due to gravity in m/sec^2.

You should include free variables in function definitions very sparingly. The values of a function's free variables should change rarely, if at all. It is perhaps best to think of them as free *constants*.

7.4 Two-Dimensional Plots

You are probably familiar with simple two-dimensional plots. Such plots show how the value of a *dependent expression* changes as the value of an *independent variable* ranges across an interval. For example, if we plot how the speed of a car varies with time, then the car's speed is the dependent expression and time is the independent variable. By convention, the independent variable is plotted along the horizontal axis and the dependent expression is plotted along the vertical axis.

Mathematica's Plot function produces such plots. Although it has many variations, in its simplest form Plot takes two parameters. The first is the dependent expression and the second specifies the interval through which the independent variable ranges. All of the variables contained in the dependent expression, with the exception of the independent variable, must have been assigned values previously.

As a concrete example, let's plot how the height of a projectile fired at 100 m/sec at an angle of 45 degrees varies with time. In this example time (t) is the independent variable, and we will let it range between 0 and 20 seconds. The dependent expression will be vertical[100, Pi/4, t], which maps time in seconds to height in meters. If we use *Mathematica* to evaluate our dependent expression,

```
In[27]:= vertical[100, Pi/4, t]

Out[27]= 50√2t − 4.9t² m
```
(7.27)

we can verify that it indeed contains t as its only symbolic variable.

We can generate our plot with

```
In[28]:= Plot[vertical[100, Pi/4, t], {t, 0, 20}]

Out[28]= (See Figure 7.2a)
```
(7.28)

This asks *Mathematica* to plot the dependent expression for values of t between 0 and 20 seconds. Note carefully the format of the range specification for the independent variable t. Similar range specifications appear in all of the visualization functions in *Mathematica*.

As you can see in Figure 7.2a, it is unnecessary to specify what range of the vertical axis is to be displayed because *Mathematica* automatically figures that out. There is an annoying problem with the plot however: It plots the behavior of the dependent expression for times greater than about 14.5 seconds, which is when the projectile hits the ground. As a result, we see the nonphysical portion of the height curve.

There are several ways to solve this problem. One way is to include an extra parameter in the Plot function call that specifies what the range of the vertical axis is to be.

```
In[29]:= Plot[vertical[100, Pi/4, t],
           {t, 0, 20},
           PlotRange->{0, 260}]

Out[29]= (See Figure 7.2b)
```
(7.29)

The third parameter specifies that the vertical axis is to range from 0 to 260, which eliminates the portion of the curve below the horizontal axis. PlotRange is an example of an *optional parameter* to Plot. In *Mathematica*, the values of optional parameters are specified by giving the name of the parameter, followed by the -> operator, followed by the value of the parameter.

A better way to solve the problem is to allow t to range only up to the time of impact, which duration will compute for us.

```
In[30]:= Plot[vertical[100, Pi/4, t],
           {t, 0, duration[100, Pi/4]}]

Out[30]= (See Figure 7.2c)
```
(7.30)

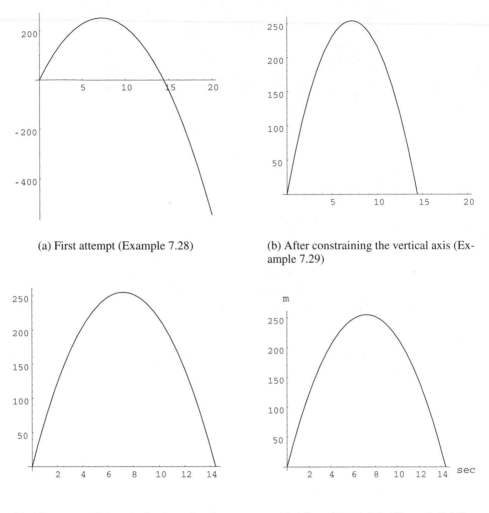

(a) First attempt (Example 7.28)

(b) After constraining the vertical axis (Example 7.29)

(c) After constraining the horizontal axis instead (Example 7.30)

(d) After adding labels (Example 7.31)

Figure 7.2 Plots of projectile height in meters as a function of time in seconds; initial angle of 45 degrees and speed of 100 m/sec

This is an improvement over the plot produced by Example 7.29. The entire horizontal axis is now used. Notice also that we do not need to specify the range of the vertical axis; *Mathematica* automatically rescales the vertical axis to accompany the portion of the curve that is being graphed.

It is easy to specify, as part of the `Plot` expression, axis labels. To illustrate, let's repeat Example 7.30.

```
In[31]:= Plot[vertical[100, Pi/4, t],
            {t, 0, duration[100, Pi/4]},
            AxesLabel->{"sec", "m"}]                          (7.31)

Out[31]= (See Figure 7.2d)
```

We have used the optional parameter AxesLabel to specify that the horizontal axis is to be labeled with sec and that the vertical axis is to be labeled with m. We will label the axes in the remainder of our plots with appropriate units.

It is also possible to specify titles for plots, but we will not do that. The information that would otherwise appear in a title will be included in the caption that accompanies each plot.

7.5 Lists

Before we consider other ways of visualizing data, we will pause to discuss the essentials of *lists* in *Mathematica*. We have already encountered lists in two different contexts. In Chapter 6, we saw that the Solve function produces lists of solutions (Example 6.34). In the last section, we used lists to specify both the range of the vertical axis (Example 7.29) and the axis labels (Example 7.31) in the Plot function. We will be using lists more extensively in the remainder of this chapter.

A list in *Mathematica* is a sequence of values. To specify a list to *Mathematica*, we enclose expressions inside of braces and separate them with commas. Thus

```
In[32]:= {2+2, 2-2, 2*2, 2/2}
                                                              (7.32)
Out[32]= {4, 0, 4, 1}
```

is a list of four integers. Each number in the result list is the value of the corresponding expression in the input list. Notice that when *Mathematica* displays a list, it encloses it in braces.

As noted above, we created two different lists in the previous section:

1. In Example 7.28, we specified that the independent variable t ranged from 0 to 20 by giving the list {t, 0, 20} as the second parameter to Plot.

2. In Example 7.31, we specified that the labels of the horizontal and vertical axes were to be sec and m, respectively, by giving the list {"sec", "m"} as the value of Plot's optional parameter AxesLabel.

The simplest way to create a list is to enumerate its components, as we have in every instance to this point. A more powerful, and often more useful, technique is to give a *rule* that *Mathematica* can use to create a list. Using this technique, we can create a long or complicated list without having to spell out each element of the list individually.

To create a list in this way, we use the `Table` function. For example, suppose that we would like to create a list of the squares of the odd integers between 1 and 21.

```
In[33]:= Table[n^2, {n, 1, 21, 2}]

Out[33]= {1, 9, 25, 49, 81, 121, 169, 225, 289, 361, 441}
```
(7.33)

This asks *Mathematica* to create a list of the values of the expression n^2 as n ranges from 1 to 21 in steps of 2. Thus, we are asking *Mathematica* to create a list consisting of 1^2, 3^2, 5^2, and so on up to 21^2.

The form of the `Table` function that we illustrate above takes two parameters. The first parameter is an expression (which we will call exp), and the second parameter is a list of the following four components:

1. a symbol (which we will call var);
2. the beginning of a numeric range (which we will call left);
3. the end of a numeric range (which we will call right); and
4. an increment value (which we will call inc).

Mathematica creates a list by first evaluating exp under the assumption that var = left, then under the assumption that var = left + inc, then under the assumption that var = left + 2 · inc), and so on for each positive n such that left + n · inc ≤ right.

7.6 Multiple-Curve Plots

Sometimes it is interesting to graph more than one dependent expression on the same plot. This makes sense only when both expressions depend on the same independent variable (so that they can share the same horizontal axis) and when the values that they produce are expressed in identical units (so that they can share the same vertical axis). Suppose, for example, that we wish to plot the vertical position of a projectile fired at 100 m/sec for initial angles of 30, 45, and 60 degrees ($\frac{\pi}{6}$, $\frac{\pi}{4}$, and $\frac{\pi}{3}$ radians).

One way to do this is to give `Plot` a *list* of dependent expressions as its first parameter. The range of the independent variable is specified in exactly the same way as before.

```
In[34]:= Plot[{vertical[100, Pi/6, t],
              vertical[100, Pi/4, t],
              vertical[100, Pi/3, t]},
             {t, 0, duration[100, Pi/3]},
             AxesLabel->{"sec", "m"}]

Out[34]= (See Figure 7.3a)
```
(7.34)

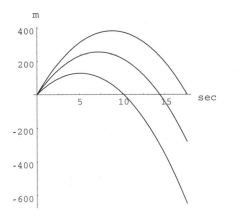

(a) Using `Plot` with a list of dependent expressions (Example 7.34)

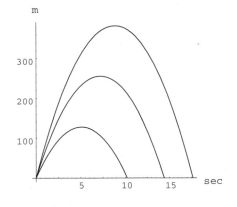

(b) Using `Show` with a list of plots (Example 7.38)

Figure 7.3   Plots of projectile height in meters as a function of time in seconds. In each plot the curves represent, from top to bottom, initial angles of 60, 45, and 30 degrees. The initial speed in each case is 100 m/sec.

The plot in Figure 7.3a shows that the projectile fired at 60 degrees goes highest and stays airborne longest, and that the projectile fired at 30 degrees stays lowest and hits the ground first. Unfortunately, this plot is displaying the nonphysical portions of two of the curves. This is because the same horizontal scale is used for all three curves, and two of the projectiles hit the ground before the third.

There are several ways to solve this problem, but the simplest is to create a list of the three curves and to then use the `Show` function to put the three curves into one plot. We will create each of the three plots just as we did in Example 7.30, but we will use assignment in each case to save the plot.

We first create the 30 degree plot, and call it `plot30`.

```
In[35]:= plot30 = Plot[vertical[100, Pi/6, t],
                    {t, 0, duration[100, Pi/6]}]

Out[35]= - Graphics -
```
(7.35)

We next create the 45 degree plot (`plot45`),

```
In[36]:= plot45 = Plot[vertical[100, Pi/4, t],
                    {t, 0, duration[100, Pi/4]}]

Out[36]= - Graphics -
```
(7.36)

and finally the 60 degree plot (`plot60`).

```
In[37]:= plot60 = Plot[vertical[100, Pi/3, t],
                  {t, 0, duration[100, Pi/3]}]

Out[37]= - Graphics -
```
(7.37)

Now that we have created and saved the three plots, we can use `Show` function to combine them.

```
In[38]:= Show[{plot30, plot45, plot60},
             AxesLabel->{"sec", "m"}]

Out[38]= (See Figure 7.3b)
```
(7.38)

The first parameter to `Show` is a list of the plots that we wish to combine into one. `Show` takes the same sorts of optional parameters as `Plot`; here we use the `AxesLabel` option to specify the labels of the axes in the combined plot. Notice that the vertical axis of the combined plot does not extend below zero, which was also true of the vertical axes of each of the three constituent plots.

7.7 Parametric Plots

None of the plots that we have produced to this point allow us to directly visualize the trajectory of a projectile. The plots have shown us only the vertical position of various projectiles as functions of time. What we would like to see is the actual curve that a projectile traces out as it moves through the atmosphere. This plot would show how horizontal and vertical position vary as functions of time. With two dependent expressions and one independent variable, we could try applying the techniques of the previous section by graphing two curves in one plot.

```
In[39]:= Plot[{horizontal[100, Pi/4, t],
               vertical[100, Pi/4, t]},
              {t, 0, duration[100, Pi/4]},
              AxesLabel->{"sec", "m"}]

Out[39]= (See Figure 7.4a)
```
(7.39)

The result, shown in Figure 7.4a, isn't very satisfactory. While we can read off the horizontal and vertical positions of the projectile for any instant of time, it really doesn't help us visualize the trajectory. The plot that we really want would graph horizontal position (on the horizontal axis) against vertical position (on the vertical axis). The problem is that this doesn't leave an axis for the independent variable.

We can solve this problem by using a *parametric* plot, which is another way to plot two dependent expressions against the same independent variable. In a parametric plot, one dependent expression is plotted relative to the horizontal axis, the second dependent expression is plotted relative to the vertical axis, and the independent variable does not appear explicitly. This is easier to illustrate with an example than to explain with words.

We can ask *Mathematica* to create a parametric plot by using the same syntax as in the multiple-curve plot of Example 7.39, but with one change: we must use the `ParametricPlot` function instead of the `Plot` function. The first parameter is a list consisting of the dependent expression to be plotted against the horizontal axis and the dependent expression to be plotted against the vertical axis.

```
In[40]:= ParametricPlot[{horizontal[100, Pi/4, t],
                          vertical[100, Pi/4, t]},
                         {t, 0, duration[100, Pi/4]},
                         AxesLabel->{"m", "m"}]

Out[40]= (See Figure 7.4b)
```

(7.40)

The resulting plot, in Figure 7.4b, shows the trajectory of a projectile fired at 100 m/sec and an angle of 45 degrees. Be careful when interpreting the plot to notice

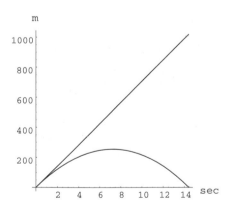

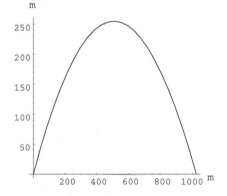

(a) Horizontal and vertical position (Example 7.39)

(b) Parametric plot with unequal scales (Example 7.40)

Figure 7.4 Two attempts to visualize the trajectory of a projectile fired at 45 degrees and 100 m/sec. On the left the straight line is horizontal position as a function of time, and the curved line is vertical position as a function of time. On the right, horizontal position is plotted against the horizontal axis and vertical position against the vertical axis over the duration of the projectile's flight.

that the two axes do not use the same scale, which means that the actual trajectory is considerably less steep than is pictured.

What is *not* shown in the plot is the time at which the projectile reaches each point on its trajectory, or even the direction in which the projectile travels. Based on our knowledge of the physics of the problem, we can supply enough of this information to allow an informed interpretation of the plot. The projectile begins at the origin, travels up and right to the apex of the curve, and then descends back to the horizontal axis, at which point it hits the ground.

7.8 Animation

Although we can now do a good job of visualizing the trajectory of a projectile fired at a given angle and speed, we still do not have a very good way of visualizing how different choices for angle and speed affect the trajectories. To do that, we need to be able to compare trajectories to one another.

One way to do this is to put more than one trajectory on the same plot. Just as we did for non-parametric plots in Example 7.38, we can do this by creating each curve individually and then combining them into one plot with the Show function.

For example, perhaps we would like to compare the trajectories of three projectiles, each fired at 100 m/sec but at differing angles of 30, 45, and 60 degrees. The first step is to create and save the three parametric plots that we want to compare.

We first create the 30 degree trajectory and save it as pplot30.

```
In[41]:= pplot30 =
         ParametricPlot[{horizontal[100, Pi/6, t],
                         vertical[100, Pi/6, t]},
                        {t, 0, duration[100, Pi/6]}]

Out[41]= - Graphics -
```
(7.41)

Next, we create the 45 degree trajectory and save it as pplot45.

```
In[42]:= pplot45 =
         ParametricPlot[{horizontal[100, Pi/4, t],
                         vertical[100, Pi/4, t]},
                        {t, 0, duration[100, Pi/4]}]

Out[42]= - Graphics -
```
(7.42)

Finally, we create the 60 degree trajectory and save it as pplot60.

```
In[43]:= pplot60 =
        ParametricPlot[{horizontal[100, Pi/3, t],
                       vertical[100, Pi/3, t]},
                      {t, 0, duration[100, Pi/3]}]

Out[43]= - Graphics -
```

(7.43)

We can now combine the three curves into a single plot by using the Show function.

```
In[44]:= Show[{pplot30, pplot45, pplot60},
            AxesLabel->{"m", "m"},
            AspectRatio->Automatic]

Out[44]= (See Figure 7.5)
```

(7.44)

The middle curve from Figure 7.5, which is the trajectory of a projectile fired at 45 degrees, also appears in Figure 7.4b. The curve in Figure 7.5 looks shallower because we have asked *Mathematica* to use the same scales on the horizontal and vertical axes. We did this by specifying Automatic to be the value of the optional parameter AspectRatio in Example 7.44. It makes sense to do this whenever the two axes are measured in the same units.

You should compare Figure 7.5 with Figure 7.3b, where we plotted three height curves against the same independent variable. The two figures both suffer from the same problem. Although they show us graphs for three different angles, they don't

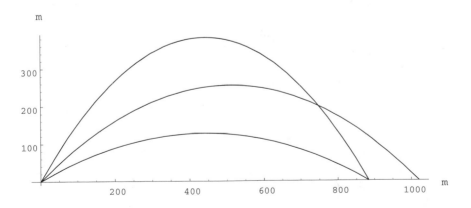

Figure 7.5   Three projectile trajectories in a parametric plot with equal scales (Example 7.44). From top to bottom, the curves represent initial angles of 60, 45, and 30 degrees. The initial speed in each case is 100 m/sec.

give us a full appreciation of what happens as the initial angle varies from 0 to 90 degrees.

Mathematica's animation capability will let us address this problem. An animation, whether in *Mathematica* or in a motion picture, consists of a sequence of discrete images called *frames*. By viewing the frames in rapid succession, we can achieve the illusion of continuous motion. The first step in producing a *Mathematica* animation, then, is to create a sequence of frames. In *Mathematica*, a frame is nothing more than a single plot.

Let's consider the problem of animating how the height curve for a projectile fired at 100 m/sec changes as the initial angle varies from 0 to 90 degrees. To do this, we will create a sequence of 13 plots in which the initial angle varies smoothly in steps of $\frac{\pi}{24}$ radians, beginning with an angle of zero radians in the first plot and ending with an angle of $\frac{\pi}{2}$ radians in the final plot.

We could do this by using the `Plot` function 13 times, much as we did three times in Examples 7.35–7.37, but this would be rather tedious. Fortunately, there is a much simpler way to create the frames. We will use the `Table` function that we discussed in Section 7.5 to create all 13 frames at once. Four of the frames appear in Figure 7.6.

```
In[45]:= Table[Plot[vertical[100, angle, t],
                {t, 0, duration[100, Pi/2]},
                PlotRange->{0, 550},
                AxesLabel->{"m", "m"}],                      (7.45)
            {angle, 0, Pi/2, Pi/24}]

Out[45]=  (See Figure 7.6)
```

Let's look at the first of the two parameters to `Table`.

```
Plot[vertical[100, angle, t],
    {t, 0, duration[100, Pi/2]},
    PlotRange->{0, 550},
    AxesLabel->{"m", "m"}]
```

The most important difference between this use of `Plot` and those from Examples 7.35–7.37 is that we do not specify a particular initial angle as the second parameter to `vertical`; we use the symbol `angle` instead. The effect of Example 7.45 is to evaluate the `Plot` expression above 13 times: the first time with `angle` = 0, the second time with `angle` = $\frac{\pi}{24}$, and so on until `angle` = $\frac{\pi}{2}$. This will result in a sequence of 13 frames in which the initial angle varies slightly from each frame to the next.

The other difference is that we have specified the value of the optional `PlotRange` parameter. Each of the frames in an animation must have identical horizontal and vertical axes. The range of the horizontal axis will be the same for each frame, but we must specify that the vertical axis should range from 0 to 550.

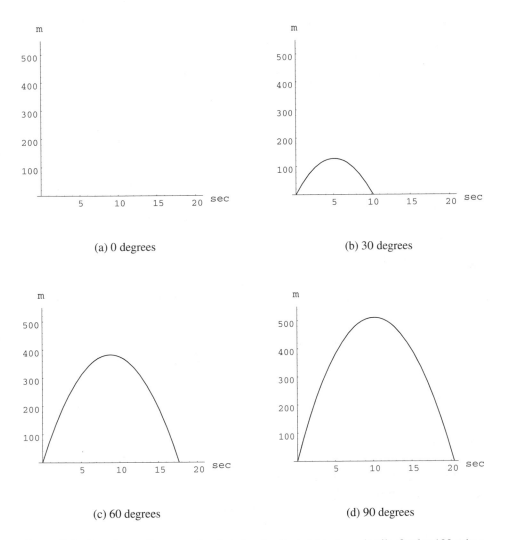

Figure 7.6   Four frames from an animation showing the height of a projectile, fired at 100 m/sec, as a function of time (Example 7.45). The animation shows a sequence of plots as the initial angle varies from 0 to 90 degrees.

If you are using a notebook interface to *Mathematica*, you will be able to select the sequence of 13 plots and then use a menu option to instruct *Mathematica* to produce an animation from them. When *Mathematica* begins running the animation, you will first see the curve for an angle to zero degrees, which is indistinguishable from the horizontal axis. You will then see the curve gradually become both higher and wider, which means that both the maximum height reached and the duration of flight increase with angle.

Whenever it runs an animation, *Mathematica* also displays an animation toolbar. You can use this toolbar to slow down or speed up the animation, to run it forwards or backwards, and to pause or single-step it.

We can animate parametric plots by using `Table` in exactly the same fashion. For example, we can animate the sequence of 100 m/sec trajectories that result as the initial angle increases from 0 to 90 degrees with

```
In[46]:= Table[
         ParametricPlot[{horizontal[100, angle, t],
                         vertical[100, angle, t]},
                        {t, 0, duration[100, Pi/2]},
                        PlotRange->{{0,1100},{0,550}},       (7.46)
                        AxesLabel->{"m", "m"}],
              {angle, 0, Pi/2, Pi/24}]

Out[46]=  (See Figure 7.7)
```

Once again, notice the two important differences between the first parameter to `Table`,

```
ParametricPlot[{horizontal[100, angle, t],
                vertical[100, angle, t]},
               {t, 0, duration[100, Pi/2]},
               PlotRange->{{0,1100},{0,550}},
               AxesLabel->{"m", "m"}]
```

and the uses of `ParametricPlot` in Examples 7.41–7.43 from which it is derived. We use the symbol `angle` in place of a numeric value, and we specify the value of the optional `PlotRange` parameter. In this case, we must specify the ranges of both the horizontal (0–1100) and the vertical (0–550) axes, since each individual plot would otherwise use conflicting values. This is different from all of our previous uses of `PlotRange`, in which we specified only the vertical range.

When you run this animation, you will see the trajectory curve becoming ever higher as the angle increases. However, you will also see the curve becoming wider until the middle of the animation, at which point it will shrink back to zero width. This reflects the fact that the distance the projectile travels is greatest when it is fired at 45 degrees.

7.9 Key Concepts

***Mathematica*'s treatment of** π**.** *Mathematica* provides the symbolic constant `Pi`, which it treats symbolically in rational number computations and numerically in floating-point computations. It is *not* a variable, as its value cannot be changed by assignment.

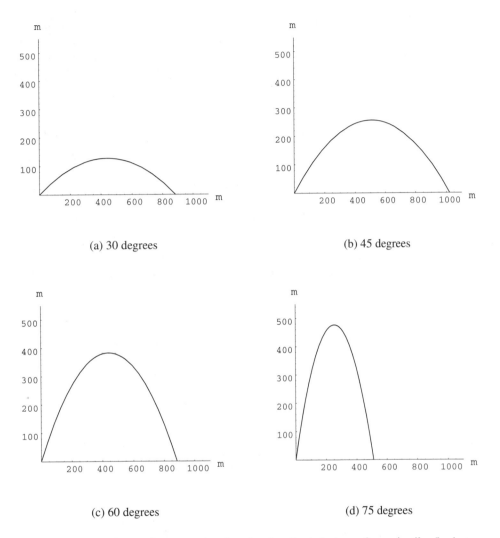

Figure 7.7   Four frames from an animation showing the trajectory of a projectile, fired at 100 m/sec, over the duration of its flight (Example 7.46). The animation shows a sequence of plots as the initial angle varies from 0 to 90 degrees.

Free variables.   When a variable appears in the body of a programmer-defined function, and it is not a parameter of a function, we call it a *free variable*. Each time a function call is evaluated, the values of the free variables are found by consulting the global variable list. This provides a second way of communicating data to a function, entirely independent of parameter passing. Free variables should be used sparingly, as they make it harder both to explain and understand exactly what a function does.

Optional parameters.   Many built-in *Mathematica* functions take optional parameters. The value of an optional parameter is specified by giving its name, followed by ->, followed by the desired value.

Lists.   Lists are sequences of values, and they are used in a number of contexts in *Mathematica*. A list can be created by explicitly specifying its members or by using the Table function to create a list from a rule.

Visualization in *Mathematica*.   *Mathematica* provides a variety of built-in functions for generating visualizations. In this chapter we used Plot to produce simple and multiple-curve, Show to combine multiple curves into a single plot, ParametricPlot to produce parametric plots, and Table to produce sequences of plots suitable for animations. Although we did everything in two dimensions, *Mathematica* also provides support for three-dimensional plotting and animation.

Power of programmer-defined functions.   Everything that we did in this chapter would have been much more difficult had we not exploited programmer-defined functions. We defined three such functions at the outset, and then used them in a number of different combinations to produce visualizations.

7.10 Exercises

7.1   Add titles to the plots from this chapter.

7.2   Experiment with constraining the two axes of the plots in this chapter to have the same scale. For which of the plots does it make sense to do this?

7.3   Create a two-dimensional plot showing how the length of the side of the square plot for each person on earth, as calculated in Chapter 2, varies with the earth's population.

7.4   Create a two-dimensional plot showing how the length of the side of the square plot for each person on earth, as calculated in Chapter 2, varies with the earth's area.

7.5   Create an animation showing how the length of the side of the square plot for each person on earth, as calculated in Chapter 2, varies with both the earth's area and its population.

7.6   Create a two-dimensional plot showing how the circumference of the earth in kilometers as calculated by Eratosthenes's method from Chapter 3 depends on the angle α of the sun. Assume a fixed distance d from Alexandria to Syene.

7.7   Create a two-dimensional plot showing how the circumference of the earth in kilometers as calculated by Eratosthenes's method from Chapter 3 depends on the distance d measured from Alexandria to Syene. Assume a fixed angle α of the sun.

7.8 Create an animation showing how the circumference of the earth in kilometers as calculated by Eratosthenes's method from Chapter 3 depends on both the angle α of the sun and the distance d from Alexandria to Syene.

7.9 Create a two-dimensional plot showing how the maximum achievable extension of the block stack from Chapter 4 depends on the number of blocks used.

7.10 Create a parametric plot showing the exact shape of a curve drawn through the upper-right corner of each block in a stack of 100 blocks, when stacked as in Chapter 4. Each block is 12 inches long and 1 inch thick.

7.11 Create an animation showing how the plot from Exercise 7.10 changes as the length of the blocks changes from 12 inches to 144 inches in increments of 12 inches.

7.12 Create a two-dimensional plot showing how the distance to the horizon as computed in Chapter 5 varies with the height of Kill Devil Hill, assuming a fixed earth radius.

7.13 Create a multiple-curve two-dimensional plot showing how the upper and lower bounds on the distance to the horizon as computed in Chapter 5 vary with the height of Kill Devil Hill, assuming a fixed earth radius. You should assume that the height of the hill is known to the nearest foot and that the radius of the earth is known to three significant digits.

7.14 Create a two-dimensional plot showing how the distance to the horizon as computed in Chapter 5 varies with the radius of the earth, assuming a fixed hill height.

7.15 Create an animation showing how the distance to the horizon as computed in Chapter 5 varies with both hill height and earth radius.

7.16 Create a multiple-curve two-dimensional plot comparing how the population of the United States will grow between 1994 and 2044 if the continuous growth rate is exactly as derived in Chapter 6, half the rate derived in Chapter 6, and twice the rate derived in Chapter 6.

7.17 Create a multiple-curve two-dimensional plot comparing the growth of a $1000 bank balance when invested for 20 years at 10% simple interest, 10% interest compounded quarterly, and 10% continuous interest, using the interest rate models from Chapter 6.

7.18 Create an animation showing how the plot from Exercise 7.17 changes as interest varies from 5% to 15%.

7.19 Create a list containing the cubes of the first 100 multiples of 13.

7.20 *Mathematica* uses the symbol I to stand for i, the imaginary unit used in complex numbers. Use *Mathematica* to find the value of $e^{i\pi} + 1$.

8

The Battle for Leyte Gulf: Symbolic Mathematics

On October 20, 1944, the United States began its invasion of the Philippines. The ensuing seven-week Battle for Leyte Gulf was the largest naval engagement in history. The United States committed 216 combat ships to the battle, but the Japanese could muster only 64. Knowing that the U.S. had superior naval power in the theater, the Japanese fleet split into three forces and employed a series of feints. The plan was to use one of the forces to draw the bulk of the U.S. fleet out of position. On October 24 this strategy worked when the U.S. Third Fleet was drawn away.

The other two forces of the Japanese fleet launched a pincer attack directly on the American invasion force, which was protected by the U.S. Seventh Fleet. The Seventh Fleet was equipped to support a marine landing, not to engage in a heavy sea battle. Early on the morning of October 25 one of the Japanese forces engaged an American battle group consisting of 16 lightly armed escort carriers, nine destroyers, and 12 destroyer escorts. The vastly superior Japanese force included four battleships, six heavy cruisers, and two light cruisers, accompanied by a screening force of destroyers.

Admiral Thomas Sprague, the commander of the American battle group, realized that his only chance was to cover the withdrawal of his escort carriers with a torpedo attack by his destroyers and destroyer escorts. The guns of the U.S. destroyers were not powerful enough to be of much use against the armor of the Japanese battleships and cruisers. The only effective weapon the destroyers had were torpedos, but they could be used only at extremely close range.

By zigzagging erratically, the destroyers were able to close to within range and fire all of their torpedos. Although the destroyers suffered heavy damage while retreating, the damage they caused—and the confusion they sowed—convinced the Japanese that they were up against a stronger force than they had anticipated. Six hours after the engagement had begun, the Japanese force withdrew.

The fact that speed and maneuverability could be used so effectively was contrary to the conventional wisdom of the day, and it caused naval architects to focus on these features for decades to come. In this chapter we will study the problem of determining the power required to give modern destroyers their speed and maneuverability. Given a trajectory for a destroyer to follow in the water, we will calculate the power required to produce that trajectory. We will proceed in four stages.

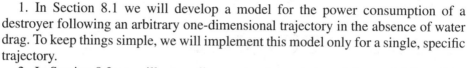

1. In Section 8.1 we will develop a model for the power consumption of a destroyer following an arbitrary one-dimensional trajectory in the absence of water drag. To keep things simple, we will implement this model only for a single, specific trajectory.

2. In Section 8.2 we will generalize our implementation of the model from Section 8.1 by developing a function that will produce a symbolic expression for power consumption when given a symbolic expression for a one-dimensional trajectory.

3. In Section 8.3 we will extend the model from Section 8.1 and the implementation from Section 8.2 to account for water drag.

4. In Section 8.4 we will use our implementation from Section 8.3 to study a one-dimensional piecewise trajectory.

We will characterize our destroyer with three constants. The mass of the destroyer, 4.5 million kilograms,

$$
\begin{array}{l}
\texttt{In[1]:= mass = 4.5*\^6} \\[2mm]
\texttt{Out[1]= } 4.5 \times 10^6 \text{ kg}
\end{array}
$$

(8.1)

determines how much force is required to produce accelerations. The power production capacity, 5 megawatts,

$$
\begin{array}{l}
\texttt{In[2]:= maxPower = 5.0*\^6} \\[2mm]
\texttt{Out[2]= } 5. \times 10^6 \text{ W}
\end{array}
$$

(8.2)

determines the maximum amount of power that the destroyer can produce at any instant in time. The drag coefficient, 5500 kg/sec,

$$
\begin{array}{l}
\texttt{In[3]:= drag = 5.5*\^3} \\[2mm]
\texttt{Out[3]= } 5500. \text{ kg/sec}
\end{array}
$$

(8.3)

determines how much force is required to maintain a constant velocity. We will use these constants throughout.

8.1 Fixed Trajectory

In this section we will determine the power consumed by our destroyer as it follows a particular one-dimensional trajectory. We will assume that the ship's east-west position is always zero and that its north-south position, which varies with time t, is given by

```
In[4]:= pos = .032 * t^2
```
$$Out[4]= 0.032\,t^2 \text{ m}$$

(8.4)

where time is measured in seconds and distance in meters. To make things simpler, we will make the substantial simplification of ignoring the effects of drag on the ship.

Before proceeding, let's visualize the motion described by pos in two different ways.

```
In[5]:= Plot[pos, {t, -100, 100},
            AxesLabel->{"sec", "m"}]
```

Out[5]= (See Figure 8.1a)

(8.5)

```
In[6]:= Plot[pos, {t, 0, 200}, AxesLabel->{"sec", "m"}]
```

Out[6]= (See Figure 8.1b)

(8.6)

The destroyer is moving back and forth on a north-south axis. Figure 8.1a shows 200 seconds in the life of the destroyer. At -100 seconds the destroyer is 320 meters north of the origin and is moving slowly south. At zero seconds it reaches the origin and reverses course, moving northward. At 100 seconds it is 320 meters away and is continuing to accelerate to the north. Figure 8.1b shows what happens in the 200 seconds after the destroyer reverses course, during which time it moves 1280 meters north.

Don't be misled by Figure 8.1 into thinking that the destroyer is not moving in a straight line. Both plots show the north-south position of the destroyer as a function of time, *not* the destroyer's trajectory. The plots contain parabolic curves instead of straight lines because the destroyer is accelerating.

Because our destroyer's motion is always along a straight line, its power consumption as a function of time can be calculated by multiplying its mass, its acceleration, and its velocity. (If the destroyer were moving in *two* dimensions, the power

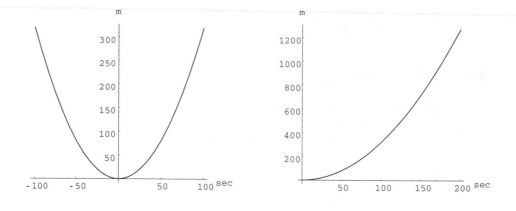

(a) 100 seconds before and after reversing course (Example 8.5)

(b) 200 seconds after reversing course (Example 8.6)

Figure 8.1   North-south position of destroyer following trajectory from Example 8.4. The position of the destroyer in meters is shown as a function of time in seconds.

calculation would be a bit more involved.) Thus, if the north-south position of the destroyer as a function of time is pos(*t*) meters, and the destroyer's mass is *m* kg, then its power requirement in watts is

$$\text{power}(t) = m \ \frac{d^2}{dt^2}\text{pos}(t) \ \frac{d}{dt}\text{pos}(t) \qquad (8.7)$$

We will have more to say about the physical principles behind this model as we implement it.

To apply this model we must differentiate the position expression pos from Example 8.4 once to obtain velocity and twice to obtain acceleration. We will use *Mathematica*'s symbolic differentiation capability to do this.

8.1.1 Differentiation

To determine power requirements, we need to find expressions for both the velocity and acceleration of the destroyer. Since velocity is the first derivative of position, we can obtain velocity by differentiating pos. Although you could probably do this quite easily by hand, let's see how to do it in *Mathematica*.

```
In[8]:= vel = D[pos, t]

Out[8]= 0.064 t m/sec
```
  (8.8)

Mathematica's built-in D function takes two parameters—a symbolic expression and a symbolic variable—and differentiates the expression with respect to the variable. We can now differentiate the expression for velocity to obtain one for acceleration.

```
In[9]:= acc = D[vel, t]

Out[9]= 0.064 m/sec²
```

(8.9)

Example 8.9 reveals that as the destroyer tracks the trajectory described by pos, it undergoes a constant acceleration of 0.064 m/sec² to the north. It is impossible for the ship to follow this trajectory forever, since a constant acceleration implies a steadily increasing speed, which cannot be sustained indefinitely. As we study the power production requirements imposed by the trajectory we have chosen, we will determine the interval of time during which the trajectory *can* be followed by our destroyer.

The next step in modeling the power requirements of the destroyer is to find the force that the destroyer's engines must exert to produce the required acceleration. Isaac Newton has already done the creative work by discovering that force equals mass times acceleration; we need only multiply mass by acceleration.

```
In[10]:= force = mass * acc

Out[10]= 288000. N
```

(8.10)

(A newton, abbreviated N, is 1 kg m/sec².) For the destroyer to maintain a constant acceleration, its engines must exert a constant force of, in this case, 288,000 newtons. Before time zero, this force serves to slow down the destroyer; afterwards, it serves to speed it up.

The final step is to find the amount of power that the engines must generate to exert this force. Power is force delivered over a distance per unit time, which amounts to the product of force and velocity. Fortunately, we have already calculated both.

```
In[11]:= pow = force * vel

Out[11]= 18432. t W
```

(8.11)

(A watt, abbreviated W, is 1 m N/sec.)

8.1.2 Power Consumption Curves

We have just developed a model for the power consumption of a destroyer moving along a fixed trajectory, and in the process have produced a symbolic expression pow that relates power consumption to time. Let's assess our model and implementation.

We will begin by plotting pow (in megawatts) for the times between -100 and 100 seconds.

```
In[12]:= Plot[pow/1*^6, {t, -100, 100},
            AxesLabel->{"sec", "MW"}]

Out[12]= (See Figure 8.2a)
```

(8.12)

(In Example 8.12, we have divided pow by 10^6 in order to scale the plot in megawatts.)

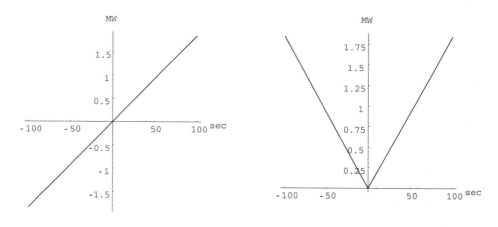

(a) Power (Example 8.12)	(b) Absolute value of power (Example 8.13)

Figure 8.2   Power requirements of destroyer following trajectory from Example 8.4 in the absence of drag. Power required in megawatts is shown as a function of time in seconds.

Figure 8.2a reveals that for negative times, the power output is negative. Negative power output means that the engines are working to *slow down* the destroyer; nevertheless, this is still power that the destroyer must generate. We can also look at the *absolute value* of power, which is more pertinent to understanding the power production demands placed on our destroyer.

```
In[13]:= Plot[Abs[pow]/1*^6, {t, -100, 100},
            AxesLabel->{"sec", "MW"}]

Out[13]= (See Figure 8.2b)
```

(8.13)

(Abs is *Mathematica*'s absolute value function.)

Figure 8.2b shows that the destroyer must generate less and less power as the ship slows down, and then more and more power after the ship starts moving in the opposite direction. Both before and after time zero, the power requirements vary linearly with time.

If you study the plots in Figure 8.2, you may wonder why it takes increasingly more power to produce the same force as the ship speeds up. The formal answer is that power is the product of force and velocity, so as velocity increases so does power. The practical answer is that the ship exerts force by pushing water out of the way with its propellers. As the destroyer gains speed the water begins rushing by the propellers, which must turn even faster to do the same amount of pushing.

8.1.3 Replacements

It is evident from the power plots that the destroyer will eventually reach its maximum power production capability, which will force it to leave the trajectory we have designed. We can easily calculate exactly when the power threshold will be reached by solving an equation.

```
In[14]:= threshold = Solve[Abs[pow] == maxPower, t]

Out[14]= {{t → −271.267}, {t → 271.267}}
```
(8.14)

There are two solutions to the equation. Approximately four and a half minutes before and after time zero, the ship will be harnessing its entire 5 megawatts of power to keep itself on track. This means that the portions of the trajectory that occur before and after these times are actually impossible for our destroyer to follow.

The solution in Example 8.14 is a list of two *replacements*. We first saw replacements in Chapter 6, which is when we began experimenting with `Solve`. The first says that the equation will reduce to an identity when t is replaced by −271.267, and the second says that the equation will reduce to an identity when t is replaced by 271.267. By virtue of the assignment expression, the value of the variable `threshold` is itself a list of replacements.

```
In[15]:= threshold

Out[15]= {{t → −271.267}, {t → 271.267}}
```
(8.15)

We have previously needed only to *create* lists; now we need to *decompose* them. We can obtain the first replacement in the list `threshold` via

```
In[16]:= threshold[[1]]

Out[16]= {t → −271.267}
```
(8.16)

and the second replacement via

```
In[17]:= threshold[[2]]

Out[17]= {t → 271.267}
```
(8.17)

How fast will the destroyer be going when it reaches its maximum power output threshold approximately 271 seconds after time zero? Recall that the expression for velocity is $.064t$, so we need to multiply .064 by the 271.267 that is contained in the second replacement of `threshold`.

```
In[18]:= .064 * 271.267

Out[18]= 17.3611 m/sec
```
(8.18)

Because the `vel` expression was so simple, it wasn't difficult to type it back into *Mathematica*, replacing t with `271.267`. Fortunately, *Mathematica* provides a more systematic way to do this that is extremely helpful when dealing with more complicated expressions. The built-in `ReplaceAll` function will apply a replacement to a symbolic expression. For example, we could have found the destroyer's velocity at its power threshold by using `ReplaceAll` to apply the replacement `threshold[[2]]` to `vel`. This will replace each occurrence of t in `vel` with 271.267.

```
In[19]:= ReplaceAll[vel, threshold[[2]]]

Out[19]= 17.3611 m/sec
```
(8.19)

This speed is approximately 38.9 mi/hr. While this is respectable, it isn't close to the top speed of a modern destroyer. Is this really the top speed that our destroyer can attain? We can answer this question and better understand the behavior of our destroyer by experimenting with other trajectories.

8.2 Arbitrary Trajectories

If we are going to investigate other trajectories for our destroyer, it would be convenient if we could get *Mathematica* to create power expressions for us. What we need is a function that takes two parameters—expressions for the trajectory of the destroyer and its mass—and returns the appropriate expression for power consumption. We could then use this function to create a power consumption formula for any combination of trajectory and mass.

8.2.1 Programmer-Defined Symbolic Functions

In Section 8.1.1 we went step by step through the process of creating a power function from a trajectory function and a mass. The steps were:

1. Differentiate position to obtain velocity.
2. Differentiate velocity to obtain acceleration.
3. Multiply acceleration by mass to obtain force.
4. Multiply force by velocity to obtain power.

In the past, whenever we needed to repeat a sequence of computations on numbers, we packaged the sequence into a function. In this case we need to repeatedly perform a sequence of computations on symbolic expressions, and it is no less natural to create a function to do the computations for us.

```
In[20]:= makePower[pos_, mass_] :=
              mass * D[D[pos,t],t] * D[pos,t]
```
(8.20)

The function `makePower` expects as parameters an expression giving the position of the destroyer (in terms of `t`) and the mass of the destroyer. It returns the result of multiplying mass by acceleration by velocity, which yields power. The function obtains velocity and acceleration by differentiating its `pos` parameter.

Let's try out the `makePower` function and see what we get.

```
In[21]:= pow = makePower[pos, mass]

Out[21]= 18432.t W
```
(8.21)

We have given `makePower` the same position expression and mass that we used in Section 8.1, and have gotten back the same power expression that we previously derived in Example 8.11.

The two parameters to `makePower` in Example 8.21 are symbolic expressions which are differentiated and multiplied. This is yet another example of the ease with which symbolic calculations can be performed in *Mathematica.*

8.2.2 Maximum Velocity in the Absence of Drag

In Section 8.1.3 we wondered whether 17.4 m/sec was really the top speed for our destroyer. Let's explore this question by experimenting with the power functions for two different trajectories. One requires the destroyer to accelerate at half,

```
In[22]:= pow1 = makePower[.016 * t^2, mass]

Out[22]= 4608.t W
```
(8.22)

and the other at one sixteenth,

```
In[23]:= pow2 = makePower[.002 * t^2, mass]

Out[23]= 72.t W
```

(8.23)

the rate from our first example.

With the first of the new trajectories, peak power is reached a little more than 18 minutes before and after time zero,

```
In[24]:= threshold1 = Solve[Abs[pow1] == maxPower, t]

Out[24]= {{t → −1085.07}, {t → 1085.07}}
```

(8.24)

at which point the destroyer's velocity will be

```
In[25]:= ReplaceAll[D[.016*t^2, t], threshold1[[2]]]

Out[25]= 34.7222 m/sec
```

(8.25)

34.7 m/sec (77.6 mi/hr). With the second of the new trajectories, peak power is reached

```
In[26]:= Solve[Abs[pow2] == maxPower, t]

Out[26]= {{t → −69444.4}, {t → 69444.4}}
```

(8.26)

over *nineteen hours* before and after time zero, at which point the ship's velocity will be 277.8 m/sec (621.4 mi/hr). Our destroyer has almost gone supersonic! Table 8.1 gives the maximum velocity attainable before the power capacity of the destroyer is exceeded for a variety of different constant accelerations. Notice that halving the acceleration quadruples the time to power threshold and doubles the velocity at power threshold.

Table 8.1 suggests that it is *acceleration*, not *velocity*, that is limited by our destroyer's peak power production capability. We can sustain a low rate of acceleration much longer than a high rate. Indeed, by accelerating slowly enough we can reach *any* velocity that we choose. We can also reach any velocity we wish by applying full power and waiting long enough. Velocity will steadily increase even as acceleration gradually declines.

Table 8.1   Effects of different accelerations in the absence of drag

Position (meters)	Acceleration (m/sec^2)	Time to threshold (seconds)	Velocity at threshold (m/sec)
$.064\,t^2$	.128	67.8	8.7
$.032\,t^2$	.064	271.3	17.4
$.016\,t^2$	.032	1085.1	34.7
$.008\,t^2$	.016	4340.3	69.4
$.004\,t^2$	.008	17361.1	138.9
$.002\,t^2$	.004	69444.4	277.8

We can demonstrate that there is no upper limit on our destroyer's velocity by looking at what happens when it follows a trajectory with *zero* acceleration.

```
In[27]:= makePower[v*t, mass]

Out[27]= 0 W
```
(8.27)

The position parameter to `makePower` describes a destroyer moving at a constant velocity with zero acceleration. The power required to follow this trajectory, once the velocity v has been attained, will be zero.

The reason that our destroyer can reach arbitrary speeds is that we have been ignoring the drag of the water, which works to slow the destroyer down. Once we account for drag, we will have a much more realistic model.

8.3 Effects of Drag

When a real ship moves through the water, the water exerts a drag force that opposes the ship's motion. This force is proportional to velocity, which means that it is larger for high velocities and smaller for low velocities. If a moving ship cuts power entirely, it will gradually coast to a stop due to the effects of drag.

The force due to drag can be modeled by multiplying velocity by a constant of proportionality called the *drag coefficient*. The drag coefficient varies from ship to ship, and its exact value depends on the construction of the ship and the density of the water. We decided in the introduction that we would use a drag coefficient of 5500 kg/sec in modeling our destroyer.

In order to move along a specific trajectory, our destroyer must now generate sufficient force to both neutralize drag and produce the desired acceleration. We can easily extend our `makePower` function to account for drag.

We will modify `makePower` to obtain a new function called `makePowerDrag`.

```
In[28]:= makePowerDrag[pos_, mass_, drag_] :=
            (mass * D[D[pos,t],t] + drag * D[pos,t]) *
                D[pos,t]
```
(8.28)

Besides changing the name of the function to reflect its new purpose, we have added `drag` as a third parameter. Furthermore, we are now multiplying the *sum* of the forces due to acceleration and drag by velocity to obtain the power requirements.

8.3.1 Maximum Velocity in the Presence of Drag

Although `makePowerDrag` is quite involved, we can easily use it to produce an improved power expression for our original trajectory.

```
In[29]:= pow = makePowerDrag[pos, mass, drag]
```

$$\text{Out}[29]= 0.064\,t\,(288000.+352.\,t)\;\text{W}$$
(8.29)

Compare this result with that obtained in Example 8.21. Let's visualize this power expression.

```
In[30]:= Plot[pow/1*^6, {t, -100, 100},
            AxesLabel->{"sec", "MW"}]
```

Out[30]= (See Figure 8.3a)
(8.30)

Although Figure 8.3a looks like a straight line, the calculations

```
In[31]:= Abs[ReplaceAll[pow, {t->100}]]
```

$$\text{Out}[31]= 2.06848 \times 10^{6}\;\text{W}$$
(8.31)

and

```
In[32]:= Abs[ReplaceAll[pow, {t->-100}]]
```

$$\text{Out}[32]= 1.61792 \times 10^{6}\;\text{W}$$
(8.32)

reveal that the power consumption at 100 seconds is 2.07 megawatts whereas the power consumption at -100 seconds is only 1.62 megawatts. This means that the destroyer is generating more power to speed up than it is generating to slow down,

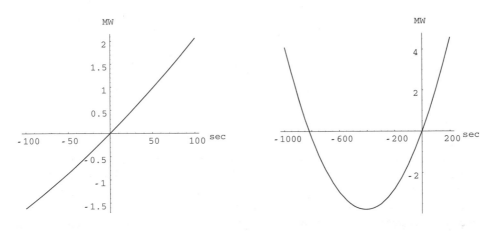

(a) Near time zero (Example 8.30)   (b) Expanded time range (Example 8.33)

Figure 8.3   Power requirements of destroyer following trajectory from Example 8.4 in the presence of drag. Power required in megawatts is shown as a function of time in seconds.

even though its acceleration never changes. This makes sense, of course, because the water's drag is helping the destroyer to slow down but making it harder to speed up.

(In both Examples 8.31 and 8.32 we have *constructed* a replacement to use with `ReplaceAll` instead of obtaining one via `Solve`. When entering a replacement from the keyboard, the right arrow symbol is created by composing a minus sign and a greater-than sign.)

We can get a better picture of what's going on by expanding the range of the plot.

```
In[33]:= Plot[pow/1*^6, {t, -1000, 200},
            AxesLabel->{"sec", "MW"}]

Out[33]= (See Figure 8.3b)
```

(8.33)

A left-to-right reading of Figure 8.3b reveals the effects of drag. At -1000 seconds, the ship is slowing down but must still supply enough power to prevent drag from slowing the ship down too quickly. At -818 seconds the ship's power output drops to zero and then becomes negative; that is, the ship must now begin using its power plant to slow the ship down—drag alone isn't sufficient. Power production peaks at -409 seconds, after which less and less power is required as the destroyer comes to a halt. After time zero, the engines begin working again to accelerate the ship forward.

When we modeled the destroyer without drag, 5 megawatts of power were required for this trajectory 271 seconds before and after time zero. Now, the power generation threshold is reached at

```
In[34]:= Solve[pow == maxPower, t]

Out[34]= {{t → −1033.03}, {t → 214.849}}
```
(8.34)

1033 seconds *before* and 215 seconds *after* time zero. Peak power production occurs earlier as we decelerate toward the origin because drag is helping us slow down; peak power production occurs earlier as we accelerate away from the origin because we must now overcome drag. When we ignored drag, the maximum velocity that we could reach was 17.4 m/sec. In the presence of drag, our maximum velocity is 13.8 m/sec.

Table 8.2 summarizes the velocities at power threshold for different constant accelerations. It is instructive to compare this with Table 8.1. Is there a top speed for our destroyer now, or can we still go as fast as we wish by accelerating sufficiently slowly?

To find out, let's repeat our experiment from Example 8.27 and calculate the power required to maintain a constant velocity v.

```
In[35]:= constantPower = makePowerDrag[v*t, mass, drag]

Out[35]= 5500. v² W
```
(8.35)

Before we took drag into account, no power at all was required to maintain a constant velocity. Now our destroyer must generate enough power to overcome drag, and the power required to do that grows with the square of the velocity. When the velocity becomes large enough, we will reach our 5-megawatt limit.

Table 8.2   Effects of different accelerations in the presence of drag

Position (meters)	Acceleration (m/sec²)	Time to threshold (sec)	Velocity at threshold (m/sec)
$.064\,t^2$	.128	63.0	8.1
$.032\,t^2$	.064	214.8	13.8
$.016\,t^2$	.032	618.1	19.8
$.008\,t^2$	.016	1519.2	24.3
$.004\,t^2$	.008	3381.9	27.1
$.002\,t^2$	.004	7139.8	28.6

```
In[36]:= Solve[constantPower == maxPower, v]
```

$$Out[36]= \{\{v \to -30.1511\}, \{v \to 30.1511\}\}$$

(8.36)

The highest velocity that our destroyer can reach, then, is a bit more than 30.1 m/sec, which is approximately 67.4 mi/hr. The entire power output of the destroyer is required to maintain this velocity.

8.4 Piecewise Trajectories

To this point, all of the trajectories that we have studied have been designed to illustrate how the maximum power production capacity of the destroyer puts a limit on its top speed. Every trajectory with which we have experimented has been impossible for our destroyer to follow, as all have ultimately required more power than our destroyer is able to supply.

In this section we will design a realistic trajectory in which the destroyer begins at rest 5000 meters south of the origin and then moves to a point 5000 meters north of the origin, where it again comes to a stop.

A function that describes such a trajectory must be defined in a *piecewise* fashion. That is, we must describe separately the pieces of the trajectory that occur before the destroyer begins moving, while it is moving, and after it stops. We will use the following position function in this section.

$$\text{move}(t) = \begin{cases} -5000 & t < -500 \\ 5000 \sin\left(\frac{\pi t}{1000}\right) & -500 \le t < 500 \\ 5000 & 500 \le t \end{cases}$$

(8.37)

This definition says that

- Before -500 seconds the north-south position of the destroyer is -5000 meters. Because this is a constant, this means that the destroyer is at rest.
- Between -500 and 500 seconds the destroyer is moving and its north-south position is $5000 \sin\left(\frac{\pi t}{1000}\right)$ meters. We have used a portion of a sine curve to describe the destroyer's motion to ensure that it both starts up and slows down gradually. This will be more evident when we plot this trajectory below.
- After 500 seconds the north-south position of the destroyer is 5000 meters. It is once again at rest, $10,000$ meters north of its original position.

Notice that this is not a particularly violent maneuver. The destroyer spends 1000 seconds moving 10,000 meters, for an average speed of 10 m/sec.

8.4.1 A Piecewise Definition

Mathematica provides a built-in function `Which` that allows us to define an expression in a piecewise fashion as we did for the function `move` above.

```
In[38]:= move = Which[t < -500, -5000,
                      t < 500, 5000*Sin[Pi*t/1000],
                      True, 5000]
```
(8.38)

This definition says that the value of move is

- -5000 when $t < -500$; and otherwise
- $5000 \sin\left(\frac{\pi t}{1000}\right)$ when $t < 500$; and otherwise
- 5000,

which is exactly what Formula 8.37 specified.

We can plot move just as easily as any other expression.

```
In[39]:= Plot[move, {t, -600, 600},
              AxesLabel->{"sec", "m"}]
```
(8.39)

```
Out[39]= (See Figure 8.4a)
```

Notice how the north-south position of the destroyer changes smoothly just after it begins moving and just before it stops. This was the point of using a portion of a sine curve to connect the two endpoints of the middle segment instead of using a straight line.

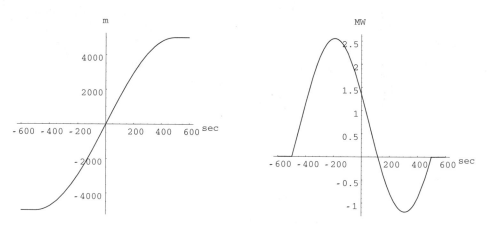

(a) Position (Example 8.39)   (b) Power required (Example 8.41)

Figure 8.4 North-south position and power requirements of destroyer following the trajectory from Example 8.38 in the presence of drag. Position is shown in meters and power is shown in megawatts as functions of time in seconds.

8.4.2 Boolean Expressions

Expressions such as the `t < -500` that appears in Example 8.38 are called *Boolean expressions*. Boolean expressions are used in *Mathematica* and other programming languages in contexts, such as `Which` expressions, where choices among alternatives must be made. Just as the value of an arithmetic expression must be a number, the value of a Boolean expression must be a truth value: either `True` or `False`. Thus, for the first 100 seconds of the time interval we used for the plots in Figure 8.4, the expression `t < -500` is true, and `move` returns the value of −5000 meters.

To this point we have seen a large variety of ways to construct arithmetic expressions. There are a comparable number of ways to construct Boolean expressions. The constants `True` and `False` are Boolean expressions, just as the constants 1 and 2 are arithmetic expressions. A common way of constructing more complicated Boolean expressions is by using relational operators. The six relational operators and their meanings appear in Table 8.3.

Table 8.3   *Mathematica*'s six relational operators

`<`	less than	`<=`	less than or equal
`>`	greater than	`>=`	greater than or equal
`==`	equal	`<>`	not equal

A Boolean expression can be written by putting a relational operator between two arithmetic expressions. For example, `2 <= 5` and `2*x < 3+y` are both Boolean expressions. The value of the first expression is `True`, and the value of the second expression depends on the values of `x` and `y`.

A `Which` expression, such as the one we used in Example 8.38, is written in the form

$$\texttt{Which}(B_1, E_1, \ldots, B_n, E_n)$$

where the B_i must be Boolean expressions. When *Mathematica* evaluates a `Which` expression, it evaluates the B_i in order until it finds one whose value is `True`. At that point, it returns the value of the corresponding E_i as the value of the `Which` expression. (It is usually a good idea to write a `Which` expression so that the last Boolean expression (B_n) is `True`, so that we can guarantee that the expression will have *some* value.

The `Which` expression is only one context in which Boolean expressions play a role, and relational operators are only one way to create Boolean expressions. We will explore Boolean expressions further in the next chapter.

8.4.3 Piecewise Power Consumption

Just as we have for every other trajectory in this chapter, we can determine the power required to follow the piecewise trajectory `move` that we defined in Example 8.38.

```
In[40]:= movePower = makePowerDrag[move, mass, drag]
```

(8.40)

We can then use `Plot` to visualize the power consumption curve.

```
In[41]:= Plot[movePower/1*^6, {t, -600, 600},
            AxesLabel->{"sec", "MW"}]

Out[41]= (See Figure 8.4b)
```

(8.41)

Figure 8.4b shows how the power requirements climb from zero to a peak of approximately 2.5 megawatts early in the motion, and then decline before peaking again as the destroyer is moving to a stop. This plot has several interesting aspects, including:

- The positive power peak is larger than the negative power peak, even though the trajectory is symmetric about the origin. This is because drag works against us as we build speed and for us as we slow down.
- Positive power is required during part of the period after time zero when the destroyer is slowing down. This is because drag would otherwise cause the destroyer to slow too abruptly.

8.5 Final Assessment

Although our model of a destroyer's power requirements raises most of the issues that would appear in a more complete model, it could be generalized in a number of ways.

The most obvious shortcoming of our model is that it works only for motion in one dimension. The model would be more useful if we could use it to compute the power required to follow an arbitrary two-dimensional path through the water. Introducing a second dimension, however, would significantly complicate the mathematics.

The tendency of the water to retard the motion of the destroyer is a more complicated phenomenon than can be accounted for with a single drag coefficient. The drag coefficient of the destroyer, for example, varies with velocity. A higher velocity may cause the destroyer to rise higher in the water, changing its drag. Similarly, higher velocities cause more turbulence, which must also be considered.

Although our model is useful if we want to decide on a trajectory and then determine power requirements, it doesn't work in reverse. What happens if we apply full power from a standing start? What happens if we suddenly cut power and coast to a stop? To answer these questions it is necessary to derive and solve differential equations.

8.6 Key Concepts

Symbolic differentiation.   The *Mathematica* D function takes two parameters. The first is a symbolic expression to be differentiated, and the second is the variable of differentiation. For example, D[2*x^3, x] calculates $\frac{d}{dx}2x^3$.

Lists.   *Mathematica* sometimes returns more than one value from a function. This is common when using Solve, because equations often have more than one solution. Multiple solutions are composed into lists, which consist of values enclosed in braces and separated by commas. Lists can also be constructed with the Table function. The components of a list can be extracted by using a subscript, which is written as an integer contained in double square brackets.

Replacements.   The *Mathematica* ReplaceAll function will substitute a value for every occurrence of a symbol that appears in a symbolic expression. For example, ReplaceAll[x+z, {x->25*y}] replaces the x in x+z with 25*y, resulting in 25*y+z.

Function composition.   Composing a complicated function from simpler pieces is the essence of programming. The makePowerDrag function, for example, is based on a number of simpler functions.

Boolean expressions.   Boolean expressions have values that are either True or False, and are used to express choices within a larger expression. One way to form a Boolean expression in *Mathematica* is to place one of the six relational operators between two arithmetic expressions.

Which expressions.   An expression whose value depends on choices that are written as Boolean expressions can be written using the *Mathematica* Which construct.

8.7 Exercises

8.1 If age is your age in years, weight is your weight in kilograms, and height is your height in meters, construct Boolean expressions that will be true only when the following statements are true.

(a) You are old enough to obtain a driver's license.

(b) You can safely climb a ladder with a load limit of 150 pounds.

(c) You can ride a roller coaster with a minimum height limit of 36 inches.

8.2 Use *Mathematica* to evaluate the following derivatives.

(a) $\frac{d}{d\theta}\tan(\theta)$

(b) $\frac{d}{dy}\sqrt{y^2 + y^3}$

(c) $\frac{d}{dx}\sin\left(\cos\left(x^2\right)\right)$

(d) $\frac{d}{dt}e^{-t}\cos\left(e^t\right)$

(e) $\frac{d}{dx}\sum_{i=1}^{10} x^i$

(f) $\frac{d}{dx}\sum_{i=n}^{10} x^i$

8.3   Use *Mathematica*'s `ReplaceAll` function to find the value of each of the derivatives from Exercise 8.2 when the variable of differentiation takes on the value 12.

8.4   Modify the trajectory in Equation 8.37 so that the destroyer returns to its starting position by tracing its trajectory in reverse.

8.5   Create a two-dimensional plot that shows the power required to follow the trajectory from Exercise 8.4.

8.6   Design a trajectory, based on the one in Equation 8.37, in which the destroyer continuously shuttles back and forth between two points 10,000 meters apart.

8.7   Create a two-dimensional plot that shows the power required to follow the trajectory from Exercise 8.6.

8.8   Create a multiple-curve two-dimensional plot that compares the power required to follow the trajectory in Equation 8.37 with and without drag.

8.9   The trajectory in Equation 8.37 begins at time −500 seconds and ends at time 500 seconds, requiring 1000 seconds to complete. Create an animation showing how the power required to follow the trajectory changes as the time required for the maneuver varies between 500 and 1500 seconds.

8.10   Approximately how fast can the maneuver in Exercise 8.9 be completed without exceeding the destroyer's power production capacity?

8.11   Create an animation showing how the power required to follow the trajectory in Equation 8.37 changes as the distance traveled varies between 5000 and 15,000 meters.

8.12   Approximately how long can the maneuver in Exercise 8.11 be made without exceeding the destroyer's power production capacity?

8.13   Create an animation showing how the power required to follow the trajectory in Equation 8.37 changes as drag varies between 0 and 10,000 kg/sec.

8.14   Approximately how high can the drag be made in Exercise 8.13 without exceeding the destroyer's power production capacity?

8.15   Create a two-dimensional plot showing the first derivative of the power consumption when the destroyer follows the trajectory in Equation 8.37.

8.16 Use the `Which` construct to implement an improved version of the `vertical` function from Chapter 7. The new version should take account of the fact that the projectile stops moving when it hits the ground.

8.17 Repeat Exercise 8.16, this time improving the `horizontal` function from Chapter 7.

8.18 Define a function that returns the horizontal distance traveled by a ballistic projectile between the time it is launched and the time it hits the ground. Your function should take as its parameter the initial speed and angle of the projectile.

8.19 Create a two-dimensional plot showing the first derivative of the function from Exercise 8.18 for a fixed initial velocity.

8.20 The local minima and maxima of a function f can be found by solving the equation

$$\frac{d}{dx} f(x) = 0 \qquad\qquad (8.42)$$

Use this approach and the function from Exercise 8.18 to determine the angle at which a ballistic projectile should be launched to maximize the horizontal distance that it travels.

9

Old MacDonald's Cow: Imperative Programming

Old MacDonald has a farm, and on that farm he has a fenced, circular pasture with a radius of 10 meters. He would like to let his cow graze inside the fence, but wants to limit her range so that she can eat only half of the grass. To do this, he plans to tie the cow to one of the fence posts with a rope. How long should the rope be so that the cow can range over only half of the pasture?

Figure 9.1 reduces this problem to geometric terms. The circle represents the pasture. It has radius r, which we know to be 10 meters. The circular arc has radius R and is centered on the point d, which lies on the circle. If the cow is tethered to a fencepost at point d with a rope of length R, the arc represents the limit beyond which the cow cannot graze. The region that is beyond the cow's reach is labeled A.

The arc intersects the circle at points a and b, which means that both ad and bd have length R. We will let X denote the area of the region bounded by the arc and the line segments ad and bd. Because the arc is part of a circle of radius R whose center is d, the ratio of X to πR^2 is the same as the ratio of θ to 2π. Thus,

```
In[1]:= X = theta/(2*Pi) * Pi*R^2

Out[1]=  R^2 theta
         ────────
            2
```

(9.1)

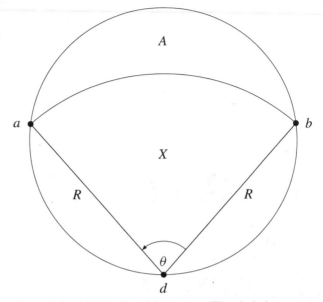

Figure 9.1   Old MacDonald's pasture. The circle has radius r, and the circular arc marks the limit of the cow's range.

Figure 9.2 is a more elaborate version of Figure 9.1. In addition to the center of the circle c, we have added the line segments ca, cb, and cd. Each of these segments is a radius of the circle and is thus of length r.

The two isosceles triangles (acd and bcd) formed by these segments are identical, and we will let Y denote the area of each. Because cd bisects the angle adb, it follows that the angles cda and bdc are both $\frac{\theta}{2}$ radians.

Figure 9.3 shows triangle acd with the perpendicular bisector ce added. Because ace is a right triangle with a hypotenuse of length r, an angle of $\frac{\theta}{2}$ radians, and a side adjacent to the angle of length $\frac{R}{2}$, it follows that

```
In[2]:= R = 2 * r * Cos[theta/2]
```
$$\text{Out[2]= } 2r \, \text{Cos}\left[\frac{\text{theta}}{2}\right]$$

(9.2)

The area Y of triangle acd is half the product of its base (R) and height ($r \sin\left(\frac{\theta}{2}\right)$), so

```
In[3]:= Y = 1/2 * R * r * Sin[theta/2]
```
$$\text{Out[3]= } r^2 \, \text{Cos}\left[\frac{\text{theta}}{2}\right] \text{Sin}\left[\frac{\text{theta}}{2}\right]$$

(9.3)

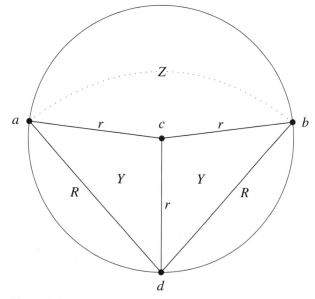

Figure 9.2   Old MacDonald's pasture with inscribed triangles

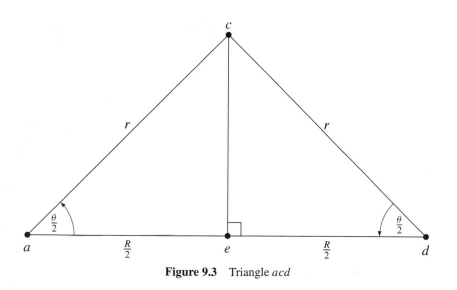

Figure 9.3   Triangle *acd*

Returning to Figure 9.2, the angles *acd* and *bcd* are both $\pi - \theta$ radians, so the angle *acb* must be $2\pi - 2(\pi - \theta)$, or 2θ radians. We will let Z denote the area of the region bounded by the circle and the line segments *ac* and *bc*. The ratio of Z to πr^2 is the same as the ratio of 2θ to 2π, so

$$In[4]:= Z = (2*theta)/(2*Pi) * Pi*r^2$$

$$Out[4]= r^2 \, theta$$

(9.4)

The area A of the region that is bounded from above by the circle and from below by the circular arc represents the portion of the pasture where the cow is prevented from grazing by the rope. Because the area that lies between the arc and the line segments ac and bc is $X - 2Y$, it follows that

$$In[5]:= A = Z - (X - 2*Y)$$

$$Out[5]= r^2 \, theta - 2r^2 \, theta \, Cos\left[\frac{theta}{2}\right]^2 + 2r^2 \, Cos\left[\frac{theta}{2}\right] Sin\left[\frac{theta}{2}\right]$$

(9.5)

This is a forbidding formula, but we can reduce it by using the *Mathematica* built-in function `Simplify` to simplify the trigonometric terms.

$$In[6]:= Simplify[A]$$

$$Out[6]= r^2 \, (-theta \, Cos[theta] + Sin[theta])$$

(9.6)

This gives us, in terms of r and θ, the area of the pasture where the cow will be unable to graze. We want this to be exactly half the area of the pasture, which is $\frac{1}{2}\pi r^2$. We can thus determine θ by solving the equation

$$\sin(\theta) - \theta \cos(\theta) = \frac{\pi}{2} \tag{9.7}$$

Once we have found the value of θ, we will be able to determine R with the equation

$$R = 2r \cos\left(\frac{\theta}{2}\right) \tag{9.8}$$

from Example 9.2.

9.1 Solving Equations in *Mathematica*

The problem of finding the roots of an equation is one of the most fundamental and enduring in mathematics. There are two different approaches for finding roots: the symbolic approach and the numerical approach. Not surprisingly, *Mathematica* provides two corresponding built-in functions for solving equations: `Solve` works symbolically and `FindRoot` works numerically.

The symbolic approach is how people usually solve equations, and it is how *Mathematica*'s built-in `Solve` function operates. With this method, an equation is manipulated until the unknown is defined in terms of the knowns. Most of high school algebra, for example, concerns symbolic techniques for solving equations, such as algebraic manipulation, the quadratic formula, and trigonometric identities.

Many, if not most, of the equations that occur in practice in science and engineering defy symbolic solution. Such equations must be solved *numerically*, which is how computers are generally used to solve equations and is how *Mathematica*'s FindRoot function operates. There are a variety of different methods for solving equations numerically, but they are all based on repeated attempts to guess the answer until an approximate solution is found. The trick is to do the guessing in such a way that an adequate approximation to the solution is quickly identified.

9.1.1 Symbolic Equation Solving

The variety of equations that you (or *Mathematica*) can solve symbolically is determined by how many equation-solving techniques you know (or *Mathematica* knows). For example, you probably learned early in your education how to solve a linear equation in one unknown by manipulating the equation until one side consisted only of the unknown. *Mathematica*'s Solve function works exactly the same way.

```
In[9]:= Solve[2*x == 4, x]

Out[9]= {{x → 2}}
```
(9.9)

Later in your education you learned how to deal with simultaneous linear equations via cancellation and substitution,

```
In[10]:= Solve[{3*x + 2*y == 7, x + y == 3}, {x,y}]

Out[10]= {{x → 1, y → 2}}
```
(9.10)

and with quadratic equations by exploiting the quadratic formula.

```
In[11]:= Solve[2*x^2 - 3*x - 9 == 0, x]

Out[11]= {{x → -3/2}, {x → 3}}
```
(9.11)

The more mathematics you know, the more kinds of equations you can solve. For example, neither you nor *Mathematica* would be able to solve

```
In[12]:= Solve[Cos[alpha] == Sin[alpha], alpha]

Out[12]= {{alpha → -3π/4}, {alpha → π/4}}
```
(9.12)

without knowing something about trigonometry.

The `Solve` function does arithmetic simplifications whenever possible, but it can work equally well with purely symbolic equations. For example, *Mathematica* can solve a quadratic equation whether the coefficients are numbers, as in Example 9.11, or symbols, as in

$$
\begin{array}{l}
\texttt{In[13]:= Solve[a*x\^{}2 + b*x + c == 0, x]} \\[2mm]
\texttt{Out[13]= } \left\{ \left\{ x \rightarrow \dfrac{-b - \sqrt{b^2 - 4ac}}{2a} \right\}, \left\{ x \rightarrow \dfrac{-b + \sqrt{b^2 - 4ac}}{2a} \right\} \right\}
\end{array}
\tag{9.13}
$$

No matter how sophisticated your mathematical knowledge becomes, there will always be equations that neither you—nor anyone else—can solve symbolically. Equation 9.7, for example,

```
In[14]:= Solve[Sin[theta] - theta*Cos[theta] == Pi/2,
              theta]

         Solve::tdep : The equations appear to involve
             transcendental functions of the variables
             in an essentially non-algebraic way.
```
(9.14)

cannot be solved symbolically; *Mathematica* admits as much by giving no answer. (*Mathematica* may sometimes fail to find a symbolic solution to an equation that *can* be solved symbolically. You should not take `Solve`'s failure to find a symbolic solution to mean that one does not exist.)

9.1.2 Numerical Equation Solving

The mere fact that an equation cannot be solved symbolically does not mean that it has no solution. For example, if we were to plot both $\sin(\theta) - \theta \cos(\theta)$ and $\frac{\pi}{2}$ on the same graph, we would see that their curves intersect. The point of intersection is the solution to Equation 9.7.

Mathematica's built-in `FindRoot` function works numerically to find floating-point approximations to the solutions of equations. For example, `FindRoot` can take the equation that baffled `Solve`

```
In[15]:= angle =
          FindRoot[Sin[theta] - theta*Cos[theta] ==
                  Pi/2,
                  {theta, 1.0}]

Out[15]= {theta → 1.9057}
```
(9.15)

and produce a floating-point approximation to its solution. The second parameter to `FindRoot` is a list consisting of the variable to solve for (`theta`) and a guess for the value of that root (1.0).

Unlike `Solve`, which returns a list of solutions, `FindRoot` returns a *single* solution. When an equation has more than one solution, the one that `FindRoot` finds depends on the initial guess that you supply.

The root that *Mathematica* found in Example 9.15 is not an exact answer—the exact answer is irrational—but it is quite close, as we can easily verify by plugging in the solution to the left side of the equation

```
In[16]:= ans1 =
            ReplaceAll[Sin[theta] - theta*Cos[theta],
                    angle]

Out[16]= 1.5708
```
(9.16)

and comparing it with the value of the right side.

```
In[17]:= ans2 = N[Pi/2]

Out[17]= 1.5708
```
(9.17)

The two sides are the same when their values are displayed to six digits. (Notice that we did not need to use a subscript for `angle` in Example 9.16, since `angle` is not a list.)

But what if we look at all 16 digits of the two results? We can do this by applying `InputForm` to `ans1` (from Example 9.16)

```
In[18]:= InputForm[ans1]

Out[18]= 1.570796621357678
```
(9.18)

and `ans2` (from Example 9.17)

```
In[19]:= InputForm[ans2]

Out[19]= 1.570796326794896
```
(9.19)

The two numbers differ after the first seven digits. By default, `FindRoot` tries to find a root that is good to six digits. We can, however, ask `FindRoot` to try to find a root that is good to 16 digits by specifying a value for the optional parameter `AccuracyGoal`.

```
In[20]:= newangle =
             FindRoot[Sin[theta] - theta*Cos[theta] ==
                         Pi/2,
                     {theta, 1.0},
                     AccuracyGoal->16]

Out[20]= {theta → 1.9057}
```
(9.20)

Even though `newangle` is the same as `angle` to six digits, it is indeed different. We can verify this by substituting `newangle` into the right-hand side of the original equation

```
In[21]:= ans3 =
             ReplaceAll[Sin[theta] - theta*Cos[theta],
                         newangle]

Out[21]= 1.5708
```
(9.21)

and looking at all 16 digits of the result.

```
In[22]:= InputForm[ans3]

Out[22]= 1.570796326794896
```
(9.22)

This is identical to the value of $\frac{\pi}{2}$ that we obtained in Example 9.19.

Unlike `Solve`, `FindRoot` will not tolerate the presence of symbolic constants in the equations it is solving. This is because all numerical techniques work by guessing a solution, plugging the solution into the equation, comparing the values of the two sides, and using the results of the comparison to come up with a better guess. This will not work if the equations contain any symbols other than the unknowns. For example, if we attempt to solve

```
In[23]:= FindRoot[a*x^2 + b*x + c == 0, {x, 1.0}]

         FindRoot::frnum : Function 1. a + 1. b + c
             is not a length 1 list of numbers at
             {x} = {1.}.
```
(9.23)

Mathematica complains because of the presence of the symbols a, b, and c.

9.2 Bisection Method

The value for `angle` that we calculated in Example 9.15 leads directly to the solution to Old MacDonald's problem. If we plug it into Equation 9.8, we find that the rope should be approximately 1.16 times the radius of the pasture, or 11.6 meters long. Although this completes the solution to the problem, we will press on because we want to shed light on how numerical equation solving works.

Although numerical equation-solving techniques have been known and used since the time of the ancient Greeks, they became much more practical with the advent of computers. In the remainder of this chapter we will study the *bisection method*, which is perhaps the oldest and simplest numerical technique for solving equations.

The bisection method is a straightforward, easily grasped technique for finding numerical solutions to equations in one unknown. Although it is reliable, it is slow compared with more sophisticated numerical techniques. *Mathematica*'s FindRoot function, to take one example, is much faster on average than the bisection method. Nevertheless, the bisection method is an excellent starting point for understanding the problem of finding roots numerically.

Suppose that we have a function f that takes one parameter and has at least one real root x, so that $f(x) = 0$. The goal of the bisection method is to find a floating-point approximation to one of the roots. In other words, the goal is to find a number x such that $f(x)$ is close to zero.

Before we can apply the bisection method, we must first find two numbers, *pos* and *neg*, such that $f(pos) > 0$ and $f(neg) < 0$. If these two numbers exist, and if f is continuous between *pos* and *neg*, there must be at least one value x that falls between *pos* and *neg* and for which $f(x) = 0$. This value x is a root of the function.

The bisection method works by narrowing the gap between *pos* and *neg* until it closes in on the correct answer. It narrows the gap by taking the average of *pos* and *neg*, which we will call *avg*. There are then two possibilities:

1. If $f(avg)$ is non-negative, we change the value of *pos* to be the same as *avg*.
2. If $f(avg)$ is negative, we change the value of *neg* to be the same as *avg*.

At this point we can still be certain that a root must lie between *pos* and *neg*. Furthermore, the distance between *pos* and *neg* will have been cut in half. If we repeat this process long enough, *pos* and *neg* will eventually be sufficiently close that their average will be a good approximation to a root.

Any equation (for example, Equation 9.7) can be rearranged into a function

```
In[24]:= cow[theta_] :=
            Sin[theta] - theta*Cos[theta] - Pi/2
```

(9.24)

whose roots are the same as those of the equation. Thus, the bisection method is generally applicable to equations in one unknown.

9.2.1 Illustration

We will use *Mathematica* to illustrate the central idea behind the bisection method by working with the function cow defined above in Example 9.24. It is not difficult to find a point at which cow is positive

```
In[25]:= cow[2.5]

Out[25]= 1.03053
```

(9.25)

and another at which cow is negative.

```
In[26]:= cow[1.5]

Out[26]= −0.679407
```

(9.26)

Thus, our initial value for *pos* is 2.5 and our initial value for *neg* is 1.5. If we plot our function between these two bounds,

```
In[27]:= Plot[cow[theta], {theta, 1.5, 2.5}]

Out[27]= (See Figure 9.4a)
```

(9.27)

we see that the graph of the function crosses the *x*-axis somewhere between 1.5 and 2.5. This is the root that we will be approximating.

By evaluating cow at the average of 1.5 and 2.5,

```
In[28]:= cow[2.0]

Out[28]= 0.170795
```

(9.28)

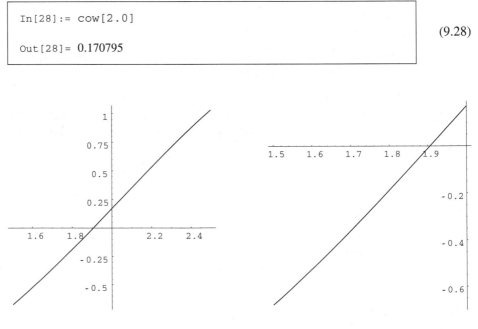

(a) Initial range (Example 9.27)   (b) Narrowed range (Example 9.29)

Figure 9.4   Using the bisection process to narrow the range in which a root must lie

we find that cow is positive at that point. This tells us that the root must lie between 1.5 and 2.0, as we can verify by plotting cow between those two bounds.

```
In[29]:= Plot[cow[theta], {theta, 1.5, 2.0}]

Out[29]= (See Figure 9.4b)
```
(9.29)

Let's evaluate cow once more at the average of 1.5 and 2.0.

```
In[30]:= cow[1.75]

Out[30]= -0.27488
```
(9.30)

The function is negative at this point, so we now know that the root lies between 1.75 and 2.0. In two steps we have cut the length of the interval in which the root must lie from 1.0 to 0.5 to 0.25. Table 9.1 shows how the interval continues to close in through 11 steps.

Table 9.1   Eleven repetitions of the bisection process

pos	neg	avg	cow(avg)
2.50000	1.50000	2.00000	0.170795
2.00000	1.50000	1.75000	-0.274880
2.00000	1.75000	1.87500	-0.0550852
2.00000	1.87500	1.93750	0.0573896
1.93750	1.87500	1.90625	0.000997637
1.90625	1.87500	1.89063	-0.0270871
1.90625	1.89063	1.89844	-0.0130550
1.90625	1.89844	1.90234	-0.00603116
1.90625	1.90234	1.90430	-0.00251738
1.90625	1.90430	1.90527	-0.000760021
1.90625	1.90527	1.90576	0.000118770

If we quit after 11 repetitions of the bisection process, the two endpoints will round to the same three digits (1.91) but not to the same four digits (1.906 versus 1.905). As a result, all that we can claim is that the root, when rounded to three digits, will be 1.91. If we want a better approximation to the root, we must continue the bisection process.

Because we are using 16-digit mantissas, the two endpoints of the interval will eventually agree to at least 15 digits. But we will never find the exact root—it is irrational, and all floating-point numbers are rational.

To keep the bisection process from continuing indefinitely, any implementation of the bisection method must include a criterion by which it can decide when it is close enough to a root. We will study this issue further in Section 9.4.

9.3 A Bisection Function

Carrying out the bisection method as outlined in the previous section, even with the assistance of *Mathematica*, would be time consuming and error prone. Each function whose root we wished to find would require a long series of tedious calculations. It would be much more convenient if we could package the bisection method into a function of three parameters so that we could invoke it as

```
In[31]:= bisection[cow, 2.5, 1.5]

Out[31]= 1.9057
```
(9.31)

The `bisection` function would take three parameters—a function, a positive guess, and a negative guess—and would narrow the guesses down to an approximate root for the function. For the remainder of this chapter, we will use *Mathematica* to implement such a function.

The body of every programmer-defined function that we have created to this point has been a pure expression. Recall from our discussion in Chapter 3 that a pure expression has no side effects; it is evaluated solely to obtain its value. Some of the many kinds of pure expressions that we have encountered so far are

- numbers (3, 3.14, 3/7, 3.14*^5);
- variables (g, t);
- arithmetic expressions (3+2, (x+5)*y);
- function calls (f[3.5], horizontal[V,theta,t]); and
- piecewise expressions (Which[x<0,-1,x==0,0,1]).

Our `bisection` function will be qualitatively different from every other programmer-defined function that we have implemented to this point: its body will be an *imperative* expression. Recall that an imperative expression is one with side effects. The kinds of imperative expressions that we have encountered so far are

- assignment expressions (x = e), which have the side effect of associating the value of the expression e with the variable x;
- delayed assignment expressions (f[x_] := e), which have the side effect of creating a function f[x] whose body is the unevaluated expression e; and
- package-loading expressions (Needs["Package`"]), which have the side effect of loading a package named Package.

Implementing the bisection method will involve arranging for a sequence of assignments to be carried out in a particular order.

As we work towards implementing `bisection`, we will survey some of the ways in which imperative expressions are commonly composed with one another in *Mathematica*. In particular, we will see how to arrange for expressions to be evaluated sequentially; how to arrange for *Mathematica* to decide, based on the value of a Boolean expression, which of two expressions to evaluate; how to arrange for *Mathematica* to evaluate a single expression multiple times; and how to arrange for *Mathematica* to encapsulate the side effects of an imperative expression so that it behaves, when viewed externally, like a pure expression.

We will be giving a deliberately concise and incomplete treatment to these ideas. Our goal is only to introduce the possibilities inherent in programming with side effects. We will devote most of the remainder of the text to a detailed investigation of this topic in the context of the C programming language.

9.3.1 Expression Sequences

It is straightforward to compose multiple expressions so that they will be executed in a sequential order. An expression sequence is created in *Mathematica* by separating two or more expressions with semicolons. Although *Mathematica* does not require it, we will also punctuate expression sequences by enclosing them with parentheses. When *Mathematica* evaluates an expression sequence, it evaluates the expressions in the specified order and then reports the value of the last expression as the value of the sequence.

For example, the sequence of arithmetic expressions

```
In[32]:= (1+1; 2+2; 3+3)

Out[32]= 6
```

(9.32)

has a value of 6 because the last expression in the sequence is $3+3$. Before reporting the value of the sequence, *Mathematica* had previously determined and discarded the values of $1+1$ and $2+2$.

Example 9.32 illustrates that it is pointless to put pure expressions into a sequence except as the last entry. Expression sequences are useful only if they contain imperative expressions whose evaluation will cause side effects. For example, the evaluation of the sequence

```
In[33]:= (x=1; x=2; y=x^2; y)

Out[33]= 4
```

(9.33)

results in x being given the value 2 and y being given the value 4, and the value of the expression sequence as a whole is 4. Thus, the expression sequence in Example 9.33 has *two* side effects.

The order in which expressions appear in a sequence can make a big difference. For example, the sequence

```
In[34]:= (x=1; y=x^2; x=2; y)

Out[34]= 1
```

(9.34)

would result in x being given the value 2 and y being given the value 1, and the value of the expression as a whole is 1. Compare this with Example 9.33, which is composed of the same four expressions in a different order.

Figure 9.5 shows a sequential composition of the ten expressions that we might evaluate to carry out two repetitions of the bisection method. They are evaluated by *Mathematica* in order from first to last. Although we have used blank lines to arrange the expressions into four groups for easier reading, this has no significance to *Mathematica*. We have also numbered the lines for easy reference; the line numbers should *not* be entered into *Mathematica*.

The first three expressions (lines 1–3) are designed to set things up by defining f, pos, and neg. Throughout the expression sequence, f is the function whose root we are finding, pos is the interval endpoint at which f is positive, and neg is the interval endpoint at which f is negative.

The next three expressions (lines 5–7) perform the first repetition of the bisection process. The variable avg is set to the average of pos and neg, and then f[avg] is computed. Since the result is positive, the value of pos is changed to be the same as that of avg.

The third group of three expressions (lines 9–11) performs the second repetition of the bisection process. A new value for avg is assigned and a new value for

```
In[35]:=

1    (f = cow;
2       pos = 2.5;
3       neg = 1.5;
4
5       avg = (pos + neg) / 2.0;
6       f[avg];
7       pos = avg;
8
9       avg = (pos + neg) / 2.0;
10      f[avg];
11      neg = avg;
12
13      avg)

Out[35]= 1.75
```

Figure 9.5   Two repetitions of the bisection process

`f[avg]` is computed. This time the result is negative, so we change `neg` instead of `pos`.

The final expression (line 13) is included so that the final value of the sequence—the one reported by *Mathematica*—will be the final value of `avg`. This is the approximation to the root of `f` that is produced by two repetitions of the bisection process.

The ten expressions in Figure 9.5 do not describe the entire process, of course. Many more repetitions are required to arrive at a good approximation.

9.3.2 Conditional Expressions

At the assignments on lines 7 and 11 of Figure 9.5 we, as creators of the expression sequence, had to make a decision. In each case we had to decide whether to change the value of `pos` via

```
pos = avg
```

or the value of `neg` via

```
neg = avg
```

We based our decision on the sign of `f[avg]` computed in the previous expression (`f[avg]` returns the values of cow(*avg*) in Table 9.1).

What this means is that there was really no way that we could have come up with all ten expressions in advance before presenting them to *Mathematica*. We had to know the result of the expression on line 6 before choosing the expression on line 7, and similarly for the expressions on lines 10 and 11.

We can address this problem by presenting the two alternatives to *Mathematica* and letting it make the choice for us. We have already seen one way in which we can arrange for *Mathematica* to make a choice. In Chapter 8 we used *Mathematica*'s `Which` expression to create conditional expressions. By using `Which`, we were able to create an expression whose value depended on the Boolean expressions that it contained. For example, the value of

```
Which[x>=0, 15, True, 20]
```

is 15 if x is non-negative and 20 otherwise.

The problem that we are faced with here is similar, and we could indeed employ a `Which` expression to solve it, but we will use an `If` expression instead. Our goal is to eventually apply, in the context of C, the concepts that we are developing here, and as we learn C we will make extensive use of a construct that is very similar to *Mathematica*'s `If` expression.

The `If` expression

```
If[x>=0, 15, 20]
```

behaves identically to the `Which` expression illustrated above. Its value is 15 if x is non-negative, and is 20 otherwise. `If` takes three parameters. If the value of the first parameter is `True`, it returns the value of the second parameter; if the value of the first parameter is `False`, it returns the value of the third parameter.

We have rewritten the expression sequence from Figure 9.5 into a version in Figure 9.6 that makes use of an `If` expression. We have replaced the expressions on lines 6–7 of Figure 9.5 with the conditional expression on lines 6–8 of Figure 9.6, and the expressions on lines 10–11 of Figure 9.5 with the conditional expression on lines 11–13 of Figure 9.6.

Each of the identical conditionals of Figure 9.6 works as follows. *Mathematica* first evaluates the Boolean expression `f[avg] >= 0`. If it evaluates to `True`, then the assignment `pos = avg` is executed; otherwise, the assignment `neg = avg` is executed.

```
In[36]:=

1    (f = cow;
2       pos = 2.5;
3       neg = 1.5;
4
5       avg = (pos + neg) / 2.0;
6       If[f[avg] >= 0,
7          pos = avg,
8          neg = avg];
9
10      avg = (pos + neg) / 2.0;
11      If[f[avg] >= 0,
12         pos = avg,
13         neg = avg];
14
15      avg)

Out[36]= 1.75
```

Figure 9.6   The bisection process with conditionals

9.3.3 Repetition

There is still a major problem with the expression sequence in Figure 9.6. After the first three expressions, which serve to set things up, we end up repeating the same pair of expressions twice before reporting the value of `avg` as an approximate root. If we wanted to obtain a better approximation to the root, we would have to repeat the pair of expressions many more times. Furthermore, we would need to devise some means of determining when we had done enough repetitions and were sufficiently close to a root.

Mathematica provides a `While` expression with which the programmer can instruct *Mathematica* to execute the same expression over and over. The expression sequence from Figure 9.6, rewritten to take advantage of a `While` expression, appears in Figure 9.7. We have replaced the two pairs of expressions on lines 5–13 of Figure 9.6 with the `While` expression on lines 5–9 of Figure 9.7.

```
In[37]:=

1     (f = cow;
2       pos = 2.5;
3       neg = 1.5;
4
5       While[Abs[pos-neg] >= 1*^-6,
6               (avg = (pos + neg) / 2.0;
7                If[f[avg] >= 0,
8                     pos = avg,
9                     neg = avg])];
10
11      avg)

Out[37]= 1.9057
```

Figure 9.7   The bisection process with repetition

`While` takes two parameters. The first is a Boolean expression that we will call the condition, and the second is an expression that we will call the body. *Mathematica* evaluates a `While` expression as follows.

1. *Mathematica* evaluates the condition.
2. If the condition evaluates to `True`, *Mathematica* evaluates the body and returns to step 1.
3. Otherwise, the evaluation of the `While` expression is finished.

The `While` expression in Figure 9.7, then, works as follows.

1. *Mathematica* evaluates the Boolean expression `Abs[pos-neg] >= 1*^-6`.
2. If the expression evaluates to `True`, *Mathematica* evaluates the two-expression sequence on lines 6–9 that makes up the second parameter to `While`, and then returns to step 1.
3. Otherwise, the evaluation of the `While` expression is finished and the value of `avg` is an approximation to the root of `f`.

Expressions such as the `While` expression Figure 9.7 are often called *loops*.

Our `While` loop repeatedly carries out the bisection process until the difference between `pos` and `neg` is less than 10^{-6}. When the `While` expression finally terminates, the value of `avg` is 1.9057, which is an approximation to the root of `cow`.

```
In[38]:= cow[avg]

Out[38]= 3.34609 × 10⁻⁷
```

(9.38)

Although f[avg] is close to zero, we can get even closer if we control the While loop differently. We will consider the problem of deciding when to terminate the loop in Section 9.4.

9.3.4 Modules

We now have a usable *Mathematica* implementation of the bisection method. By assigning appropriate values on lines 1–3 of Figure 9.7, we can solve any equation we please. The implementation still has a significant shortcoming, however. It changes the values of the global variables f, pos, neg, and avg as it calculates a root.

This is a serious problem. You probably do not expect the values of variables to change when you invoke, for example, FindRoot. If we want to make our implementation of the bisection method more generally applicable, we need to change it so that it has no side effects. In other words, we need to modify the program in Figure 9.7 so that it is a pure, instead of an imperative, expression.

Fortunately, we do not need to significantly modify the code in Figure 9.7. We merely need to arrange for *Mathematica* to treat the four variables f, pos, neg, and avg as being *local* to the bisection method implementation instead of *global*.

Figure 9.8 shows how we can use *Mathematica*'s Module construct to achieve this end. Module takes two parameters: a list of variables and an expression. Notice that the first parameter to Module on line 1 of Figure 9.8 is a list of the four variables to which the expression sequence of Figure 9.7 assigns, and that the second parameter on lines 2–12 is exactly that expression sequence.

```
In[39]:=

1    Module[f, pos, neg, avg,
2          (f = cow;
3             pos = 2.5;
4             neg = 1.5;
5
6             While[Abs[pos-neg] >= 1*^-6,
7                   (avg = (pos + neg) / 2.0;
8                    If[f[avg] >= 0,
9                        pos = avg,
10                       neg = avg])];
11
12            avg)]

Out[39]= 1.9057
```

Figure 9.8   The bisection process encapsulated into a module

When *Mathematica* evaluates a Module expression, it evaluates the second parameter as if it had been typed directly into *Mathematica*, but with one exception. All variables in the list that appears as the first parameter are treated as local. This means that the effects of the assignments on lines 2–12 of the module will not be visible after the `Module` has been evaluated. In other words, the assignments are to local versions of the variables `f`, `pos`, `neg`, and `avg`, *not* to the global variables of the same name.

By encapsulating the side effects caused when its second parameter is evaluated so that they are no longer visible, the `Module` expression in Figure 9.8 qualifies as a pure expression. This means that we can evaluate the expression anywhere in a *Mathematica* notebook without having to worry that the values of existing variables will inadvertently be changed.

```
In[40]:=

1    bisection[function_, positive_, negative_] :=
2
3        Module[f, pos, neg, avg,
4                (f = function;
5                 pos = positive;
6                 neg = negative;
7
8                 While[Abs[pos-neg] >= 1*^-6,
9                       (avg = (pos + neg) / 2.0;
10                       If[f[avg] >= 0,
11                           pos = avg,
12                           neg = avg])];
13
14                avg)]
```

Figure 9.9   A bisection method function

9.3.5 The Finished Function

All that remains is to package the `Module` expression of Figure 9.8 as a function so that it is more convenient to use. In Figure 9.9 we have completed the task of converting our implementation into a function of three parameters. We can now use it to compute roots of equations, as in

$$
\boxed{
\begin{array}{l}
\texttt{In[41]:= bisection[cow, 2.5, 1.5]}\\[6pt]
\texttt{Out[41]= 1.9057}
\end{array}
}
\qquad (9.41)
$$

If our `bisection` function is called as in Example 9.41, lines 3–14 of Figure 9.9 will be exactly as in Figure 9.8 once the formal parameters are replaced with the actual parameters. This means that the loop will eventually terminate with `avg` having a value of 1.9057. At that point, the value `avg`, which is the final expression in the body of `bisection`, will be returned as the result of the function.

9.3.6 Programming Styles

Different programming languages facilitate different styles of programming. An experienced programmer is typically the master of more than one style, and as a result is able to exploit the nuances of more than one programming language.

Mathematica facilitates at least *three* different styles of programming: functional programming, imperative programming, and rule-based programming. Most of the book to this point has focused on functional programming, which is programming with pure expressions. Except for a limited use of assignment expressions, we worked entirely with pure expressions in Chapters 2–8.

In this chapter we have focused on imperative programming, which is programming with imperative expressions. We did this in preparation for our transition to C in Chapter 10, because C facilitates only the imperative style of programming.

We will not discuss rule-based programming in this book. While rule-based programming is a powerful and, in some ways, unique aspect of *Mathematica*, the issues involved are a bit more subtle than with functional and imperative programming. Discussing rule-based programming at this point would entail a rather lengthy diversion.

9.4 Assessment

In this section we will assess our implementation of the bisection method. We defer to Chapter 14 an assessment of the method itself.

There is plenty of room for improvement in our `bisection` function. The convergence criterion—the test that we use to decide whether or not to continue the `While` expression—is seriously flawed. In the current implementation we continue the bisection process until the difference between `pos` and `neg` is less than 10^{-6}. In this section we will explore some reasons why this is a bad idea and then develop a better criterion.

9.4.1 Poor Approximation

The most obvious problem is that our `bisection` function does not generally make use of all the digits that are available in *Mathematica*'s floating-point mantissas. For example, compare the 16-digit solutions obtained with `FindRoot`

```
In[42]:= InputForm[FindRoot[cow[theta], {theta, 1.0},
                  AccuracyGoal->16]]

Out[42]= {theta → 1.905695729309883}
```

(9.42)

and `bisection`.

```
In[43]:= InputForm[bisection[cow, 2.5, 1.5]]

Out[43]= 1.905695915222167
```

(9.43)

These two solutions agree only in their first seven digits. The answer produced by `bisection` cannot hope to be meaningful beyond its first seven digits because of the way the bisection process is cut off when `pos` and `neg` become within 10^{-6} of each other.

9.4.2 Infinite Repetition

If the root we are looking for is sufficiently large, it may be impossible for `pos` and `neg` to ever get within 10^{-6} of each other. For example, suppose that we create a function whose root is at 2×10^{20},

```
In[44]:= bigRoot[x_] := x - 2.*^20
```

(9.44)

and then use `bisection` to find the root.

```
In[45]:= bisection[bigRoot, 3.*^20, 1.*^20]

Out[45]= $Aborted
```

(9.45)

As `bisection` runs, `pos` and `neg` will converge until `pos` is 2.0×10^{20} and `neg` is $1.999999999999999 \times 10^{20}$. Because of roundoff error they will never become equal, and because 16-digit mantissas are in use they can never get any closer. Since their difference is greater than 10^{-6}, the `While` loop will never terminate, the function will never return an answer, and we will have to interrupt the computation. This possibility—which is called an *infinite loop*—must be carefully avoided when writing functions containing loops.

9.4.3 Abysmal Approximation

If the root we are looking for is small, the loop may terminate while we are still orders of magnitude away from that root. For example, if we create a function whose root is at 1×10^{-20},

```
In[46]:= smallRoot[x_] := x - 1.*^-20
```

(9.46)

and then use `bisection` to find the root,

```
In[47]:= bisection[smallRoot, 1.0, 0.0]

Out[47]= 9.53674 × 10⁻⁷
```
(9.47)

our answer is over 13 orders of magnitudes larger than the actual root—not a very good approximation.

9.4.4 No Repetition

If the initial guesses are sufficiently close together, the loop may not run even once, which will mean that the value returned by bisection will be undefined. For example, here

```
In[48]:= bisection[smallRoot, 2.*^-8, 0]

Out[48]= avg$12
```
(9.48)

we end up with a symbolic answer.

9.4.5 Bad Guesses

The bisection method requires that the initial values for pos and neg satisfy the inequalities f[pos] > 0 and f[neg] < 0. If they do not, as in

```
In[49]:= bisection[bigRoot, 1.*^20, 3.*^20]

Out[49]= 3. × 10²⁰
```
(9.49)

the bisection process will converge to something that is not even remotely a root.

9.4.6 An Improved Implementation

The central problem with the bisection function is that we are relying on the *absolute* difference between pos and neg to determine whether or not the loop should continue. This ignores the number of digits in the floating-point mantissas and the magnitude of the roots.

A much better approach is to let the loop continue until the difference between pos and neg stops getting smaller. When this happens, the loop cannot possibly make any more progress and we may as well stop it. A revised version of bisection using this idea appears in Figure 9.10.

By continuing until the difference between pos and neg stops decreasing, we guarantee that every digit of the floating-point mantissas will be exploited. This approach also adapts nicely to the variations in mantissa size between different

```
In[50]:=

1    bisection[function_, positive_, negative_] :=
2
3      Module[f, pos, neg, avg, delta,
4
5           (f = function;
6            pos = positive;
7            neg = negative;
8
9            If[f[pos]<0, Return["Positive guess is bad"]];
10           If[f[neg]>0, Return["Negative guess is bad"]];
11
12           delta = Abs[pos-neg] * 2;
13
14           While[Abs[pos-neg] < delta,
15                 (avg = (pos + neg) / 2.0;
16                  delta = Abs[pos-neg];
17                  If[f[avg] >= 0,
18                      pos = avg,
19                      neg = avg])];
20
21           avg)]
```

Figure 9.10   Improved bisection function

versions of *Mathematica*. The absolute difference between pos and neg will be large when the root is large and small when the root is small, which is exactly what we want.

We have implemented the change by replacing the termination condition for the While loop that begins on line 14. This condition compares the difference between pos and neg to delta, which is the difference between pos and neg that held during the *previous* repetition of the loop. We have included delta as one of the variables introduced on line 3, we have added the assignment on line 12 to give delta a value *before* the loop is executed so that there will be a basis for the first termination test, and we have added the assignment on line 16 so that the value of delta will be continually updated.

We have also added the two conditional expressions on lines 9 and 10. These conditionals verify that the values for pos and neg are appropriate. If they are not, we return from the function immediately with a warning message. (The evaluation of a Return expression forces the enclosing function to return immediately with the parameter of the Return becoming the function's result.)

Our improved implementation yields close approximations for each of Examples 9.43–9.49, which we explored earlier in this section. For example, the results obtained by bisection for the root of the cow function

```
In[51]:= InputForm[bisection[cow, 2.5, 1.5]]

Out[51]= 1.905695729309883
```

(9.51)

now differs only in the last digit from the result produced by `FindRoot` in Example 9.42.

One problem remains with this convergence test: it doesn't behave well for roots at zero. We will explore this problem in the exercises.

9.5 Key Concepts

Solving equations.   Equations can be solved either symbolically or numerically. People generally solve equations symbolically by doing algebraic transformations. Computers can often solve equations numerically, which involves repeatedly guessing at a solution until a sufficiently good answer is found.

Bisection method.   The bisection method is a simple but effective numerical technique for solving equations in one unknown. The key idea is to determine two guesses that are known to bracket a root and to then narrow the distance between them.

Pure versus imperative expressions.   Pure expressions are evaluated to obtain a value; imperative expressions are evaluated to cause a side effect. Most of the computing that we did prior to this chapter was with pure expressions, but in this chapter we made extensive use of imperative expressions.

Composing imperative expressions.   Common ways to compose imperative expressions are with expression sequences, conditional expressions, repetition expressions, and modules. An expression sequence executes two or more simpler expressions in order, a conditional expression chooses among alternative expressions, a repetition expression repeatedly executes an expression, and a module hides the side effects of an expression.

9.6 Exercises

9.1   Which of the following equations can *Mathematica* solve for x using `Solve`? Using `FindRoot`?

(a) $\cos(x) = x$

(b) $\sqrt{ax} = x$

(c) $\sin^2(x) + \cos^2(x) = 1$

(d) $e^x = e^{-x}$

(e) $ax^5 + 5 = 0$

(f) $x^5 + 5 = 0$

9.2   Suppose that when the bisection method is used to find a root for a function f, values for *pos* and *neg* are chosen by mistake so that both $f(pos)$

and $f(neg)$ are positive. Under what circumstances will a root still be found? What will happen if a root is not found?

9.3 Repeat Exercise 9.2 for the case where initial values for *pos* and *neg* are chosen so that $f(pos)$ is negative and $f(neg)$ is positive.

9.4 If the bisection method is used with a function that has multiple roots, what determines which of the roots will actually be found?

9.5 How can the bisection method go wrong when the equation being solved involves a discontinuous function, as with $\frac{1}{x} - 1$?

9.6 If the difference between the initial values of *pos* and *neg* is 10.0, how many repetitions of the bisection method will be required to guarantee that the root found is within 0.00001 of an actual root?

9.7 We were unable to solve the equation from Example 6.27 using `Solve`. Find a solution using `FindRoot`. (To make things easier for `FindRoot`, write all of the constants in the equation as floating-point numbers.)

9.8 Use the `bisection` function from Figure 9.9 to solve the equation from Example 6.27. Compare your result with the one obtained in Exercise 9.7.

9.9 Use the `bisection` function from Figure 9.10 to find the length of rope that Old MacDonald should use if he wants his cow to be able to range over only one quarter of his pasture.

9.10 Use the `bisection` function from Figure 9.10 to determine how high a hill must be so that the distance to the horizon from the top of that hill will be 75,000 feet as discussed in Chapter 5.

9.11 Repeat Examples 9.45–9.49 using the `bisection` function from Figure 9.10.

9.12 Repeat Exercise 8.20 using the `bisection` function from Figure 9.10 to solve the equation.

9.13 Use the `bisection` function from Figure 9.10 to find a root of the function $(x - 2)(x - 4)(x - 6)$. Find an initial value for `pos` that is greater than 6 and an initial value for `neg` that is less than 2 such that the root at 2 is found. Find a second pair in the same range so that the root at 6 is found. Is there an initial value for `pos` greater than 6 and an initial value for `neg` less than 2 so that the root at 4 is found?

9.14 Modify the bisection function from Figure 9.9 so that the While loop that begins on line 8 continues so long as the absolute value of the ratio of pos-neg and pos+neg is greater than 10^{-6}.

9.15 Repeat Examples 9.43–9.48 using the version of bisection from Exercise 9.14. How do the results compare with those obtained by the version of bisection from Figure 9.9? From Figure 9.10?

9.16 Modify the bisection function from Figure 9.10 by adding the expression

```
Print[InputForm[{pos, neg, avg, f[avg]}]]
```

as the first expression of the expression sequence on lines 15–19 that makes up the body of the While. This will display the values of pos, neg, avg, and f[avg] at the beginning of each repetition of the While loop.

Use the modified implementation to create versions of Table 9.1 for the functions from Examples 9.45–9.48.

9.17 Use the modified implementation of the bisection function from Exercise 9.16 to investigate the difficulty that the convergence test exhibits when finding a root at zero.

9.18 Modify the bisection function from Figure 9.10 as follows.

- Add a variable count to the list of local variables on line 3.
- Add an assignment to initialize count to be zero immediately after the expression that initializes delta on line 12.
- Add as the first expression of the body of the While loop that begins on line 14 an assignment expression that increases the value of count by one.
- Replace avg with count on line 21 so that the function will now return the number of iterations required to find a root.

Use the modified implementation to create a plot showing how the number of repetitions required to find the root of the cow function depends on the difference between the initial guesses.

9.19 Use the version of bisection from Exercise 9.18 to create a plot showing the number of repetitions required to find the root of the equation $x - 10^{-n} = 0$ as n varies from 1 to 100. Use 1 as the initial value of pos and 0 as the initial value of neg.

9.20 Demonstrate that the version of bisection from Exercise 9.18 does not take advantage of the situation when one of the initial guesses happens to be an exact root.

10

Introduction to C

This chapter marks a transition. To this point we have used *Mathematica* to implement our models. For the remainder of the book we will use the programming language C instead. We will deviate in this chapter from our approach of organizing each chapter around a problem. Instead, we will set the stage for the rest of the book by comparing *Mathematica* and C, and we will begin our study of C by examining two simple C programs that solve problems with which you are already familiar.

The field of computational science is so broad, and the collection of available programming languages so varied, that no single language can be appropriate for every problem. As we stressed in Chapter 1, we have chosen to focus on *Mathematica* and C for two reasons. First, each is widely used for scientific programming. Second, each is representative of one of the two classes of programming languages that are most commonly used by computational scientists. *Mathematica* and C are far from being the only choices, however. We could just as easily have written this book using Fortran, for example, as our conventional language; we *have* written a version of this book that uses Maple as its computer algebra system.

10.1 *Mathematica* Background

Mathematica is a system for doing numerical, symbolic, and graphical mathematics. It is also known, somewhat inaccurately, as a computer algebra system. *Mathematica* provides an interface that, among other things, supports writing *Mathematica* programs using notebooks, has an easily accessible help feature, and facilitates saving notebooks to files.

Mathematica is also the name for the programming language that we have studied to this point and which is used within the *Mathematica* system. Among other things, it allows the manipulation of numerical, symbolic, and graphical data and supports the definition of programmer-defined functions.

Although experimental computer algebra systems have existed for more than 25 years, they have been widely used only for the last ten. The first commercial version of *Mathematica*, for example, was released in 1988. Other computer algebra systems include Maple and Macsyma.

Mathematica is a copyrighted product of Wolfram Research Inc. The interface presented by the *Mathematica* system varies somewhat depending on the graphical capabilities of the computer on which it is running. The DOS version, for example, is necessarily very different from the Macintosh version. The *Mathematica* programming language, however, is identical across platforms.

10.2 C Background

C is an example of what is traditionally called a high-level programming language, but it is more accurately called a conventional programming language. Conventional languages were first developed in the early 1950s, when it became clear that writing large programs in low-level machine languages was too tedious and error prone. The most successful of the conventional languages from those days, COBOL and Fortran, are still with us today, albeit in significantly different forms.

C was developed in 1970 by Dennis Ritchie at AT&T's Bell Laboratories, where it was first used to implement the UNIX operating system on the PDP-11 computer. C gradually spread beyond Bell Labs and the PDP-11. It gained in popularity, and in 1988 its design was standardized by the American National Standards Institute (ANSI) and later by the International Standards Organization (ISO).

C is not a copyrighted product as *Mathematica* is, and as a result many vendors have marketed implementations of C. As with *Mathematica*, it is important to distinguish between the programming systems that support the use of C and the language C itself. The programming systems provided by the various vendors differ tremendously, much more than do the various *Mathematica* systems. Thanks to the ANSI and ISO standards, however, the language itself is identical across all of these systems.

10.3 An Example C Program

Figures 10.1–10.2 show a C implementation of the bisection method that we discussed in Chapter 9. It is instructive to compare this program with the *Mathematica* implementation in Figure 9.9. The differences will certainly stand out at first reading, but there are also many similarities. The similarities between *Mathematica* and C will pay off as we begin studying C.

The captions of Figures 10.1–10.2 briefly describe the program. The caption of Figure \10.2 also includes, in parentheses, the name of the file (`bisect1.c`) containing the program that appears on the diskette included with this book. We

```
1    #include <stdio.h>
2    #include <math.h>
3
4
5    /* Old MacDonald's cow function from Chapter 9. */
6
7    double cow (double theta)
8    {
9      return(sin(theta) - theta*cos(theta) - 3.141592654/2.0);
10   }
11
12
13   /* Uses the bisection method to find an approximate root
14      for "f", where f(pos) > 0 and f(neg) < 0. */
15
16   double bisection (double f (double), double pos, double neg)
17   {
18     double avg, delta;
19     delta = fabs(pos-neg) * 2;
20
21     while (fabs(pos-neg) < delta) {
22       avg = (pos + neg) / 2.0;
23       delta = fabs(pos-neg);
24       if (f(avg) >= 0) {
25         pos = avg;
26       }
27       else {
28         neg = avg;
29       }
30     }
31
32     return(avg);
33   }
```

Figure 10.1   C implementation of bisection method (continued in Figure 10.2)

will similarly caption every C program that we present in this book. We will also occasionally refer by name to programs that are included on the diskette but do not appear in figures. The line numbers contained in the two figures are not a part of the program, and do not appear in the version on the diskette. We have included them—as we will in all of our example programs—for ease of reference.

C programs consist of a mixture of comments and code. The comments in C programs serve to make things easier for human readers and are otherwise irrelevant. We have included the comments for bisect1.c in Figures 10.1–10.2 to show you that C comments begin with the two characters /* and end with the two characters */.

For the remainder of the book we will omit the comments from programs when they appear in figures. They *will* be present, however, in the versions on the diskette.

```
34   /* Prompts for the positive and negative approximations to
35       the root of the function cow defined above, then
36       computes and displays an approximation to cow's root. */
37
38   void main ()
39   {
40     double positive, negative, root;
41     printf("Enter positive guess: ");
42     scanf("%lf", &positive);
43     printf("Enter negative guess: ");
44     scanf("%lf", &negative);
45     root = bisection(cow, positive, negative);
46     printf("The root is %.10g\n", root);
47   }
```

Figure 10.2   C implementation of bisection method (`bisect1.c`)

We make this omission for two reasons. First, it will save a considerable amount of space. Second, the explanations we will give in the book for each part of each program will be far more comprehensive than the comments. When you write programs of your own, you should take inspiration from the versions of our programs that appear on the diskette and be generous with comments. A program without comments is like an unalphabetized phone directory: it is full of information that is difficult to understand.

Notice that `bisect1.c` contains such formatting features as blank and indented lines. This is done for the benefit of the human reader but is of no further consequence. C ignores white space in a program except as it serves to separate words from one another. We will use a consistent formatting style throughout this book, and as you become familiar with it you will be able to understand much about a program simply from its appearance.

It is not our goal in this chapter to undertake a detailed explanation of how `bisect1.c` works. In fact, we will not have discussed every aspect of C present in Figures 10.1–10.2 until we complete Chapter 15. For now it is important only to notice that, whereas the *Mathematica* implementation consisted only of the `bisection` function, the C implementation consists of the function `cow` whose root is to be found (lines 7–10), the `bisection` function (lines 16–33), and a `main` function (lines 38–47). We will return to this point in Section 10.4.2.

10.4 Interpreters versus Compilers

When you begin using C, the first difference between *Mathematica* and C that you will notice, and the one that will most profoundly color your impressions of the two languages, is that *Mathematica* is an *interpreted* language whereas C is a *compiled* language. Even though there are more important differences between the two languages, and even though the interpreted/compiled distinction has no bearing on what programs can be written with the two languages, the distinction will crucially influence how you go about developing programs. Consequently, we will be taking a significantly different approach to writing C programs than we have taken to writing *Mathematica* programs.

10.4.1 *Mathematica* is Interpreted

Interpreted languages such as *Mathematica* facilitate an interactive style of program development with which you should already be familiar. When you type a command, *Mathematica* reads it, evaluates it, and immediately displays the result. This ability to interactively enter a command and immediately see its result is the most important aspect of an interpreted language. It makes it easy to develop and test programs incrementally, one piece at a time.

Think back to how we developed the implementation of the bisection method in Chapter 9. We began by entering a sequence of assignment expressions, the last two of which were assignments to the variables pos and neg. We then selectively and repeatedly evaluated these two assignments and watched as pos and neg narrowed down to an approximate root. This interactive experiment helped us understand how the bisection method worked, and helped convince us that our implementation efforts were on the right track.

From this point we focused on packaging the implementation as a function. We embedded the assignments to pos and neg into a conditional expression so that *Mathematica* could decide which to execute. We embedded the conditional inside of a loop so that *Mathematica* could take care of doing the repetition for us. We encapsulated the loop-based program inside of a module so that we could hide its side-effects. Finally, we embedded everything inside a function. In the end, after interactively experimenting with our program at each step of its development, we had a function called bisection that we could use to solve a function by interactively entering the function call

$$
\begin{array}{|l}
\hline
\texttt{In[1]:= bisection[cow, 2.5, 1.5]} \\
\\
\texttt{Out[1]= 1.9057} \\
\hline
\end{array}
\qquad (10.1)
$$

and observing the result. Experimenting with different initial guesses, or even with a different function, was as easy as typing in different parameters to bisection.

10.4.2 C is Compiled

Compiled languages such as C, in contrast, do *not* facilitate this interactive style of program development. Your experience with writing C programs will vary somewhat depending on what kind of computer and what version of C you are using, but it will be roughly as follows.

You will use a text editor to create a file containing a complete C program. You will then use a C compiler to translate your program into a form that your computer can execute. If your program contains syntax errors the translation will fail, the compiler will tell you where the problems are, and you will return to the text editor to repair them. Otherwise the compiler will produce a machine language translation, in a file separate from your program, that you can run.

The translation process does not produce any results. To obtain results, you must run the program. If the translated program produces the wrong results, you will correct the program using the text editor, recompile the program, and run it again. You will repeat this process until your program behaves as desired.

The bisection program in Figures 10.1–10.2, when compiled and run, will behave as illustrated below.

```
Enter positive guess: 2.5
Enter negative guess: 1.5
The root is 1.905695729
```
(10.2)

The program first prints out a message asking for the positive guess, which the user then types in. (Throughout this book, we will distinguish user input in C interaction examples by highlighting it with **boldface**.) Next the program asks for the negative guess, which the user types in. Finally, the program calculates and displays the result.

Besides calculating the result with the function bisection—which is *all* that the *Mathematica* program must do—the C program must also take care of reading in input values and writing out the result. None of this is automatic: code to produce the output and deal with the input shown in Example 10.2 is explicitly included in the main function.

The C program also contains an explicit definition of the function cow whose root is to be found. This was not part of the *Mathematica* implementation, in which the function was supplied as a parameter to the bisection function from the keyboard as shown in Example 10.1. Because there is no way in C to read a function from the keyboard in the way that numbers can be read, we had no choice but to incorporate the function into the program. As a result, if we want to find the root of a different function, we will have to modify bisect1.c.

Because a C program must be complete and self-contained, C affords no way for a programmer to experiment directly with individual expressions as is possible with *Mathematica*. This is not to say that incremental program development is impossible in C. In fact, it is as important to adopt a careful and deliberate approach to programming in C as it is in *Mathematica*. The process is different in C, however, as we will study in Chapter 13.

10.5 Differences Between *Mathematica* and C

While the fact that *Mathematica* is interpreted and C is compiled certainly affects the process of program development, it has no bearing on what can and cannot be done with the two languages. In this section we will focus on some of the more essential differences between *Mathematica* and C.

10.5.1 Intended Purposes

C is a general-purpose, conventional programming language. It can be used to implement all kinds of applications, including operating systems, Web browsers, and text editors. It can also be used to do low-level communication with laboratory devices. It can even be used to implement other programming languages. In fact, portions of *Mathematica* itself are implemented in C!

C can also be used to do numerical computations, which is how we will use it for the remainder of the text. Unlike *Mathematica*, however, C provides no ready support for symbolic or graphical computations. This does not mean that it is impossible to do symbolic computations in C—the fact that *Mathematica* is implemented in C is evidence to the contrary—but it does mean that doing symbolic computations in C is rather difficult.

Mathematica is a much higher-level language than C. It is designed to do numerical, symbolic, and graphical mathematics. No one would ever try to implement a text editor, for example, in *Mathematica*. In this book, of course, we are interested only in numerical, symbolic, and graphical mathematics, which is one reason *Mathematica* has been the focus of our attention to this point.

10.5.2 Support for Numerical Computing

It is only in the area of numerical computing that *Mathematica* and C can be considered competitors. In a nutshell, standard numerical programs are often easier to produce in *Mathematica*, but the corresponding C versions are usually faster. We will now explore some of the reasons why.

Both *Mathematica* and C provide libraries of functions for doing low-level numerical computations. For example, each provides functions for exponentiation, trigonometry, and inverse trigonometry. *Mathematica* goes further than C, however, by providing implementations of more advanced numerical methods for solving equations, integrating functions, and the like. As a C programmer, you will either have to implement these methods on your own or locate C implementations, written by others, that you can borrow. Although public-domain libraries containing C implementations of standard numerical methods have grown up over the years, C places the burden on the programmer to track down and exploit them.

There is a payoff to using C, however. A careful C implementation of a numerical method can be substantially faster than the corresponding *Mathematica* implementation, even when that *Mathematica* implementation is in the form of a built-in function. Everything else being equal, compiled languages are typically more efficient than interpreted languages. In the final analysis, the design of C was optimized for speed, while the design of *Mathematica* was optimized for ease of use.

10.5.3 Expressions and Statements

As you saw in Chapters 2–9, *Mathematica* is an *expression-oriented* programming language. Expressions are the only kind of syntactic construct in *Mathematica*. Numbers, arithmetic expressions, Boolean expressions, function calls, assignments, conditionals, and while loops are all examples of expressions. Expressions can be arbitrarily composed via nesting so long as they produce appropriate values. For example, an assignment expression can be added to a conditional expression so long as they both have numbers as values; the resulting expression will itself have a numerical value.

As you will see in Chapters 10–18, C is a *statement-oriented* programming language. C has two kinds of syntactic constructs: statements and expressions. The body of a function in C is made up of statements, and statements are in turn composed of expressions. Statements and expressions are *not* interchangeable in C.

C expressions are similar to *Mathematica* expressions: all have values, some have side effects, and they can be nested to produce new expressions. C statements do not have values, almost always cause side effects, and can be nested to produce new statements. In C, numbers, arithmetic expressions, Boolean expressions, and function calls are expressions; assignments, conditionals, and while loops are statements. (Strictly speaking, assignments are expressions, but we will treat them as statements.)

The distinction between statements and expressions will become clearer after we have looked at a few examples. What is important to realize right now is that, when programming in C, you will have to be aware of what things are expressions and what things are statements, because expressions and statements have different syntactic niches.

10.5.4 Creating Applications

Mathematica programs are useful only if you have access to *Mathematica*. For example, if you want to share a *Mathematica* program that you have written with a friend, your friend must also have access to *Mathematica*. In other words, the program is useless in the absence of the interpreter.

C is different. Once a C program has been compiled to produce an executable version, the compiler is no longer needed. The executable will run on any computer system of the type on which the original program was compiled. *Mathematica* itself is a convenient example of a compiled C program that is distributed and used independently of any implementation of C.

10.5.5 Choosing Between *Mathematica* and C

Mathematica is the appropriate choice for moderate-scale numerical, symbolic, and graphical computations. The performance advantages of C are relevant only if you are working with large amounts of data or with extremely complicated calculations. The flexibility advantages of C are relevant only if you are doing extremely specialized calculations.

In the final analysis, though, intrinsic merit is not always the deciding factor in determining which languages are used to solve a given problem. Conventional languages such as C have a longer history and are familiar to more users. If your task is to modify an existing program, you will not even have a choice about what language to use. As a result, it is wise to be familiar with both a traditional conventional language and a modern computer algebra language.

10.6   Learning C

The differences between *Mathematica* and C derive mostly from aspects that are beyond the programmer's control, such as how numbers are represented and what library functions are available. For numerical computations, the *Mathematica* and C programming languages are remarkably similar.

During the first part of this book, we focused almost entirely on programming with pure expressions. The only imperative expression that we made any use of in *Mathematica* was the assignment expression, except in Chapter 9 when we touched on more advanced kinds of imperative expressions in preparation for our transition to C.

Fortunately, almost everything you learned about programming with pure numerical expressions in *Mathematica* will carry over to C. Consequently, we will be able to focus on programming with statements—all of which exist as imperative expressions in *Mathematica*—from the outset. Furthermore, should you be ultimately interested in investigating programming with imperative expressions in *Mathematica*, much of what you will learn about C in the remainder of the book will apply to *Mathematica*.

Just as we did with *Mathematica*, we will make no attempt to exhaustively cover every aspect of C. Our goal is to introduce you to a subset of C in which you can write a variety of useful programs and which gives you the background and insight necessary to learn the remainder as time and interest dictate.

10.7   Eratosthenes's Problem

Figure 10.3 shows a C implementation of the problem inspired by Eratosthenes that we discussed in Chapter 3. When compiled and run, this program prompts for and reads the angle of the sun as measured at Alexandria, prompts for and reads the distance from Alexandria to Syene, calculates the circumference of the earth, and reports the result. A typical interaction would be

```
Enter angle at Alexandria (degrees): 7.2
Enter distance to Syene (stadia): 5000
The earth's circumference is 39375 km
```

(10.3)

```
1    #include <stdio.h>
2
3    void main (void)
4    {
5      double angle, distance, circ;
6
7      printf("Enter angle at Alexandria (degrees): ");
8      scanf("%lf", &angle);
9
10     printf("Enter distance to Syene (stadia): ");
11     scanf("%lf", &distance);
12
13     circ = (360/angle) * distance * 0.1575;
14
15     printf("The earth's circumference is %g km\n", circ);
16   }
```

Figure 10.3   C implementation of Eratosthenes's problem (`circum.c`)

In this section we will examine `circum.c` in enough detail to explain the purpose of each of its components. We will follow up in Chapter 11 by revisiting the material in this section in considerably more detail. In preparation for all of this, we first sketch the program's overall structure.

- The first line accesses a library file that declares the input/output functions used in the program.
- The rest of the program consists of the implementation of one function, which is called `main`. The function begins on line 3 and ends on line 16.
- The function consists of the declaration of three variables on line 5 and a sequence of six statements on lines 7–15. The statements are executed in order.
- The statements on lines 7, 10, and 15 display information to the user, and the statements on lines 8 and 11 read information from the user. The remaining statement, the assignment on line 13, performs the actual computation and saves the result in the variable `circ`. It could be carried over almost unchanged to a *Mathematica* notebook.
- The statements are composed from a variety of arithmetic expressions.

We will explore the program from the inside out, beginning with the declarations, expressions, and statements before considering the `main` function and the library inclusion.

10.7.1 Variable Declarations

A variable in *Mathematica* can take on any value. For example, it can be assigned a floating-point value in one statement and a rational value in the next. Any given variable in a C program, in contrast, can take on only one type of value, and that type must be specified before the variable can be used.

All of the variables that are used in a C function must be declared at the beginning of that function. A declaration consists of a type name followed by a comma-separated list of variable names terminated by a semicolon. All of the variables in the list are taken by C to be of the specified type.

The declarations for the `main` function of `circum.c` appear on line 5 of Figure 10.3:

```
double angle, distance, circ;
```

As we will see in Chapter 11, C provides three types of floating-point numbers with varying characteristics, and the `double` type is one of them. Thus, the declaration above specifies that the three variables `angle`, `distance`, and `circ` are to contain floating-point numbers of type `double`.

10.7.2 Expressions

A variety of expressions appear in `circum.c`, including numbers, variables, arithmetic expressions, and function calls. You should be familiar with each of these kinds of expressions from your experience with *Mathematica*.

- There is both an integer constant (360) and a floating-point constant (0.1575) on line 13. Although `circum.c` doesn't illustrate this, floating-point numbers in C can also be written using a form of scientific notation that is similar to—but not identical to—that used in *Mathematica*.
- Each of the three variables declared on line 5 is subsequently used at two different points of the program.
- The compound arithmetic expression that appears on line 13,

```
(360/angle) * distance * 0.1575
```

uses division and multiplication and is written exactly as it would be written in *Mathematica*.
- The library function `printf` is used to write information to the display on lines 7, 10, and 15. The library function `scanf` is used to read information from the keyboard on lines 8 and 11.

10.7.3 Statements

Two different kinds of statements appear among the six in the body of `main`: one assignment statement and five function calls (to the `printf` and `scanf` library functions).

The statement on line 13 is an assignment statement. As with a *Mathematica* assignment expression, a C assignment statement associates the value of the expression on its right-hand side with the variable on its left-hand side. Thus, the assignment statement

```
circ = (360/angle) * distance * 0.1575;
```

associates the value of the expression $(360/\text{angle}) * \text{distance} * 0.1575$ with the variable `circ`. This is almost exactly the way this assignment would be written in *Mathematica*. The only difference is that the C assignment statement is terminated with a semicolon.

The statements on lines 7, 10, and 15 all call the `printf` library function. The calls on lines 7 and 10 display the character strings that are passed as parameters. The call on line 15 is a bit more complicated, as it incorporates the value of the variable `circ` into the string before displaying it; the value of `circ` is substituted for the symbol `%g`.

The statements on lines 8 and 11 call the `scanf` library function. In both cases, a floating-point number typed by the user is read from the keyboard. On line 8 the number read is stored into the variable `angle`; on line 11, the number read is stored into the variable `distance`.

10.7.4 Main Function

Every C program must define a function `main`, as is the case in `circum.c`. When a compiled C program is run, its execution begins with the `main` function. When the `main` function finishes, the program terminates.

The `main` function in Figure 10.3 begins on line 3 and ends on line 16. Line 3 contains the header of the function,

```
void main (void)
```

which specifies the name of the function, the number and types of its parameters, and the type of result it returns. We will discuss function headers in more detail in Chapters 12 and 13; for now it suffices to know that this header says that `main` takes no parameters and returns no results.

The body of `main` is enclosed by the matching braces on lines 4 and 16, and consists of a sequence of variable declarations followed by a sequence of statements. When `main` is called, C uses the declarations to create a table of local variables, as illustrated in Figure 10.4. The variables in the table have no particular initial values.

angle	7.2
distance	5000
circ	

angle	7.2
distance	5000
circ	39375.0

(a) After `scanf` on line 11   (b) At end

Figure 10.4   Variable table for `main` function of `circum.c`

The statements in `main` are then executed from top to bottom; the table is used to record and look up the values of the variables as necessary. Figure 10.4a shows the state of the table immediately after the `scanf` statement on line 11 has been executed. At this point, `angle` and `distance` have been given values via `scanf`, while `circ` has yet to be given a value.

When the end of the `main` function is reached, the variable table is discarded and the program terminates. Figure 10.4b shows the values of the variables just before the table is discarded. By this point, all of the variables have been given values.

10.7.5 Libraries

Just as every variable must be declared before it is used, so must every function. The inclusion directive on line 1 of `circum.c`,

```
#include <stdio.h>
```

makes the declarations of the C library functions contained in the file `stdio.h`—in particular `printf` and `scanf`—available to the program. This is similar in nature to the *Mathematica* `Needs` function. Files such as `stdio.h` that contain function declarations are called *header* files. If you use a function without including the header file in which the function is declared, the compiler will complain.

10.8 Kitty Hawk Problem

Figure 10.5 shows a C implementation of the Kitty Hawk problem that we discussed in Chapter 5. This program is structurally almost identical to `circum.c`. When

```
1    #include <stdio.h>
2    #include <math.h>
3
4    void main (void)
5    {
6      double radius, height, distance;
7
8      printf("Enter radius of earth (ft): ");
9      scanf("%lf", &radius);
10
11     printf("Enter height of hill (ft): ");
12     scanf("%lf", &height);
13
14     distance = sqrt(2*radius*height + height*height);
15
16     printf("Distance to horizon is %g ft\n", distance);
17   }
```

Figure 10.5 Kitty Hawk problem (`horizon.c`)

compiled and run, it will prompt for two values, perform a computation, and display the result. A typical interaction would be

```
Enter radius of earth (ft): 2.09e7
Enter height of hill (ft): 66
Distance to horizon is 52524.3 ft
```
(10.4)

($2.09e7$ stands for 2.09×10^7.)

This program makes use of a library function, `sqrt`, that `circum.c` did not use. As a result, we must include *two* header files at the beginning of the program: `stdio.h` and `math.h`. This is because `stdio.h` contains only headers for input and output functions, while `math.h` contains only headers for math functions. Since `sqrt` computes the square root of its parameter, its header is found in `math.h`.

10.9 Key Concepts

Mathematica **and C.**   *Mathematica* is a computer algebra system for doing numerical, symbolic, and graphical mathematics. It is also the name for the interpreted programming language that is used within the *Mathematica* system. As a high-level language, it is specialized for mathematical applications. C is a conventional compiled language for doing general-purpose programming. It is a much lower-level language than *Mathematica*, and can be used for systems programming and applications programming in addition to numerical programming. Numerical programs written in C are faster, but *Mathematica* provides more built-in functions for doing numerical calculations.

Interpreted versus compiled languages.   *Mathematica* is an interpreted language, whereas C is a compiled language. *Mathematica*, like all interpreted languages, provides an interface that allows the programmer to interactively enter expressions and observe results. C, like all compiled languages, provides a compiler that translates C programs into a directly executable form. The programmer must manage more input/output details in a compiled language than in an interpreted language. Compiled programs, however, run faster than interpreted programs.

Simple C programs.   The simplest C programs consist of a single `main` function containing variable declarations and a sequence of output, input, and assignment statements. Any library functions that are used must be declared in an included header file. The statements are composed of expressions and are executed in sequence.

10.10 Exercises

10.1   Which of the problems that we studied in Chapters 2–9 could be as easily implemented in C as in *Mathematica*?

10.2   Write a *Mathematica* expression that adds an assignment expression to a conditional expression. What is its value? What are its side effects?

10.3   Consider the following list of problems to which a programming language could be applied. Which are clearly best done in *Mathematica*? In C? Which could reasonably be done in either?

(a) Calculate the escape velocity for Jupiter.

(b) Implement a spreadsheet program.

(c) Solve the problems from a calculus textbook.

(d) Prove an algebraic identity.

(e) Determine the time required for a sky diver to reach terminal velocity.

(f) Implement a compiler for a programming language.

10.4   If you were stranded on a desert island and had your choice of using either *Mathematica* or C for all of your programming needs, which would you choose?

10.5   *Mathematica*'s interpretive interface tends to make it faster to develop new programs from scratch. The C compiler makes it faster to run a program once it has been written. Under what circumstances is fast development time more important than fast running time? Under what circumstances is the reverse true?

10.6   Find out how to edit, compile, and run C programs on your computer. Compile and run `bisect1.c`, `circum.c`, and `horizon.c`.

10.7   Modify the messages displayed by `bisect1.c` when it prompts the user for the positive and negative guesses.

10.8   Modify `bisect1.c` so that it prompts first for the negative guess and then for the positive guess.

10.9   Modify `bisect1.c` so that it uses 3.14 as the value for π instead of 3.141592654. How much of a difference does this make?

10.10   Modify `bisect1.c` so that it finds a root of the equation $\cos(\theta) = \theta$.

10.11   Modify `circum.c` so that it reads `angle` in radians.

10.12   Modify `circum.c` so that it reads `distance` in kilometers.

10.13   Modify `circum.c` so that it displays its result in stadia.

10.14   Modify `horizon.c` so that it uses the simplified distance formula given as Formula 5.22.

10.15   Modify `horizon.c` so that it uses interval arithmetic to display upper and lower bounds on the distance to the horizon, assuming that both `radius` and `height` can be off by at most 1 percent.

10.16   Repeat Exercise 10.15 using your solution to Exercise 10.14 as the starting point.

10.17   Write a C program that solves the population density problem from Chapter 2. It should prompt for both the earth's population and the earth's land area in square feet, and should report the length in feet of the side of the square that each person would receive if the earth's surface were divided up evenly.

10.18   Modify your solution to Exercise 10.17 so that it prompts for the earth's land area in square miles.

10.19   Write a C program that solves the continuous interest problem from Chapter 6. It should prompt for a beginning balance, an annual interest rate, and a period of years and should report the balance that results at the end of the period. You will need to include `math.h` as on line 2 of Figure 10.5 and use the math library function `exp`, which is analogous to the *Mathematica* function `Exp`.

10.20   Modify your solution to Exercise 10.19 so that it prompts for a period of days.

11

Robotic Weightlifting: Straight-Line Programs

Industrial robots are widely used to perform repetitive tasks on assembly lines. They look nothing like the popular conception of robots, which is to say that they look nothing like people. The *animatronics* industry, however, deliberately produces robots that resemble humans. Such robots are commonly seen at large amusement parks, and have even been used as advertising gimmicks at trade shows. In this chapter we will consider a problem that might arise in designing a robot that can enter weightlifting competitions.

The *squat* is a weightlifting exercise designed to increase strength between the waist and knees. To perform a squat, a lifter places a weight on his upper back, about two inches below his shoulder. He then bends his ankles, knees, and hips as if sitting down, all the while keeping the weight directly over his ankles. Maintaining this alignment is critical, for otherwise the lifter is likely to keel over. When his thighs reach a near-horizontal position, the lifter reverses the process and stands back up. This exercise is diagrammed in Figure 11.1.

Imagine that we have a robot with controllable ankle, knee, and hip joints, and that we wish to program it to do the squat exercise. To do this we must solve a *kinematics* problem and a *dynamics* problem. The kinematics problem involves deciding how the three joints should bend over time to ensure that the weight will move straight down until the robot's thighs are parallel to the ground. The dynamics problem involves deciding what forces to apply to the joints to make them behave as we want.

In this chapter we will investigate the kinematics of our weightlifting robot. We will develop a C program to determine the x- and y-coordinates of the ankle, knee, hip, and shoulder joints as our robot performs the squat exercise.

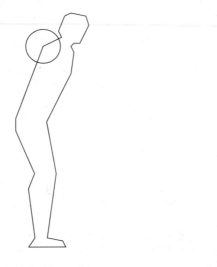

(a) Upright position   (b) Squatting position

Figure 11.1   The squat exercise

11.1 Trigonometry of a Link Diagram

We will begin with a diagram that distills the problem and makes some convenient simplifications. Figure 11.2 is a *link diagram* that shows all of the joints and rigid body parts that play a role in our problem.

The link diagram is a stick figure that represents the robot. The shaded circles represent the ankle, knee, hip, and shoulder joints, while the solid lines that connect the circles represent the rigid shin, thigh, and torso body parts. Our stick figure can rotate freely at its joints, but the connecting body parts are rigid and cannot bend.

We will use the position of the ankle joint, which is fixed, as the origin of an x-y coordinate system in which x increases toward the right and y increases toward the top. We will make the slight simplification of assuming that the weight is exactly at the shoulder instead of a few inches below. This means that, during the squat exercise, the x-coordinate of the shoulder must always be zero to ensure that it remains directly over the ankle.

If we view the kinematics of the squat as a strictly geometric problem, there are any number of ways to bend the ankles, knees, and hips so that the weight moves down while remaining directly over the ankles. If we were to take into consideration how a real robot is constructed, however, we would discover that many geometrically plausible joint configurations would cause serious damage. Even for a robot, there are good ways and bad ways to exercise.

We will assume that our robot lifts most effectively when the knees bend freely, the ankles bend half as much as the knees, and the hips bend as necessary to keep the weight directly over the ankles. We will also assume that the robot's thigh length is 67%, and its shin length 59%, of the length of its torso.

As a result of these assumptions and the constraints of the problem, the only quantities that the eventual user of our model will need to provide are the length of the torso (torso) and the angle formed by the thigh with the shin (kneeAngle). To complete our model we must derive expressions for the

- lengths of the thigh (thigh) and shin (shin);
- angles formed by the shin with the floor (ankleAngle) and by the torso with the thigh (hipAngle); and
- coordinates of the knee (kneeX, kneeY), hip (hipX, hipY), and shoulder (0, shoulderY);

To derive expressions for these elements, we will exploit the trigonometric properties of right triangles. We have included in Figure 11.2 six dotted reference lines that indicate the triangles of interest, and have labeled one angle in each triangle (α, β, γ). All of the reference lines are parallel to either the x- or y-axis.

The lengths thigh and shin are, by assumption, directly proportional to the length torso.

$$\text{thigh} = .67\,\text{torso} \qquad (11.1)$$

$$\text{shin} = .59\,\text{torso} \qquad (11.2)$$

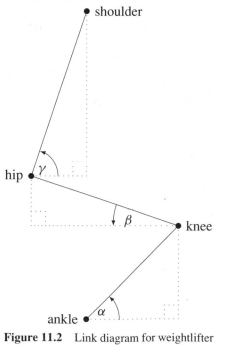

Figure 11.2 Link diagram for weightlifter problem

The angles ankleAngle and α are identical, and by our assumption both are half of kneeAngle.

$$\alpha = \text{ankleAngle} = \frac{1}{2}\text{kneeAngle} \qquad (11.3)$$

Since the ankle is at the origin, the coordinates of the knee are

$$\text{kneeX} = \text{shin}\cos(\alpha) \qquad (11.4)$$
$$\text{kneeY} = \text{shin}\sin(\alpha) \qquad (11.5)$$

The angle kneeAngle is divided by one of the reference lines. The upper portion is β and the lower portion is identical to α, so

$$\beta = \text{kneeAngle} - \alpha \qquad (11.6)$$

Using the coordinates of the knee computed above, the coordinates of the hip are

$$\text{hipX} = \text{kneeX} - \text{thigh}\cos(\beta) \qquad (11.7)$$
$$\text{hipY} = \text{kneeY} + \text{thigh}\sin(\beta) \qquad (11.8)$$

Using the coordinates of the hip, the y-coordinate of the shoulder is

$$\text{shoulderY} = \text{hipY} + \text{torso}\sin(\gamma) \qquad (11.9)$$

We're not done yet, because γ is an unknown. The equation

$$\text{hipX} + \text{torso}\cos(\gamma) = 0 \qquad (11.10)$$

holds because the x-coordinate of the shoulder is zero. It follows that

$$\gamma = \arccos\left(\frac{-\text{hipX}}{\text{torso}}\right) \qquad (11.11)$$

Finally, the angle hipAngle is also divided by one of the reference lines. The upper portion is γ and the lower portion is identical to β, so

$$\text{hipAngle} = \gamma + \beta \qquad (11.12)$$

Given values for kneeAngle and torso, our model makes it a simple matter to calculate the lengths of the shin and torso, the angles of the ankle and hip, and the coordinates of each joint.

11.2 Components of a Straight-Line Program

The definitions that we have derived lead directly to the C program `lifter.c` that begins in Figure 11.3 and concludes in Figure 11.4.

When `lifter.c` is compiled and run, it will interact with the user as illustrated in Figure 11.5. It begins by prompting the user for the knee angle, next prompts the user for the torso length, and finally computes and displays the lengths of the

```
1    #include <stdio.h>
2    #include <math.h>
3
4    void main (void)
5    {
6       int shin, thigh, torso;
7       double kneeX, kneeY, hipX, hipY, shoulderY;
8       double ankleAngle, kneeAngle, hipAngle;
9       double alpha, beta, gamma;
10      double pi;
11
12      printf("Enter knee angle (deg): ");
13      scanf("%lf", &kneeAngle);
14
15      printf("Enter torso length (mm): ");
16      scanf("%d", &torso);
17
18      pi = acos(-1);
19
20      ankleAngle = kneeAngle / 2;
21      thigh = 0.67 * torso;
22      shin = 0.59 * torso;
23
24      alpha = (ankleAngle / 180) * pi;
25      kneeX = shin * cos(alpha);
26      kneeY = shin * sin(alpha);
27
28      beta = (kneeAngle / 180) * pi - alpha;
29      hipX = kneeX - thigh * cos(beta);
30      hipY = kneeY + thigh * sin(beta);
31
32      gamma = acos(-hipX/torso);
33      hipAngle = ((gamma + beta) / pi) * 180;
34      shoulderY = hipY + torso * sin(gamma);
```

Figure 11.3 Weightlifter problem implementation (continued in Figure 11.4)

shin, thigh, and torso and the angles and coordinates for the ankle, knee, hip, and shoulder. The knee angle that we specified in Figure 11.5 is for a lifter standing straight up, which is why all of the *x*-coordinates are zero.

Although it is quite a bit longer, the overall structure and composition of lifter.c is essentially identical to that of circum.c (Figure 10.3), which we studied in Chapter 10. It is instructive to compare the two programs.

• Both programs begin by including header files. While circum.c includes only stdio.h, lifter.c also includes math.h.

```
35        printf("\nTorso = %d mm\n", torso);
36        printf("Thigh = %d mm\n", thigh);
37        printf("Shin  = %d mm\n", shin);
38
39        printf("\nAnkle angle = %g deg\n", ankleAngle);
40        printf("          X = %g mm\n", 0.0);
41        printf("          Y = %g mm\n", 0.0);
42
43        printf("\nKnee angle = %g deg\n", kneeAngle);
44        printf("         X = %g mm\n", kneeX);
45        printf("         Y = %g mm\n", kneeY);
46
47        printf("\nHip angle = %g deg\n", hipAngle);
48        printf("        X = %g mm\n", hipX);
49        printf("        Y = %g mm\n", hipY);
50
51        printf("\nShoulder X = %g mm\n", 0.0);
52        printf("         Y = %g mm\n", shoulderY);
53    }
```

Figure 11.4   Weightlifter problem implementation (`lifter.c`)

```
Enter knee angle (deg): 180
Enter torso length (mm): 686

Torso = 686 mm
Thigh = 459 mm
Shin = 404 mm

Ankle angle = 90 deg
        X = 0 mm
        Y = 0 mm

Knee angle = 180 deg
        X = 0 mm
        Y = 404 mm

Hip angle = 180 deg
        X = 0 mm
        Y = 863 mm

Shoulder X = 0 mm
        Y = 1549 mm
```

Figure 11.5   User interaction with `lifter.c`

- Besides the inclusions, the two programs are composed entirely of a single main function. The `main` function of `lifter.c` is by far the longer of the two, extending from line 4 to line 53.

- Whereas `circum.c` declares three variables and contains six statements, `lifter.c` consists of the declarations of 15 different variables on lines 6–10 and the sequence of 31 statements on lines 12–52.

- The simpler `circum.c` contains five input/output statements and a single assignment statement. In contrast, the first four and the last 14 of the statements in `lifter.c` perform input/output, while the remaining 13 statements are assignments.

The extra length of `lifter.c` relative to `circum.c` is due to the greater complexity of the weightlifting problem. Nevertheless, its extra length doesn't make it any harder to understand. If you understand what `circum.c` does, you should be able to figure out what `lifter.c` does.

In Chapter 10 we dissected `circum.c` by exploring the five components from which it was constructed: types, expressions, statements, the `main` function, and library inclusions. We will repeat that analysis here by dissecting `lifter.c`. Our goal in Chapter 10 was to give you a solid, though not overly detailed, understanding of how simple C programs work. Our goal here is to fill in the details that we omitted in our discussion of `circum.c`.

11.3 Types

We saw in Chapter 10 that any given variable in a C program can take on only one type of value, and that the type must be declared before the variable can be used. All three variables in `circum.c` were of type `double`. In fact, C provides 11 different types of numbers. The six most important for our immediate purposes are identified in Table 11.1. In this section we will investigate the differences among the three types of integers and three types of floating-point numbers listed in Table 11.1.

Table 11.1   Six of C's numerical types

Type name	Description
`short`	Short integer
`int`	Integer
`long`	Long integer
`float`	Single-precision floating-point number
`double`	Double-precision floating-point number
`long double`	Extended-precision floating-point number

11.3.1 Characteristics of Numerical Types

Mathematica provides floating-point numbers, rational numbers, and integers. C provides only floating-point numbers and integers.

The three types of integers (short, int, long) differ in the minimum and maximum values that they permit. Similarly, the three types of floating-point numbers (float, double, long double) differ in the mantissa length, minimum exponent, and maximum exponent that they permit.

As in *Mathematica*, the exact limits on the permissible values of the different types of numbers in C vary among different types of computers. Fortunately, there is no need to guess what these limits are or to discover them by trial and error. The program limits.c (from the diskette) will determine and print characteristics of the six types of numbers we are studying in this section.

It is not important to our discussion that you understand how limits.c works, so we have not included it in a figure. What *is* important is that you make a point of compiling and running limits.c on your computer. As an illustration, the output of limits.c when compiled by the GNU C compiler and run on a Sun workstation appears in Figure 11.6.

The int and long types are identical, as is typically the case on 32-bit machines such as the Sun workstation. The range of the short type is much more restrictive. On less expensive 16-bit machines, the short and int types are usually identical, with the long type having the larger range of values.

Although the ranges of the integer types may appear to be arbitrary, they are not. The size of the int type is generally matched to the width of the computer's data paths. The int range from Figure 11.6 is equal to $-2^{31} \ldots 2^{31} - 1$, for a total of

```
Minimum short = -32768
Maximum short = 32767

Minimum int = -2147483648
Maximum int = 2147483647

Minimum long = -2147483648
Maximum long = 2147483647

Minimum positive float = 1.17549e-38
Maximum positive float = 3.40282e+38

Minimum positive double = 2.22507385850720e-308
Maximum positive double = 1.79769313486232e+308

Minimum positive long double =
    3.36210314311209350626267781732175e-4932
Maximum positive long double =
    1.18973149535723176508575932662801e+4932
```

Figure 11.6   Numerical limits on a Sun workstation

2^{32} different numbers. The `short` range from Figure 11.6, which is usually the `int` range on a 16-bit computer, is equal to $-2^{15} \dots 2^{15} - 1$, for a total of 2^{16} different numbers.

By matching the size of its integers to the capabilities of the computer, C can provide extremely fast integer arithmetic. To do arithmetic on its much larger integers, *Mathematica* must partition them into smaller machine-sized pieces, perform operations on the pieces, and combine the individual results to obtain an answer.

On our Sun workstation, a `float` has a six-digit mantissa and an exponent between -38 and 38. By comparison, a `double` has a 15-digit mantissa and an exponent between -308 and 308, and a `long double` has a 33-digit mantissa and an exponent between -4932 and 4932.

Many computers have special hardware for doing floating-point arithmetic, and the characteristics of the C floating-point numbers are designed to match the capabilities of that hardware. *Mathematica* does the same thing with its machine-precision floating-point numbers, which we used throughout Chapters 2–9. As a result, on the same computer *Mathematica*'s floating-point numbers and C's double-precision floating-point numbers will typically have the same characteristics.

In *Mathematica*, as we saw in Chapter 4, arithmetic on small integers is faster than arithmetic on floating-point numbers, which is in turn much faster than arithmetic on large integers. In C, where *all* integers are fairly small, arithmetic on integers is always much faster than arithmetic on floating-point numbers.

For the remainder of this text we will use the `int` type when we need an integer and the `double` type when we need a floating-point number. No matter what kind of computer you are using, the `int` type will be tuned to the capabilities of that machine. The six-digit mantissas of the `float` type are not quite long enough for our purposes, and `long doubles` cannot be used with such math library functions as `cos` and `sqrt`.

Nevertheless, there are circumstances where other choices would be appropriate. For some applications on 16-bit machines, the range of the `int` type is too restrictive, making `longs` more attractive. `Floats` take half the storage space of `doubles` and, depending on the available floating-point hardware, can also be faster. This can be an important consideration in applications that must either store a large number of floating-point numbers or do a large number of floating-point operations.

11.3.2 Variable Declarations

We saw in Chapter 10 that all of the variables that are used in a C function must be declared at the beginning of that function. The declarations for the `main` function of `lifter.c` appear on lines 6–10 of Figure 11.3. Recall that a declaration consists of a type name followed by a comma-separated list of variable names terminated by a semicolon, and that all of the variables in the list are taken by C to be of the specified type.

To make our program more readable, we have chosen to group our declarations according to the purposes of the variables. The names of the variables are taken directly from the trigonometric model. Thus, we first declare three body part variables of type `int`, then five coordinate variables of type `double`, then three degree variables of type `double`, then three radian variables of type `double`, and finally a variable of type `double` that will contain the value of π. This organization is entirely for the benefit of our human readers.

11.3.3 Type Conversions

There are a number of circumstances, three of which we will discuss later in this chapter, when C will convert one type of number into another type of number.

Converting an integer into a floating-point number can cause roundoff error if the integer contains more digits than are allowed in the mantissa of the floating-point number. As Figure 11.6 reveals, this is never a problem on our Sun workstation when we convert an `int` to a `double`, but could be a problem when we convert an `int` to a `float`.

When a floating-point number is converted into an integer, any fractional part is discarded. If the resulting number is too big, an integer overflow will result.

11.4 Expressions

Four different kinds of expressions—numbers, variables, arithmetic expressions, and function calls—appear in `lifter.c`, just as they did in `circum.c`. We discuss them next, highlighting differences between *Mathematica* and C.

11.4.1 Numbers

Integer constants, and floating-point constants without exponents, are written in C exactly as they are in *Mathematica*. Our program contains three integer constants $(-1, 2, 180)$ and three floating-point constants (0.67, 0.59, 0.0). Floating-point constants with exponents are written differently. Whereas the exponent is delimited with `*^` in *Mathematica*, it is delimited with `e` in C. For example, the *Mathematica* constant `2.5*^7` is written in C as `2.5e7`. Some examples of C floating-point constants with exponents appear in Figure 11.6.

11.4.2 Variables

Whereas *Mathematica* provides direct support for both numerical and symbolic calculations, C supports only numerical calculations. A *Mathematica* variable, until it is given a value via assignment, stands for itself and can be used as a mathematical unknown. A C variable, in contrast, always has a value.

Unfortunately, the initial value of a freshly declared C variable is not predictable. This means that you must be sure to initialize each variable before you use it. In `lifter.c`, all of our variables are given values before they are used. The variables `kneeAngle` and `torso` obtain values from the keyboard via calls to the `scanf` function on lines 13 and 16; all of the other variables are initialized via assignment.

11.4.3 Arithmetic Expressions

C provides the four arithmetic operators +, -, *, and /, but it does *not* provide an exponentiation operator. (C *does* provide the library function pow for doing exponentiation, however.) Otherwise, arithmetic expressions are written exactly as they are in *Mathematica*, and obey the same precedence rules.

An operation on a pair of integers always produces an integer result, and an operation on a pair of floating-point numbers always produces a floating-point result. When an operation involves both an integer and a floating-point number, the integer is first converted to floating-point form and a floating-point result is produced. Examples of this in lifter.c include line 20, where an integer constant divides a floating-point variable; line 21, where a floating-point constant multiplies an integer variable; and line 32, where an integer variable divides a floating-point value.

Recall that when two integers are divided in *Mathematica*, the quotient is expressed as a fraction in lowest terms. C, in contrast, does not permit fractions. When two integers are divided in C, the quotient is expressed as an integer with any remainder discarded. For example, the quotient 7/3 would be 2. The only way to represent a fraction such as $\frac{7}{3}$ in C is to approximate it with a floating-point number such as 2.33333.

The behavior of C when overflow, underflow, or division by zero occurs varies from computer to computer and according to whether integers or floating-point numbers are involved. The program errors.c (on the diskette) will give you an idea of how your implementation of C behaves. The results that it displays for both floating-point and integer arithmetic on our Sun workstation appear in Figure 11.7. You, of course, should compile and run it on your own computer, as your results may vary.

On our workstation, floating-point overflow and division by zero both result in the special symbol Inf, which stands for infinity, while underflow results in zero. The symbol Inf is not something that you can use in a program or provide as input, but is simply how C reports numbers that are too large for it to represent.

Errors in integer arithmetic are handled much less gracefully. For example, adding 1 to the largest possible integer yields the smallest possible integer, while subtracting 1 from the smallest possible integer yields the largest possible integer. This wrap-around effect means that every integer expression, except for division by zero, yields an integer result. This result will be wrong, however, if overflow or

```
Floating-point arithmetic:
 1.79769e+308 * 1.79769e+308 = Inf
 2.22507e-308 * 2.22507e-308 = 0
 1 / 0 = Inf

Integer arithmetic:
 2147483647 + 1 = -2147483648
 -2147483648 - 1 = 2147483647
 1 / 0 = Arithmetic Exception (core dumped)
```

Figure 11.7 Arithmetic results of errors.c on a Sun workstation

underflow occurred. Division by zero results in an arithmetic exception that causes the program to stop dead in its tracks.

11.4.4 Function Calls

Function calls are written in C as they are written in *Mathematica*, with one exception. C uses standard mathematical notation by using parentheses to delimit function parameters, whereas *Mathematica* uses square brackets. The program `lifter.c` contains calls to five different built-in functions: `printf` does formatted output, `scanf` does formatted input, `acos` computes arccos, `cos` computes cosine, and `sin` computes sine. We will discuss these and related library functions in Section 11.7.

C functions expect to receive a fixed number of parameters of particular types, and they produce a result of a particular type. To give two examples, `cos` expects one `double` as its parameter and produces a `double` as its result, whereas `pow` expects two `doubles` as its parameters and produces a `double` as its result.

The compiler will complain if you pass the wrong number of parameters to a function. If you pass an `int` where a `double` is expected, or a `double` where an `int` is expected, C will perform the appropriate conversion. An example of this appears in `lifter.c` on line 18, where an `int` constant is passed where a `double` is expected.

Passing a parameter of the appropriate type is no guarantee that the parameter is sensible. For example, the `sqrt` function expects a non-negative `double` parameter, and the `acos` function expects a `double` between −1 and 1. The library functions will behave in different ways on different machines when given bad parameters. The program `errors.c`, which we used earlier to explore arithmetic errors, also illustrates the error behavior of several built-in functions. The results obtained on our workstation appear in Figure 11.8. The special symbol `NaN` stands for "not a number."

```
Built-in functions:
  sqrt(-1) = NaN
  acos(2) = 0
  log(0) = -Inf
```

Figure 11.8   Behavior of three built-in functions on erroneous input

11.5 Simple Statements

There are two different kinds of statements among the 31 in the body of `main`: function calls and assignment statements.

11.5.1 Function Calls as Statements

Lines 12–16 and 35–52 contain 18 different function calls involving the C library functions `scanf` and `printf`. Each of these function calls is an expression that is being used as a statement.

A C expression can always be used where a statement is required, but when it is evaluated its value is discarded. Thus, it serves no purpose to use an expression as a statement unless the expression has some side effect. The side effect of a `printf` function call is to write information to the display, and the side effect of a `scanf` function call is to read information from the keyboard.

Lines 12 and 13 prompt the user for, and then read in, the knee angle of the lifter. Similarly, lines 15 and 16 prompt for and read in the torso length of the lifter. Lines 35–52 display the results calculated by the program. We will describe the behavior of the `scanf` and `printf` functions in more detail in Section 11.7.

11.5.2 Assignment Statements

The statements on lines 18–34 are all assignment statements. We saw in Chapter 10 that, just as in *Mathematica*, a C assignment statement associates the value of the expression on its right-hand side with the variable on its left-hand side. For example, the assignment statement

```
kneeX = shin * cos(alpha);
```

on line 25 associates the value of the expression `shin * cos(alpha)` with the variable `kneeX`.

(Technically, assignments in C are expressions, and just as in *Mathematica* the value of an assignment is the value of the subexpression on the right-hand side. Because we will never use C assignments in contexts where their values matter, we can treat them as statements.)

The type of the value of the right-hand-side expression must match the type of the left-hand-side variable. If they do not match and both are numerical types, C will convert the value of the expression into the type of the variable as discussed in Section 11.3.3. Examples of this occur on lines 21 and 22, where an expression of type `double` is assigned to a variable of type `int`.

11.6 Main Function

You know from Chapter 10 that every C program must define a function called `main`, which is where execution begins and ends when the program is run. The `main` function in Figures 11.3–11.4 begins on line 4 and ends on line 53.

Line 4 contains the header of the function,

```
void main (void)
```

which specifies the name of the function, the number and types of its parameters, and the type of result it returns. We will discuss function headers in more detail in Chapters 12 and 13; as you already know, this header says that `main` takes no parameters and returns no results.

The body of `main` is enclosed by the matching braces on lines 5 and 53, and consists of a sequence of variable declarations followed by a sequence of statements. When `main` is called, C uses the declarations to create a table of local variables, as illustrated in Figure 11.9. The variables in the table have no particular initial values.

The statements in `main` are then executed from top to bottom, with the table being used to record and look up the values of the variables as necessary. Figure 11.9a shows the state of the table immediately after the assignment on line 22 has been executed. (We have rounded the 15-digit mantissas of the `doubles` to save space.) At this point, `torso` and `kneeAngle` have been given values via `scanf`, while `shin`, `thigh`, `ankleAngle`, and `pi` have been given values via assignment.

When the end of the the `main` function is reached, the variable table is discarded and the program terminates. Figure 11.9b shows the values of the variables just before the table is discarded. By this point, all of the variables have been given values.

shin	404
thigh	459
torso	686
kneeX	
kneeY	
hipX	
hipY	
shoulderY	
ankleAngle	90.00
kneeAngle	180.00
hipAngle	
alpha	
beta	
gamma	
pi	3.14

(a) Immediately after assignment on line 22

shin	404
thigh	459
torso	686
kneeX	0.00
kneeY	404.00
hipX	0.00
hipY	863.00
shoulderY	1549.00
ankleAngle	90.00
kneeAngle	180.00
hipAngle	180.00
alpha	90.00
beta	90.00
gamma	90.00
pi	3.14

(b) Immediately before end

Figure 11.9   Variable table for `main` function of `lifter.c`

11.7 Libraries

Our program makes use of five library functions: `printf` and `scanf` from `stdio.h`, and `sin`, `cos`, and `acos` from `math.h`. This is why `lifter.c` includes the header files `stdio.h` and `math.h` in its first two lines. Unlike *Mathematica*, C does not provide any built-in mechanism for obtaining information about what library functions are available. Although online information is usually available, the means of accessing it vary radically among different types of computers. In this section we will identify and explain some of the immediately useful functions from `math.h` and `stdio.h`. Appendix C contains a summary of some of the C library's more useful functions.

11.7.1 Math Functions

The math library header file (`math.h`) declares a variety of functions, including

- trigonometric functions (`sin`, `cos`, `tan`);
- inverse trigonometric functions (`asin`, `acos`, `atan`);
- exponentiation functions (`exp`, `pow`, `sqrt`);
- logarithmic functions (`log`, `log10`);
- absolute value (`fabs`).

All of these functions except for `pow` take one `double` parameter and return a `double` result; `pow` takes two `double` parameters and returns a `double` result.

The trigonometric functions work with radians and their behavior should be apparent from their names. The other functions behave as follows:

- `exp(x)` $= e^x$
- `pow(x,y)` $= x^y$
- `sqrt(x)` $= \sqrt{x}$
- `log(x)` $= \ln x$
- `log10(x)` $= \log_{10} x$
- `fabs(x)` $= |x|$

11.7.2 An Output Function

The standard I/O library header file (`stdio.h`) declares, among a variety of other functions, `printf` and `scanf`. While these two functions are quite complicated in their most general forms, they are easy to use for basic output and input. In this section we will describe a simple form of `printf`, and in the next section we will describe a simple form of `scanf`. We will discuss more of the nuances of `printf` and `scanf` as they arise in our example programs in succeeding chapters.

The function `printf` writes information to the display. It takes one or more parameters, the first of which must be a character string and is called the *format string*. In its simplest form, as on line 12 of `lifter.c`,

```
printf("Enter knee angle (deg): ");
```

`printf` writes the format string to the display. The string is displayed exactly as given. To produce a newline in the output, the two-character sequence "`\n`" must be included in the format string. For example

```
printf("Enter knee angle (deg):\n");
```

would result in the termination of the displayed string with a newline.

A slightly more involved version of printf is used to write the values of expressions to the display. Consider, for example, line 36 of lifter.c.

```
printf("Thigh = %d mm\n", thigh);
```

Notice that there are two parameters and that the format string contains a percent sign. The percent sign and the d that follows are called a *conversion specification*. When printf encounters the conversion specification in the format string, it replaces it in the output with the value of the second parameter. For example, if the value of thigh were 484, the output produced would be

```
Thigh = 484 mm
```

Besides telling printf where to insert the value of the extra parameter, the conversion specification also tells printf how to format the value. An illustration of the four most useful conversion specifications—at least for our purposes—appears in Figure 11.10.

The %d specification is used for formatting integers. For floating-point numbers we have our choice of %f, %e, and %g. The %f specification displays the value without an exponent, placing six digits after the decimal point. The %e specification displays the value in scientific notation, with one digit before and six after the decimal point. The %g specification displays the value using a six-digit mantissa, including an exponent as necessary. All of these format specifications have variants, which we will explore in succeeding chapters, that allow you to control such things as mantissa length and column width. In most programs, however, it suffices to use %d for int values and %g for double values.

It is critical that the type of a displayed value match the type expected by the conversion specification. If it does not, garbage will be displayed. The last two examples in Figure 11.10 show the results of formatting a floating-point number with %d and an integer with %g. C will not automatically convert a value to the

Example	Result
printf("%d", 123456789);	123456789
printf("%f", 12.3456789);	12.345679
printf("%f", .00000123456789);	0.000001
printf("%e", 12.3456789);	1.234568e+01
printf("%e", .00000123456789);	1.234568e-06
printf("%g", 12.3456789);	12.3457
printf("%g", .00000123456789);	1.23457e-06
printf("%d", 12.3456789);	1076408572
printf("%g", 123456789);	3.21193e-273

Figure 11.10   Output conversion specifications

appropriate type in a `printf` expression, and most C compilers will not even warn you that you have made a mistake.

11.7.3 An Input Function

The function `scanf` reads information from the display. As with `printf`, a complete understanding of `scanf` requires mastering many nuances. We will simplify things in this section by considering a form that takes two parameters. The first is a format string containing a single conversion specification, and the second is a variable name preceded by an ampersand (&) character.

For example, line 13 of `lifter.c` is

```
scanf("%lf", &kneeAngle);
```

The conversion specification `%lf` (that's the letter ell, not the digit one) tells `scanf` to read a `double` from the keyboard, and the second parameter `&kneeAngle` tells `scanf` to store the value it reads into the variable `kneeAngle`. Similarly, line 16

```
scanf("%d", &torso);
```

reads an `int` from the keyboard and stores it into `torso`.

This behavior of `scanf` is illustrated by the first two examples in Figure 11.11. The first column shows the input entered by the user, the second column shows the code executed by C, and the third column shows the output that is produced. (Throughout Figure 11.11, n is an `int` variable and x is a `double` variable.)

As with `printf`, it is essential that you use the conversion specification that is appropriate for the type of the variable into which you are reading. Failure to do this will yield surprising results. The third and fourth examples of Figure 11.11 show what can happen when reading into an `int` using `%lf` and when reading into a `double` using `%d`.

It is beyond the scope of this book to explain the purpose of the ampersand that appears before `kneeAngle` and `torso`. If you leave it off, the compiler will warn you and your program will generally crash when run. This is illustrated by the fifth example of Figure 11.11.

It is important to understand how `scanf` will behave if the user makes a mistake and enters an unexpected value at the keyboard. When `scanf` is called, it begins by reading and discarding any leading white space characters (spaces, tabs, and newlines) that have been entered from the keyboard. It then reads characters until it finds one that cannot be part of the type of number for which it is looking. Finally, it converts the characters that it has read into a number, which it stores.

Suppose that we write a program to read first an `int`, then another `int`, and finally a `double` from the keyboard, but that the user mistakenly enters a floating-point number followed by two integers instead:

```
1.2e8 34 45
```

The behavior of `scanf` in this circumstance is illustrated by the last three examples of Figure 11.11. The three calls behave as follows.

Input	Code	Output
12345	scanf("%d", &n); printf("%d", n);	12345
12.345	scanf("%lf", &x); printf("%g", x);	12.345
12345	scanf("%lf", &n); printf("%d", n);	1086856320
12.345	scanf("%d", &x); printf("%g", x);	2.74533e-313104
12345	scanf("%d", n); printf("%d", n);	Crash!!
1.2e8 34 56	scanf("%d", &n); printf("%d", n);	1
.2e8 34 56	scanf("%d", &n); printf("%d", n);	1
.2e8 34 56	scanf("%lf", &x); printf("%g", x);	.2e8

Figure 11.11   Using scanf, with int n and double x

- The first call is expecting an int. It stops reading characters as soon as it encounters the decimal point, which cannot be part of an integer. As a result, it stores 1 into n.
- The second call, which is also expecting an int, picks up where the first call left off. It doesn't read *any* characters because it encounters the decimal point right away. As a result it cannot produce a number to store and so leaves n unchanged.
- The third call is expecting a double. It isn't hindered by the decimal point and so reads in .2e8, which it stores into x. The 34 and 56 that were entered from the keyboard remain unread.

This behavior of scanf means that the user of lifter.c must be careful to type nothing except white space and properly formatted numbers. By better exploiting the capabilities of scanf and by using some of the other functions defined in stdio.h, it is possible to deal more gracefully with erroneous input. We will not explore this issue here.

Unfortunately, it is extremely challenging to write programs that can deal gracefully with arbitrary user input errors. As things stand now, more than half of the statements in lifter.c are devoted to input and output. If the program were modified to make it deal sensibly with anything the user might type, it could easily become several times longer. The output of lifter.c echoes the knee angle and torso lengths, which are supplied as input, so that the user can verify that they were entered correctly.

11.8   Assessment

Our implementation contains a subtle flaw. We chose to represent the lengths of the robot's shin, thigh, and torso as integers by storing these values in variables of type int. While this means that the user must enter the measured length of the torso to the nearest millimeter, the real problem is that the computed lengths of the thigh and shin (67% and 59% of the torso length, respectively) are subject to quite a bit of roundoff error. This error can be quite large when the torso length is small.

In the sample run of our program in Figure 11.5, we specified a torso length of 686 mm, and the program computed a thigh length of 459 mm and a shin length of 404 mm. In reality, 67% of 686 is 459.62 and 59% of 686 is 404.74. In converting these floating-point values into integers, C has not only introduced roundoff error, it has introduced more than was strictly necessary by *truncating* instead of *rounding*.

There are two ways to deal with this problem. The best solution is to represent the three measurements as floating-point numbers. (We chose to use integers so that we could illustrate their declaration and use in C.) A different solution is to change the program so that the floating-point numbers are rounded instead of truncated. The time-honored way of rounding a positive floating-point number is to add 0.5 before C gets a chance to truncate, as in

```
thigh = 0.67 * torso + 0.5;
```

The truncated result will then be properly rounded: 0.67 * 686 + 0.5 is 460.12, which truncates to 460.

11.9   Key Concepts

Numerical types.   *Mathematica* provides rational and floating-point numbers. C provides integers and floating-point numbers, and allows the programmer to choose among several different sizes of each kind of number. The range of integers in C is much narrower than in *Mathematica*. We will exclusively use integers of type int and floating-point numbers of type double.

Type declarations.   The type of value that every variable can contain must be explicitly declared in a C program. This is different from *Mathematica*, in which a variable may contain a rational number at one point and a floating-point number later.

Expressions.   Arithmetic expressions are written in C just as they are in *Mathematica*. The only significant difference is that there is no exponentiation operator in C. (There is, however, the exponentiation function pow in the math library.)

Statements.   C functions are written as sequences of statements. In this chapter we examined two kinds of statements: assignment statements and function calls.

Main function.   Every C program is composed of a collection of functions. Every program must contain a function called `main`, which serves as the starting point of the program.

Libraries.   Like *Mathematica*, C provides a number of built-in functions. All of these functions are contained in libraries. Before a function from a library can be used, the header file for that library must be included into the program. For our purposes, the two most useful libraries will be the math library `math.h` and the input/output library `stdio.h`.

Declaration before use.   All variables and functions in a C program must be declared before they can be used. This principle is evident throughout the weightlifting program. The variables used in the `main` function are declared at the beginning of the function, and the built-in functions used in the `main` function are declared by including the appropriate library header files.

11.10 Exercises

11.1   Modify `lifter.c` so that the output is labeled with "millimeters" and "degrees" instead of "mm" and "deg."

11.2   Modify the output statements of `lifter.c` so that all of the angles are displayed together, all of the *x*-coordinates are displayed together, and all of the *y*-coordinates are displayed together.

11.3   Modify `lifter.c` so that it prompts for and displays all results in centimeters.

11.4   Modify `lifter.c` so that it prompts for and displays all angles in radians.

11.5   Modify `lifter.c` so that it prompts for the torso length in centimeters instead of millimeters. It should still display the final results in millimeters.

11.6   Modify `lifter.c` so that it prompts for the knee angle in radians instead of in degrees. It should still display the final results in degrees.

11.7   Modify `lifter.c` so that it uses 3.14159 as the value for `pi`. How much of a difference does this make in the results?

11.8 Modify `lifter.c` so that it rounds instead of truncates when converting floating-point values into integers. How much of a difference does this make in the results?

11.9 Modify `lifter.c` so that `thigh` is obtained by multiplying `torso` by `(67/100)` and `shin` is obtained by multiplying `torso` by `(59/100)`. Explain the result.

11.10 Try running `lifter.c` with an extremely small torso length, such as 1 centimeter. Then modify `lifter.c` so that `torso`, `thigh`, and `shin` are all variables of type `double`. You will also have to change the `printf/scanf` expressions that manipulate those variables. Rerun `lifter.c` with the same value for `torso`. Explain why the output is so very different.

11.11 We developed the model in Section 11.1 so that everything could be determined from torso and kneeAngle. Develop a model that defines everything in terms of shin and ankleAngle.

11.12 Modify `lifter.c` so that it prompts for `shin` and `ankleAngle` and determines everything else using the model from Exercise 11.11.

11.13 Write a C program that prompts the user for a positive integer n and displays the sum of the first n integers, which is

$$\frac{1}{2}n^2 + \frac{1}{2}n \qquad (11.13)$$

Your program should make no use of floating-point variables or constants. Be sure that your program works for odd values of n.

11.14 Write a C program that prompts the user for a positive integer n and displays the sum of the first n integers, using Formula 11.13. Your program should do its calculations, and display its results, using floating-point numbers.

11.15 For how large a value of n can you obtain the sum of the first n integers using your solution to Exercise 11.13? Using your solution to Exercise 11.14?

11.16 Write a C program that solves the population density problem from Chapter 2 using interval arithmetic. It should prompt for upper and lower bounds on both the earth's population and the earth's land area in square feet, and should report upper and lower bounds on the side of the square that each person would receive if the earth's surface were divided up evenly.

11.17 Write a C program that solves Eratosthenes's problem from Chapter 3 using interval arithmetic. It should prompt for upper and lower bounds on both the distance from Alexandria to Syene and the angle of the sun at Alexandria, and should display upper and lower bounds on the circumference of the earth.

11.18   Write a C program that solves the horizon distance problem from Chapter 5 using interval arithmetic. It should prompt for upper and lower bounds on both the earth's radius and the hill's height, and should display upper and lower bounds on the distance to the horizon from the top of the hill.

11.19   Write a C program that does comparisons of simple, compound, and continuous interest as discussed in Chapter 6. It should prompt for an initial balance, an annual interest rate, a duration in years, and the number of compounding intervals per year and should print out the final balances under all three forms of interest.

11.20   Modify your solution to Exercise 11.18 so that all of the variables and constants are integers, all of the `scanf` statements read integers, and all of the `printf` statements write integers. Explain the behavior of the resulting program.

12

Sliding Blocks: Conditionals and Functions

Figure 12.1 shows a block sitting on a ramp. The ramp is inclined from the horizontal at an angle of θ radians, and the block's initial position is L meters from the lower end of the ramp. The block is initially prevented from moving, but is released at time zero. If the friction between the block and the ramp is high enough, the block will remain in place. Otherwise, it will accelerate down the ramp until it reaches the end, at which point it will abruptly stop.

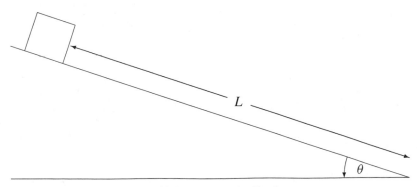

Figure 12.1   Block on inclined ramp

In this chapter we will develop a series of models and implementations for determining the position of the block on the ramp at any given time. In our first model we will ignore friction and assume that the ramp is of infinite length. In our second model we will take friction into account, and in our third and final model we will assume that L is finite.

12.1 An Infinite Ramp without Friction

In this section we will develop our first model and implementation of the sliding block problem. We will assume that there is no friction and that the ramp is infinite.

12.1.1 Dividing Gravity into Components

In the absence of the ramp, the force due to gravity would accelerate the block straight down. In the presence of the ramp, however, only a fraction of that force serves to accelerate the block down the ramp. Figure 12.2 shows how the force due to gravity, mg, can be divided into a component parallel to the ramp and a component orthogonal to the ramp. (We are using m to stand for the mass of the block.)

Only the component of mg that is parallel to the ramp, $mg \sin \theta$, will accelerate the block down the ramp. If the ramp is horizontal, $\sin \theta$ will be zero and there will be no acceleration. If the ramp is vertical, $\sin \theta$ will be 1 and the block will be in free fall.

If an object of mass m begins at rest at time zero and is then subjected to a constant force ma, its position at time t will be $\frac{1}{2}at^2$. Accordingly, the position of the block on the ramp as a function of time is given by

$$\text{position}(t) = \begin{cases} 0 & t < 0 \\ \frac{1}{2}g \sin \theta t^2 & t \geq 0 \end{cases} \tag{12.1}$$

where position is measured relative to the block's starting location.

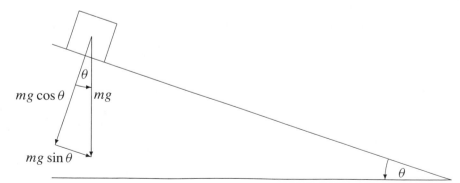

Figure 12.2   Components of the block's acceleration, where $mg \cos \theta$ is perpendicular to the ramp and $mg \sin \theta$ is parallel to the ramp

12.1.2 Conditional Statements

Any implementation of this model will have to make a choice. If t is negative, the position of the block will be zero, independent of the value of θ. Otherwise, the position of the block on the ramp will be given by the expression $\frac{1}{2}g\sin\theta t^2$.

We have already encountered two different situations where a *Mathematica* program was required to make a choice. In Chapter 8 we used the *Mathematica* `Which` construct to create a conditional expression to describe the trajectory of a destroyer through the water. In Chapter 9 we used the *Mathematica* `If` construct to create the conditional expression needed to implement the bisection method.

C provides two different conditional constructs: one for creating conditional *expressions* and the other for creating conditional *statements*. In practice, the conditional expression is rarely used—it has an extremely obscure syntax—and we will not use it in this book. We will instead focus on the conditional statement.

Figure 12.3 shows a program that implements the model that we have developed in this section. It prompts for the angle of the ramp (in degrees, which it then converts to radians) and the elapsed time (in seconds). It then calculates and displays the position of the block (in meters).

We will focus on the conditional statement that appears on lines 15–20 of Figure 12.3, which calculates the position of the block. Depending on the value of the variable `time`, C chooses which of two assignment statements to execute. If

```
1    #include <stdio.h>
2    #include <math.h>
3
4    void main (void)
5    {
6      double angle, theta, time, pos;
7
8      printf("Enter the angle of the ramp (degrees): ");
9      scanf("%lf", &angle);
10     theta = acos(-1) * angle/180;
11
12     printf("Enter the time (seconds): ");
13     scanf("%lf", &time);
14
15     if (time < 0) {
16       pos = 0;
17     }
18     else {
19       pos = 1./2. * 9.8 * sin(theta) * time * time;
20     }
21
22     printf("Position of the block is %g meters\n", pos);
23   }
```

Figure 12.3 Implementation of first model (`block1.c`)

`time` is negative, C executes the assignment on line 16; otherwise, it executes the assignment on line 19.

Conditional statements can be written in a variety of ways, as we will see as we implement increasingly more sophisticated models in this chapter. The conditional in `block1.c`, which chooses between two alternatives, is an `if` statement in its most fundamental form. The general form of an `if` statement is:

```
if (CONDITION) {
    STATEMENTS
}
else {
    STATEMENTS
}
```

Specifically, it consists of

1. The keyword `if`.
2. A Boolean expression enclosed in parentheses. We will discuss Boolean expressions in Section 12.1.3.
3. A sequence of statements, enclosed in braces. (Although Figure 12.3 contains only one statement in this position, there can be any number.) These statements are executed if the Boolean expression is true.
4. The keyword `else`.
5. A sequence of statements, enclosed in braces. These statements are executed if the Boolean expression is false.

12.1.3 Relational Operators

The Boolean expression that controls the `if` in Figure 12.3, `time < 0`, compares the values of two arithmetic expressions. C provides six relational operators for comparing numbers. They appear in Table 12.1.

Just as in *Mathematica*, a Boolean expression in C can be formed by placing a relational operator between two arithmetic expressions. The six relational operators are exactly the same as in *Mathematica*.

Note carefully the difference between the C assignment operator = and equality operator ==. Just as in *Mathematica*, it is a very common mistake to use the assignment operator in a Boolean expression in place of the equality operator.

Table 12.1   C relational operators

<	less than	<=	less than or equal
>	greater than	>=	greater than or equal
==	equal	!=	not equal

Unfortunately, this often yields a C program that compiles with no error messages but does not behave as intended.

Figure 12.4 shows the result of running `block1.c`. We have specified an angle of 45 degrees and a duration of five seconds. During that time, the block moves approximately 87 meters down the ramp.

```
Enter the angle of the ramp (degrees): 45
Enter the time (seconds): 5
Position of the block is 86.6206 meters
```

Figure 12.4   Interaction with `block1.c`

12.2 An Infinite Ramp with Friction

In this section we will extend the model and implementation from the previous section by accounting for the friction between the block and the ramp. We will continue to assume that the ramp is infinite.

12.2.1 Accounting for Friction

In Section 12.1 we resolved the force due to gravity into a component parallel to the ramp, $mg \sin \theta$, and a component orthogonal to the ramp, $mg \cos \theta$. We used the parallel component to determine how the block accelerated down the ramp. We will now use the orthogonal component to help us model the effects of friction.

Friction counteracts the acceleration of the block down the ramp by producing a force that accelerates the block *up* the ramp. The magnitude of the force up the ramp due to friction never exceeds the magnitude of the force down the ramp due to gravity, for otherwise the block would move up the ramp! In fact, the force due to friction exactly counteracts the force due to gravity up to a maximum value, at which point gravity wins and the block begins moving.

The maximum value that the force due to friction can reach is directly proportional to the component of the force due to gravity that is orthogonal to the ramp. The constant of proportionality is called the *coefficient of friction*, which we will denote by μ. The maximum value of the force due to friction is thus $-mg\mu \cos \theta$.

The coefficient of friction depends on the materials from which the ramp and the block are made. A small coefficient of friction (close to zero) means that the ramp and block are slippery, while a high coefficient of friction means that the ramp and block are rough.

Once the block begins moving, the overall acceleration of the block will be the difference between the acceleration due to gravity and the acceleration due to friction, or

$$g \sin \theta - g\mu \cos \theta \qquad (12.2)$$

Putting all of this together, the position of the block as a function of time is given by

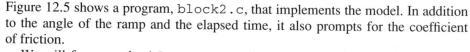

$$
\text{position}(t) = \begin{cases} 0 & t < 0 \\ 0 & \mu\cos\theta > \sin\theta \\ \frac{1}{2}g(\sin\theta - \mu\cos\theta)t^2 & \text{otherwise} \end{cases} \tag{12.3}
$$

Any implementation of this model will have to choose among the three alternatives in Equation 12.3. The first alternative reflects the fact that the block is restrained prior to time zero. The second alternative reflects the fact that the block will not move unless the gravitational force down the ramp exceeds the maximum possible friction force up the ramp. The third alternative characterizes the block in motion, when gravity has overcome friction.

12.2.2 Multi-Way Conditional Statements

Figure 12.5 shows a program, `block2.c`, that implements the model. In addition to the angle of the ramp and the elapsed time, it also prompts for the coefficient of friction.

We will focus on the `if` statement, on lines 20–28 of Figure 12.5, that calculates the position of the block. If `time` is negative, C executes the assignment on line 21. Otherwise, if `factor` is negative, it executes the assignment on line 24. Only when neither of the two conditions holds does it execute the assignment on line 27.

The `if` statement from `block2.c` is an example of a three-way conditional. The general form of a multi-way `if` statement is:

```
if (CONDITION) {
    STATEMENTS
}
else if (CONDITION) {
    STATEMENTS
}
else if (CONDITION) {
    STATEMENTS
}
else {
    STATEMENTS
}
```

Although it diagrams a four-way conditional, any number of `else if` parts may be included between the `if` part and the `else` part.

Only one of the statement sequences is executed. The conditions are tested in order, beginning with the `if` and proceeding through the `else ifs`. When the first true condition is found, the associated statement sequence is executed and the

```
1    #include <stdio.h>
2    #include <math.h>
3
4    void main (void)
5    {
6      double angle, theta, time, pos, mu, factor;
7
8      printf("Enter the angle of the ramp (degrees): ");
9      scanf("%lf", &angle);
10     theta = acos(-1) * angle/180;
11
12     printf("Enter the coefficient of friction: ");
13     scanf("%lf", &mu);
14
15     printf("Enter the time (seconds): ");
16     scanf("%lf", &time);
17
18     factor = sin(theta) - mu * cos(theta);
19
20     if (time < 0) {
21        pos = 0;
22     }
23     else if (factor < 0) {
24        pos = 0;
25     }
26     else {
27        pos = 1./2. * 9.8 * factor * time * time;
28     }
29
30     printf("Position of the block is %g meters\n", pos);
31   }
```

Figure 12.5   Implementation of second model (`block2.c`)

remainder of the conditional is skipped. Only if none of the conditions is true is the statement sequence associated with the `else` executed.

12.2.3 Boolean Operators

The assignments on lines 21 and 24 of Figure 12.5 are identical. Therefore, it is possible to collapse the three-way `if` into a more compact two-way `if`. An alternative way of implementing the conditional appears in Figure 12.6. We have combined the first two cases from Figure 12.5 into one. The Boolean expression on line 1,

```
(time < 0 || factor < 0)
```

```
1        if (time < 0 || factor < 0) {
2          pos = 0;
3        }
4        else {
5          pos = 1./2. * 9.8 * factor * time * time;
6        }
```

Figure 12.6 Improved version of `if` from `block2.c` (excerpt of `block3.c`)

is true when `time` is negative *or* when `factor` is negative (or both). The symbol `||` is an example of a *Boolean operator*. Boolean operators are used to compose simple Boolean expressions into more complicated ones.

C has three Boolean operators, which appear in Table 12.2. The `and` and `or` operators can be written *between* any two Boolean expressions. For example, the expression

```
(time < 0 && factor < 0)
```

will be true only when *both* `time` and `factor` are negative.

Table 12.2   C Boolean operators

&&	and
||	or
!	not

The `not` operator is written *in front of* a single Boolean expression, and serves to reverse its sense. For example

```
(! (time < 0))
```

will be true when `time` is *not* less than zero.

Figure 12.7 shows the result of running `block3.c`. We have specified the same angle and duration as in Figure 12.4, along with a coefficient of friction of 0.8. The block moves only approximately 17 meters in five seconds, as compared with 87 meters in the absence of friction.

```
Enter the angle of the ramp (degrees): 45
Enter the coefficient of friction: .8
Enter the time (seconds): 5
Position of the block is 17.3241 meters
```

Figure 12.7   Interaction with `block3.c`

12.3 A Finite Ramp with Friction

We will now consider the most general case of our problem, in which there is a ramp of finite length.

12.3.1 Accounting for Finite Length

Our model requires only a minor extension to account for the finite length of the ramp.

$$\text{position}(t) = \begin{cases} 0 & t < 0 \\ 0 & \mu \cos\theta > \sin\theta \\ \frac{1}{2}g(\sin\theta - \mu\cos\theta)t^2 & \frac{1}{2}g(\sin\theta - \mu\cos\theta)t^2 < L \\ L & \text{otherwise} \end{cases} \qquad (12.4)$$

The first two cases are the same as in Equation 12.3. The remaining two cases give the position of the block before and after it has reached the bottom of the ramp.

It would appear that implementing this model would require using a four-way `if` statement. As we will see below, however, a more streamlined solution requires only a two-way `if` statement followed by a one-way `if` statement.

12.3.2 One-Way Conditionals

An implementation that takes the third and final model into account appears in Figure 12.8. We have made several changes. We now prompt for the length of the ramp. We have also added an `if` statement on lines 30–32 to enforce the limit on the length of the ramp. If the calculated position would be beyond the end of the ramp, this `if` statement sets `pos` back to the `limit`.

The `if` statement that we have added is different from the others we have seen in that the `else` part is omitted. If the condition is false, the `if` statement has no effect. The form of a one-way conditional is:

```
if (CONDITION) {
    STATEMENTS
}
```

In fact, any of the `if` statements that we have seen in this chapter can be written without an `else` part, if desired.

12.3.3 Precedence

To this point, we have seen three kinds of operators in C: arithmetic operators, relational operators, and Boolean operators. Because it is possible to write Boolean expressions that mix all three kinds of operators, it is helpful to know the precedence rules that apply to them.

```
1    #include <stdio.h>
2    #include <math.h>
3
4    void main (void)
5    {
6      double angle, theta, time, pos, mu, factor, limit;
7
8      printf("Enter the angle of the ramp (degrees): ");
9      scanf("%lf", &angle);
10     theta = acos(-1) * angle/180;
11
12     printf("Enter the coefficient of friction: ");
13     scanf("%lf", &mu);
14
15     printf("Enter the length of the ramp (meters): ");
16     scanf("%lf", &limit);
17
18     printf("Enter the time (seconds): ");
19     scanf("%lf", &time);
20
21     factor = sin(theta) - mu * cos(theta);
22
23     if (time < 0 || factor < 0) {
24       pos = 0;
25     }
26     else {
27       pos = 1./2. * 9.8 * factor * time * time;
28     }
29
30     if (pos > limit) {
31       pos = limit;
32     }
33
34     printf("Position of the block is %g meters\n", pos);
35   }
```

Figure 12.8   Implementation of third model (`block4.c`)

We last discussed precedence in Section 3.4.1, where we discussed the precedence of the *Mathematica* arithmetic operators. The C arithmetic, relational, and Boolean operators are given in Table 12.3 in order of decreasing precedence. Except for the unary minus and negation operators, all are left associative.

According to this table, for example, the expression

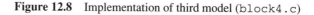

x*y+4 <= 2*z || 3 == 5

is equivalent to the fully parenthesized expression

(((x*y)+4) <= (2*z)) || (3 == 5)

Table 12.3   Precedence of C arithmetic, relational, and
Boolean operators

– and !	right associative unary operators
* and /	left associative
+ and –	left associative
< and <= and > and >=	left associative
== and ! =	left associative
&&	left associative
||	left associative

As always, if you are in doubt about how the precedence rules work, use parentheses
to group expressions explicitly.

Figure 12.9 shows the result of running `block4.c`. We have specified the same
angle, duration, and coefficient of friction as in Figure 12.7, along with a ramp length
of 10 meters. The program reports the block's position at the end of five seconds as
10 meters because it has reached the end of the ramp; on an infinite ramp, the same
block moved 17 meters.

```
Enter the angle of the ramp (degrees): 45
Enter the coefficient of friction: .8
Enter the length of the ramp (meters): 10
Enter the time (seconds): 5
Position of the block is 10 meters
```

Figure 12.9   Interaction with `block4.c`

12.4 Programmer-Defined Functions

Suppose that before developing the implementation (`block4.c`) of our final model,
we had discovered a C library header file `block.h` that provided a function called
`position` that behaved as in Equation 12.4. We would have then been able to
implement our program as illustrated in Figure 12.10. Having included `block.h`
on line 3, all that would have been required would be to read in the four relevant
values, solve the problem with the call on line 22 to the handy `position` function,
and print out the result.

Unfortunately, the C library provides no such header file and no such function.
If we wish to write our `main` function as in Figure 12.10, we must implement
our own `position` function. The reasons for dividing a C implementation into
functions should be familiar to you; they are identical to the reasons for dividing a
Mathematica implementation into functions that we saw beginning in Chapter 5.

```
1    #include <stdio.h>
2    #include <math.h>
3    #include <block.h>
4
5    void main (void)
6    {
7      double angle, theta, time, pos, mu, limit;
8
9      printf("Enter the angle of the ramp (degrees): ");
10     scanf("%lf", &angle);
11     theta = acos(-1) * angle/180;
12
13     printf("Enter the coefficient of friction: ");
14     scanf("%lf", &mu);
15
16     printf("Enter the length of the ramp (meters): ");
17     scanf("%lf", &limit);
18
19     printf("Enter the time (seconds): ");
20     scanf("%lf", &time);
21
22     pos = position(theta, mu, limit, time);
23
24     printf("Position of the block is %g meters\n", pos);
25   }
```

Figure 12.10   Postulating a `position` function in the C library (`block5.c`)

Figure 12.11 shows an implementation of our model that is divided into a `main` function and a `position` function. The `main` function is exactly the same as in Figure 12.10, so we have elided it.

The implementation of `position` is composed of a header (on lines 4–5) and a body (on lines 6–23). The implementation of `position`, of course, is not the only example of a function implementation in the program. The `main` function also consists of a header and a body.

12.4.1 Function Headers

The header of a C function serves as the interface between the function body and the remainder of the program by specifying information that is relevant to both. It gives

- the name of the function;
- the number of parameters that must be passed in each call to the function;
- the type required of each of the values in a function call; and
- the type of value (if any) that is returned by the function.

```
1    #include <stdio.h>
2    #include <math.h>
3
4    double position (double theta, double mu,
5                           double limit, double time)
6    {
7      double pos, factor;
8
9      factor = sin(theta) - mu * cos(theta);
10
11     if (time < 0 || factor < 0) {
12       pos = 0;
13     }
14     else {
15       pos = 1./2. * 9.8 * factor * time * time;
16     }
17
18     if (pos > limit) {
19       pos = limit;
20     }
21
22     return(pos);
23   }
24
25
26   void main (void)
27   {
28     . . . . .
29   }
```

Figure 12.11   Using a programmer-defined function (`block6.c`)

The header also gives one category of information,

- the name of each formal parameter,

that is relevant only to the function body and *not* to the remainder of the program.

A header begins with the return type of the function, followed by its name. Thus, `position` returns a value of type `double`. The `main` function doesn't return a value, which is indicated by using the keyword `void` in place of a return type.

Following the function name in a header is a parenthesized list giving information about the function's parameters. The `position` function expects four parameters, each of type `double`. The `main` function expects no parameters, as indicated by the keyword `void`. You might have expected that the parameter list of a function expecting no parameters would be empty. This would be more logical, but compatibility with older versions of C necessitates the use of the keyword `void`.

The parameter list also gives the names of the formal parameters. Thus, the four parameters to `position` are `theta`, `mu`, `limit`, and `time`. This information is of absolutely no significance except within the function body.

12.4.2 Function Bodies

As we learned in Chapter 11, when the `main` function is called C creates a table to keep track of the local variables. It then executes the sequence of statements that make up the body, using the table to keep track of variable values. It returns when it reaches the end of the body. Because `main` doesn't take any parameters or return a result, this is a special case of what happens when a function such as `position` is called.

When the `position` function is called, C creates a table to keep track of the values of the formal parameters as well as the local variables. Each formal parameter is initialized with the value of the corresponding actual parameter, but the local variables are not initialized. For example, if `position` is called with

```
position(1.05, 0.95, 200.0, 12.3)
```

then C creates a table such as appears in Figure 12.12a. Once the variable table is created, the sequence of statements that make up the body of the function is executed. After the conditional statement on lines 11–16 is executed, the table appears as in Figure 12.12b.

The formal parameters and local variables used in a function are the private property of that function. Except by passing the initial values for the formal parameters in a function call, no other function has access to a function's formal parameters or local variables. Similarly, the values stored in a function's table of variables have no effect on the values of the variables that are used in other functions, even if they have the same name.

Line 22 is a `return` statement, which is a type of statement that we have not previously seen. A `return` statement consists of the keyword `return` followed by an expression. When C encounters a `return` statement, it terminates the execution

theta	1.05
mu	0.95
limit	200.0
time	12.3
factor	
pos	

theta	1.05
mu	0.95
limit	200.0
time	12.3
factor	0.395
pos	292.62

(a) Immediately after call   (b) After first conditional

Figure 12.12   Table of variables for call to `position` in `block6.c`

of the function and returns the value of the expression as the value of the function call. The function's variable table is discarded at this point. If and when the function is called again, a new variable table is created.

Every function that returns a value must have a `return` statement. We will see numerous examples of such functions in this as well as in succeeding chapters. Once we begin writing programs with conditionals and loops, we will see examples of functions with more than one `return` statement.

If a function does not return anything, as has been the case with every `main` function that we have written to this point, the `return` statement may be omitted. If a `return` statement is used in such a function, it must consist only of the keyword `return`.

12.4.3 Function Prototypes

A C program may contain any number of functions, exactly one of which must be called `main`. Our example program `block6.c` consists of two functions, `position` and `main`.

C requires that a function must be declared before it can be called. One way to do this is to organize the program so that the implementation of each function appears before the point where it is first called. This is what we have done with `block6.c` in Figure 12.11, where we have placed the implementation of `position` before that of `main`, from which `position` is called.

The necessity to order functions in this way is inconvenient in large programs, and it can even be impossible if a program contains two functions each of which calls the other. Fortunately, there is an alternative. All that C requires is that a function be *declared* before it is called; it can be *implemented* anywhere. Figure 12.13 contains a version of our program in which `position` is declared (on lines 4–5) and `main` implemented before `position` is implemented.

The declaration of `position` on lines 4–5 is called a *prototype*. The prototype of a function is simply its header, terminated with a semicolon. It gives the name of the function, the number and types of its parameters, and the type of its return value. C makes no use of the formal parameter names in a prototype, and they can in fact be omitted.

We have made use of function prototypes before now, though you have probably been unaware of it. The library header files `stdio.h` and `math.h` that we have been including at the beginning of each program contain prototypes for the library functions.

A prototype details everything that C needs to know in order to compile a call to a function. By adding a comment to each prototype, we can also arrange for a prototype to detail everything that a *programmer* needs to know in order to use a function. A commented prototype for `position` appears in Figure 12.14.

By reading this commented prototype, a programmer can learn everything necessary to call `position` without being forced to read the code. We will make extensive use of commented prototypes in Chapter 13.

```
1     #include <stdio.h>
2     #include <math.h>
3
4     double position (double theta, double mu,
5                         double limit, double time);
6
7
8     void main (void)
9     {
10      . . . . .
11    }
12
13
14    double position (double theta, double mu,
15                        double limit, double time)
16    {
17      double pos, factor;
18
19      factor = sin(theta) - mu * cos(theta);
20
21      if (time < 0 || factor < 0) {
22        pos = 0;
23      }
24      else {
25        pos = 1./2. * 9.8 * factor * time * time;
26      }
27
28      if (pos > limit) {
29        pos = limit;
30      }
31
32      return(pos);
33    }
```

Figure 12.13   Using a prototype (`block7.c`)

```
/* Returns the distance in meters moved by a block "time"
   seconds after it is released on a ramp of length "limit"
   meters that is inclined "theta" radians from the horizontal.
   The coefficient of friction between the ramp and the block
   is "mu". */

double position (double theta, double mu,
                    double limit, double time);
```

Figure 12.14   Commented prototype of `position`

Large C programs often consist of more than one file. It is possible for a programmer to create a header file containing prototypes for functions that are implemented in one file but used in another. We will explore this issue further beginning in Chapter 15.

12.5 Assessment

In our implementations of the three models of the sliding block problem, we have directly realized the position function as defined in Equations 12.1, 12.3, and 12.4. This gives us a way to determine the position of the block as a function of ramp angle, coefficient of friction, and ramp length. It does not, however, afford us a very good way of appreciating the accelerating motion of the block over time. It would be more interesting if our programs printed out a sequence of positions or created a graph or animation of the block's motion. We have not yet developed enough expertise in C to take either approach.

The models that we have developed in this chapter are all oversimplified. The coefficient of friction between the block and the ramp will actually drop once the block begins moving. To describe the motion of the block on the ramp more accurately, we must take into account both the static and dynamic coefficients of friction. We will do this as an exercise.

12.6 Key Concepts

Conditional statements.   It is possible for a C program to choose among alternative statement sequences through the use of `if` statements. In its various forms, an `if` statement permits one-way, two-way, and multi-way choices.

Boolean expressions. `If` statements are controlled with Boolean expressions, which are composed using relational and Boolean operators. The six relational operators are `<` (less than), `<=` (less than or equal to), `>` (greater than), `>=` (greater than or equal to), `==` (equal to), and `!=` (not equal to). The three Boolean operators are `&&` (and), `||` (or), and `!` (not).

Programmer-defined functions.   C programs are composed of one or more programmer-defined functions, exactly one of which must be called `main`. A function consists of a header and a body. The header gives information needed to use the function (name of function, number and types of parameters, type of return value), while the body consists of the sequence of declarations and statements that implement the function.

Function prototypes.   A function prototype contains all of the information needed to use a function, but does not actually implement the function. It consists of the function's header together with an explanatory comment.

12.7 Exercises

12.1   If age is your age in years, weight is your weight in pounds, and height is your height in inches, construct Boolean expressions in C that will be true only when the following statements are true.

(a) You are old enough to obtain a driver's license but too young to retire.

(b) You are not a teenager.

(c) You are either younger than 20 and less than 150 pounds, or are older than 40 and more than 6 feet.

(d) You are neither old enough to vote, tall enough to bump your head on a five-foot door frame, nor the right weight to box in the 132–140 pound division.

12.2   Develop a model of the sliding block problem to account for a block that stubbornly refuses to go faster than 2 m/sec. Once this block's velocity reaches 2 m/sec, it continues at that speed until it reaches the end of the ramp. The velocity at time t of a sliding block with coefficient of friction μ on a ramp tilted at an angle θ is

$$gt\,(\sin\theta - \mu\cos\theta) \tag{12.5}$$

12.3   Implement a new version of the sliding block program based on the model that you developed in Exercise 12.2.

12.4   Develop a model of the sliding block problem that takes better account of the physics of friction as discussed in Section 12.5. It should make use of two coefficients of friction, μ_s and μ_d. The static coefficient of friction, μ_s, applies before the block starts moving and determines when motion begins. The dynamic coefficient of friction, μ_d, applies after the block starts moving, and together with gravity determines the acceleration of the block. The dynamic coefficient of friction is never larger than the static coefficient of friction.

12.5   Implement a new version of the sliding block program based on the model that you developed in Exercise 12.4.

12.6   Implement a C program based on the ballistic trajectory problem from Chapter 7. It should prompt the user for the direction and magnitude of the projectile's initial velocity and for a time in seconds. It should calculate and report the x- and y-coordinates of the projectile at that time. Assume that the projectile stops when it hits the ground.

12.7   Implement a C program that prompts for and reads the coefficients of the quadratic equation $ax^2 + bx + c = 0$ and displays all of the real roots.

12.8 Repeat Exercise 12.7, but display all of the roots, whether they are real or complex.

12.9 Implement a C program that prompts for an angle θ in radians and reports either the value of $\sin\theta$ or $\cos\theta$, depending on which has the smaller absolute value.

12.10 Implement a C program that prompts for a positive floating-point number x. If x is not positive, your program should complain. Otherwise, it should report the largest of $2\sqrt{x}$, x^2, and 2^x.

12.11 Implement a C program that computes the position of the destroyer from Chapter 8 when it follows the trajectory described by Equation 8.37. Your program should prompt for a time, and should report the position of the destroyer at that time.

12.12 Implement a C program that computes the power required by the destroyer from Chapter 8 when it follows the trajectory described by Equation 8.37. Your program should prompt for the mass and drag coefficient of the destroyer and for a time t. It should report the power that must be produced by the destroyer at time t.

12.13 Repeat Exercise 10.17. Your program should include a programmer-defined function that takes as parameters the earth's population and land area and returns the length of the side of the square that each person would receive if the land area were divided evenly.

12.14 Reimplement the solution to Eratosthenes's problem from Figure 10.3. Your program should include a programmer-defined function that takes as parameters the angle of the sun in Alexandria and the distance to Syene and returns the circumference of the earth.

12.15 Reimplement the solution to the horizon distance problem from Figure 10.5. Your program should include a programmer-defined function that takes as parameters the height of the hill and the radius of the earth and returns the distance to the horizon from the top of the hill.

12.16 Repeat Exercise 11.16 by reusing, with no modifications, the programmer-defined function you created in Exercise 12.13.

12.17 Repeat Exercise 11.17 by reusing, with no modifications, the programmer-defined function you created in Exercise 12.14.

12.18 Repeat Exercise 11.18 by reusing, with no modifications, the programmer-defined function you created in Exercise 12.15.

12.19   Repeat Exercise 11.19. Your program should include three programmer-defined functions to compute the three forms of interest.

12.20   Implement a C program that prompts for a beginning balance, an annual interest rate, the number of compounding intervals per year, and three investment periods. It should compare the results of an annual interest investment for the first investment period, a compound interest investment for the second investment period, and a continuous interest investment for the third investment period. It should identify which of the three options is superior. You should reuse, with no modifications, the programmer-defined functions from Exercise 12.19.

13

Rod Stacking: Designing with Functions

A number of cylindrical steel rods of various radii are stacked horizontally in a bin. The rods on the bottom are lying on the floor and braced by the sides of the bin. Every other rod is supported from below by exactly two rods. Assuming that we know the radius of each rod as well as the center coordinates of the rods on the bottom, what are the center coordinates of the remaining rods?

In this chapter we will develop a C program to solve the instance of this problem that is illustrated in Figure 13.1, which shows a way in which eight rods might be stacked. Three of the rods are supported by the floor, with the remaining five stacked on top. The radius of each rod, and the center coordinates of the three bottom rods, are given in Table 13.1.

This is a more difficult problem than any we have encountered to this point. This means that solving the problem will require a more intricate model, and that coding the model will require a more complicated implementation, than has been the case in any earlier chapter. We will use this opportunity to focus on the process by which one designs and then implements a C program that cannot be conveniently realized as a single `main` function.

13.1 Decomposing the Problem

The key insight required to develop a model of the rod-stacking problem is to realize that the problem can be divided into a collection of identical subproblems. Developing a model for solving the subproblem will yield a model for solving the whole problem.

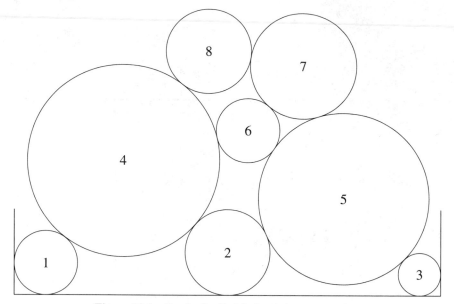

Figure 13.1   Stack of cylindrical rods in cross section

Table 13.1   Radii and center coordinates of rods from Figure 13.1

| | | Center coordinates | |
Rod	Radius	x	y
1	3.00	3.00	3.00
2	4.00	20.00	4.00
3	2.00	38.00	2.00
4	9.00		
5	8.00		
6	3.00		
7	5.00		
8	4.00		

The subproblem is illustrated in Figure 13.2. We need to be able to determine the coordinates of a rod that is supported by two other rods whose center coordinates are known. Once we know how to do that, we can determine the coordinates of all the rods in any stack. For example, the stack in Figure 13.1 could be solved by determining the coordinates of rod 4, then rod 5, and so on through rod 8. By dealing with the rods in this order, we would always be deriving the unknown coordinates of a supported rod from the known coordinates of its two supporting rods.

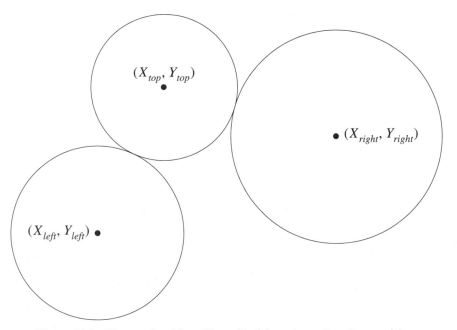

Figure 13.2   Three-rod problem. The radii of the rods are R_{left}, R_{right}, and R_{top}.

13.1.1 Three-Rod Model

We will turn our focus to the problem of determining the position of a circular rod that is supported by two other circular rods. Figure 13.3 shows the same three rods that we saw in Figure 13.2, but with two reference triangles added.

The upper triangle connects the centers of the three rods. We have labeled its sides a, b, and c, and one of its angles α. The lower triangle is a right triangle whose hypotenuse connects the centers of the two supporting rods. We have labeled its sides c, d, and e, and one of its angles β. Side d is parallel to the x-axis, and side e is parallel to the y-axis.

We will assume that we know the radii of the left (R_{left}), right (R_{right}), and top (R_{top}) rods; the coordinates of the left rod (X_{left}, Y_{left}); and the coordinates of the right rod (X_{right}, Y_{right}). Our goal is to find the coordinates of the top rod (X_{top}, Y_{top}). This will involve a bit of trigonometry.

We begin by determining the lengths of the sides of the two triangles. Because the left and top rods are tangent,

$$a = R_{left} + R_{top} \tag{13.1}$$

and because the right and top rods are tangent,

$$b = R_{right} + R_{top} \tag{13.2}$$

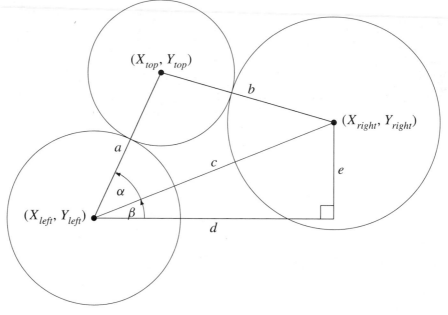

Figure 13.3   Three rods with triangles

The coordinates of the left and right rods are fixed, which means that we can easily calculate c, d and e.

$$c = \sqrt{(X_{right} - X_{left})^2 + (Y_{right} - Y_{left})^2} \tag{13.3}$$

$$d = X_{right} - X_{left} \tag{13.4}$$

$$e = Y_{right} - Y_{left} \tag{13.5}$$

Now let's turn our attention to determining the sines and cosines of the angles α and β. Because β is involved in a right triangle,

$$\cos \beta = \frac{d}{c} \tag{13.6}$$

$$\sin \beta = \frac{e}{c} \tag{13.7}$$

Dealing with α is slightly more difficult as it involves using the law of cosines.

$$\cos \alpha = \frac{a^2 + c^2 - b^2}{2ac} \tag{13.8}$$

$$\sin \alpha = \sqrt{1 - \cos^2 \alpha} \tag{13.9}$$

Figure 13.4 is similar to Figure 13.3, with the rods removed, the line c removed, and a reference line added to form a new triangle. One of the angles of this triangle is $\alpha + \beta$. Since this triangle is a right triangle, it follows that

$$X_{top} = X_{left} + a \cos(\alpha + \beta) \tag{13.10}$$

$$Y_{top} = Y_{left} + a \sin(\alpha + \beta) \tag{13.11}$$

All that remains to complete our model is to determine the sine and cosine of the angle $\alpha + \beta$. Two standard trigonometric identities do the trick.

$$\sin(\alpha + \beta) = \sin \alpha \cos \beta + \sin \beta \cos \alpha \tag{13.12}$$

$$\cos(\alpha + \beta) = \cos \alpha \cos \beta - \sin \alpha \sin \beta \tag{13.13}$$

No unknowns remain, so our model is complete. Notice that we never did determine the values of α and β, nor do we need to. Because of this, our eventual implementation will not need to make any use of the trigonometric functions in the C library.

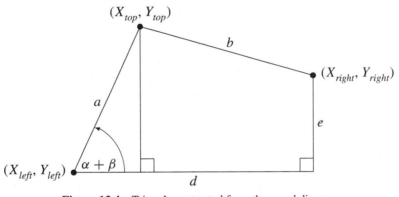

Figure 13.4   Triangles extracted from three-rod diagram

13.2 Design

The rod-stacking problem is too complicated to implement as a program containing only a single `main` function. Although it would be technically possible to do so, the resulting program would be excessively long, hard to understand, and difficult to test. A much better approach is to divide the implementation into a collection of programmer-defined functions.

In this section we will design an implementation to solve the rod-stacking problem. It is important to carefully plan how a program will be organized before writing even a single line of code. Large computer programs are the most complicated artifacts that have ever been constructed by humankind. It is just as important to design a program before implementing it as it is to design a suspension bridge before building it.

Like any engineering design problem, designing a program is a mixture of art and science. The portion that is art can only be acquired by experience. The portion that is science is the topic of countless textbooks, university courses, and business seminars. As computing technology has evolved, approaches to software design have proliferated, and these approaches make up part of the field of *software enginering*. The essence of design, however, is to answer the following two questions.

- How should the program's data be represented?
- What programmer-defined functions are needed?

The first question is the more important of the two and must be answered first. In fact, the ability to design appropriate data representations is what distinguishes an accomplished programmer. The decision of how to represent a program's data has repercussions throughout the program.

To this point in the text we have not needed to raise this issue because all of our data have been conveniently representable with integers or floating-point numbers. The program we are designing in this chapter is different, and it would benefit from a more inspired choice of data representation. Nevertheless, our design will make use of no data other than numbers. This is because our goal in this chapter is to illustrate the role that functions play in program design and implementation.

We will represent the *x*- and *y*-coordinates of the centers of the rods using numbers of type `double`, and we will represent the radii of the rods using numbers of type `double`. We will revisit this decision at two different points in the text. In Section 13.4 we will discuss how our implementation could have been simplified with a different data representation, and in Chapter 16 we will revise the design and implementation that we are about to produce to take advantage of that representation.

All of this means that our focus here will be on answering the second question. We will start by describing a simple but powerful way of beginning with a program description and deciding what programmer-defined functions are required. We will then illustrate this design approach by applying it to the rod-stacking problem.

13.2.1 Functional Decomposition

The functional decomposition design process begins with descriptions of what a program is to do and how its data should be represented. It produces a collection of commented function prototypes whose implementation will yield the desired program. As we explain below, it is an iterative process.

The first step is to produce a commented prototype for the `main` function. For all of the programs in this text, the prototype of `main` will always be

```
void main (void);
```

and the comment will be the description of what the entire program is supposed to do. This is because what the `main` function accomplishes, from beginning to end, is exactly what the program accomplishes.

The next step is to carefully consider the problem of *implementing* main. The goal is to identify functions which—if only they existed—would make implementing main easy. Sometimes the problem is so simple that no subsidiary functions are required. Other times, the functions we need already exist, perhaps as C library functions.

It is often the case, though, that the functions we wish we had do not already exist. In this case, we must design commented prototypes for each of these functions, taking care that the prototypes we produce are exactly what we will need to implement main.

Next we must turn our attention to the problem of implementing these newly designed functions. As we consider each function, we may well uncover the need for still other functions, for which we must also create prototypes. The design process continues until we have created prototypes for all of the programmer-defined functions required to implement the program. For this design process to work, the functions for which we create prototypes must become successively simpler, so that we ultimately arrive at functions that can be implemented directly in terms of library functions.

13.2.2 Designing main

We will apply the functional decomposition process to designing an implementation for the rod-stacking problem. We always begin with main because it is the only function that is guaranteed to be a part of every program. Its commented prototype appears in Figure 13.5.

Next we must consider what would be involved in implementing main. The derivation of our model in Section 13.1 suggests that main would be easy to implement if we had available a function that would solve the three-rod problem. We could then make a sequence of calls to this function, successively determining and displaying the center coordinates of each rod.

Suppose that we had such a function, called threeRod. It would need to take as parameters the center coordinates and radii of the two supporting rods and the radius of the supported rod, and would return as its result the center coordinates of the supported rod.

Unfortunately, a C function can return only one value, not two. Since threeRod must return *two* coordinates, we must identify a different approach. Suppose instead that we had one function called threeRodX, which would calculate the *x*-coordinate of the supported rod, and another called threeRodY, which would

```
/* Displays a table showing the x-y coordinates of the centers
   of rods 4--8 from Figure 13.1, using the raw data from
   Table 13.1. */

void main (void);
```

Figure 13.5 Prototype for main

calculate the *y*-coordinate of the supported rod. The prototypes for these two functions are shown in Figure 13.6.

If we had these two functions, the implementation of `main` would be straightforward. All that remains is to design `threeRodX` and `threeRodY`.

```
/* Returns the x-coordinate of the center of a supported rod
   of radius rTop. The left supporting rod has center
   coordinates (xLeft, yLeft) and radius rLeft. The right
   supporting rod has center coordinates (xRight, yRight)
   and radius rRight. */

double threeRodX (double xLeft, double yLeft,
                  double xRight, double yRight,
                  double rLeft, double rRight, double rTop);


/* Returns the y-coordinate of the center of a supported rod
   of radius rTop. The left supporting rod has center
   coordinates (xLeft, yLeft) and radius rLeft. The right
   supporting rod has center coordinates (xRight, yRight)
   and radius rRight. */

double threeRodY (double xLeft, double yLeft,
                  double xRight, double yRight,
                  double rLeft, double rRight, double rTop);
```

Figure 13.6   Prototypes for `threeRodX` and `threeRodY`

13.2.3 Designing `threeRodX`

Everything that is required to implement `threeRodX` was laid out in Section 13.1.1, where we derived the model for the three-rod problem. To find the *x*-coordinate of the supported rod, we must find the lengths of the sides (a, b, c, d, e) and the sines and cosines of the two angles (α, β) that are labeled in Figure 13.3.

Equations 13.1–13.5 tell us that a, b, d, and e can all be calculated via simple arithmetic on the parameters to `threeRodX`, but that a calculation based on the distance formula is required to calculate c. It would be helpful to have a function, such as the one whose prototype appears in Figure 13.7, to do the distance calculation for us.

The sine and cosine of β can be obtained by division (Equations 13.6 and 13.7), but obtaining the sine and cosine of α requires an application of the law of cosines (Equations 13.8 and 13.9). Let's invent a function to do this for us; its prototype appears in Figure 13.7.

```
/* Returns the distance between points (x1,y1) and (x2,y2). */

double distance (double x1, double y1, double x2, double y2);


/* Returns the cosine of one of the angles of a triangle. The
   adjacent sides are of length adj1 and adj2, while the
   opposite side is of length opp. */

double cosine (double adj1, double adj2, double opp);


/* Returns the square of x. */

double sq (double x);
```

Figure 13.7   Prototypes for three helping functions

The remainder of the calculations required to compute the *x*-coordinate of the supported rod involve only simple arithmetic.

13.2.4 Designing `threeRodY`

The calculations required to determine the *y*-coordinate of a supported rod are identical, except for the last step, to those required to determine the *x*-coordinate. This means that the implementations will be virtually identical, and that any functions that help in implementing `threeRodX` will also help in implementing `threeRodY`.

13.2.5 Designing `distance`

To compute the distance between two points we need to square two numbers and take a square root. We can obtain a square root function from the C math library, but it would be helpful to create a function to do squaring for us such as the one whose prototype is in Figure 13.7.

13.2.6 Designing `cosine`

To apply the law of cosines, we need to square three numbers, do some simple arithmetic, and compute a square root. The function `sq` that we designed to help implement `distance` will also help us here.

13.2.7 Designing `sq`

The implementation of `sq` will require only multiplication. We have now completed our design and can implement it.

13.3 Implementation

Although a good design makes it much easier to produce a correct program, it is important to take an equally disciplined approach to implementation. We produced our design by beginning with `main` and working down to the simplest programmer-defined functions. In doing the implementation we will reverse the process. By implementing and testing the simplest functions first and working back towards `main`, we can test each function as we write it.

13.3.1 Implementing `sq`

Of the six functions that we need to implement, `sq` is the only one that does not depend on any of the others. For this reason, we will implement and test it first. Once we have convinced ourselves that it works, we will be able to use it with confidence when we implement the other functions.

Figure 13.8 shows an implementation of `sq`. We have added a function body to the prototype that we developed during the design phase. Because `sq` is such a simple function, its implementation turns out to be only one line long.

Figure 13.8 also contains a *driver* for `sq`. A driver is a `main` function constructed for the express purpose of testing some other function. For this reason, drivers are always straightforward. The one on lines 10–16 of Figure 13.8 prompts the user for a parameter to pass to `sq` and then displays the result of calling `sq` with that parameter.

The implementation of `sq` together with its driver constitutes a complete C program, which we can compile and run to test `sq`. Nevertheless, the driver will not be a part of the final program, which will have a very different `main` function. When we are through testing `sq` we will remove its driver from the developing program.

```
1    #include <math.h>
2    #include <stdio.h>
3
4    double sq (double x)
5    {
6       return(x*x);
7    }
8
9
10   void main (void)
11   {
12      double x;
13      printf("Enter double: ");
14      scanf("%lf", &x);
15      printf("sq(%g) = %g\n", x, sq(x));
16   }
```

Figure 13.8   Implementation of `sq` with driver (`rod1.c`)

13.3.2 Implementing `cosine`

Neither `cosine` nor `distance` depends on any programmer-defined function other than `sq`, which means that we could implement either one next. We have arbitrarily chosen `cosine`, and its implementation appears in Figure 13.9.

Notice what we did in moving from the program in Figure 13.8 to the program in Figure 13.9: We retained the implementation of `sq`, but eliminated its driver. We added an implementation of `cosine`, and we added a driver for testing `cosine`. The result is a complete program that can be used for testing `cosine`.

Suppose that we turn up a problem while testing `cosine`. We would then have to examine the program to find the programming mistake. Because we have already tested `sq` and convinced ourselves that it is correct, we would be able to focus our attention on the implementation of `cosine` as the most likely source of the error. If we had not previously convinced ourselves that `sq` was correct, we would have to consider that the mistake might be in either `cosine` or `sq`.

This might not seem like such an important consideration since `cosine` and `sq` are so simple, but as programmer-defined functions become more complicated it becomes increasingly important to be able to isolate the source of an error to a single function body. The tactic of working back from the simplest functions to the main function, using drivers to test each function as we write it, makes this possible.

```
1    #include <math.h>
2    #include <stdio.h>
3
4    double sq (double x)
5    {
6       return(x*x);
7    }
8
9
10   double cosine (double adj1, double adj2, double opp)
11   {
12      return(((sq(adj1) + sq(adj2) - sq(opp)) / (2*adj1*adj2)));
13   }
14
15
16   void main (void)
17   {
18      double a1, a2, op;
19      printf("Enter three doubles: ");
20      scanf("%lf%lf%lf", &a1, &a2, &op);
21      printf("cosine(%g, %g, %g) = %g\n",
22             a1, a2, op, cosine(a1, a2, op));
23   }
```

Figure 13.9 Implementation of `cosine` with driver (`rod2.c`)

13.3.3 Implementing `distance`

Figure 13.10 shows the next step in the evolution of our program. We have retained sq and cosine, eliminated the driver that we used to test cosine, and added implementations of distance and its driver. Now we have a complete program that we can use to test distance.

```
1    #include <math.h>
2    #include <stdio.h>
3
4    double sq (double x)
5    {
6       return(x*x);
7    }
8
9
10   double cosine (double adj1, double adj2, double opp)
11   {
12      return((sq(adj1) + sq(adj2) - sq(opp)) / (2*adj1*adj2));
13   }
14
15
16   double distance (double x1, double y1,
17                    double x2, double y2)
18   {
19      return(sqrt(sq(x1-x2) + sq(y1-y2)));
20   }
21
22
23   void main (void)
24   {
25      double x1, y1, x2, y2;
26      printf("Enter a point (two doubles): ");
27      scanf("%lf%lf", &x1, &y1);
28      printf("Enter another point (two doubles): ");
29      scanf("%lf%lf", &x2, &y2);
30      printf("distance(%g, %g, %g, %g) = %g\n",
31             x1, y1, x2, y2, distance(x1, y1, x2, y2));
32   }
```

Figure 13.10   Implementation of `distance` with driver (`rod3.c`)

While testing can increase our confidence that a program is correct, it can never prove that a program contains no mistakes. Test cases should be devised with an eye toward revealing mistakes. It is pointless to employ a test case unless you know what answer you expect the program to produce in response.

Major mistakes in a function will usually be revealed by the simplest of test cases. To test the `distance` function, for example, we might use (0,0) and (3,4) as input values, expecting to see 5 as the result.

To find the subtle mistakes in a function, it is a good idea to devise test cases that describe boundary situations. We might, for example, test `hypot` with input values of (0,0) and (0,0), expecting zero as the result. These input values describe two points with a distance of zero between them.

13.3.4 Implementing `threeRodY`

The three helping functions that we have just implemented were quite straight-forward. The implementation of `threeRodY`, shown in Figure 13.11, is consider-ably more involved. Our program at this point consists of `sq`, `cosine`, `distance`, `threeRodY`, and a driver. In the interests of brevity, we have eliminated all but `threeRodY` from the figure.

The implementation makes use of nine local variables that are inspired by the derivation of the model that we did in Section 13.1. The body of the function consists of nine assignment statements that give values to these variables, followed by a return statement that combines the values of several of the variables to compute the y-coordinate of the supported rod. The order of the assignment statements is not arbitrary. The assignments on lines 8–12 must come before the assignments on lines 14–17, which use the values of a, b, c, d, and e.

```
1    double threeRodY (double xLeft, double yLeft,
2                       double xRight, double yRight,
3                       double rLeft, double rRight, double rTop)
4    {
5       double a, b, c, d, e;
6       double cosA, sinA, cosB, sinB;
7
8       a = rLeft + rTop;
9       b = rRight + rTop;
10      c = distance(xLeft, yLeft, xRight, yRight);
11      d = xRight - xLeft;
12      e = yRight - yLeft;
13
14      cosA = cosine(a, c, b);
15      sinA = sqrt(1 - sq(cosA));
16      cosB = d/c;
17      sinB = e/c;
18
19      return(yLeft + a * (sinA*cosB + sinB*cosA));
20   }
```

Figure 13.11   Implementation of `threeRodY` (`rod4.c`)

13.3.5 Implementing `threeRodX`

There's not much difference between `threeRodY` and `threeRodX`, which is shown in Figure 13.12. The only difference other than the name of the function is the return statement on line 19.

Because these two functions are so similar, because they take exactly the same parameters, and because they don't depend on each other, it makes sense to test `threeRodY` and `threeRodX` simultaneously with similar drivers. The same test cases can be used with each, and a mistake uncovered in one is likely to be a mistake in the other.

Coming up with test cases, however, presents a "chicken and egg" problem. If we just make up some coordinates and radii and feed them to the two functions, we will have no way to judge whether the results are correct. Any calculations we do to determine what the answers should be are as likely to contain mistakes as the program itself. We must resolve this dilemma, however. The implementations of `threeRodY` and `threeRodX` are our most complicated ones, so they are the most likely to contain mistakes.

A good first step is to use graph paper to sketch several representative three-rod problems. Although we will not be able to predict *exactly* what the answer should be, we will be able to predict *approximately* what it should be. This will reveal major problems in the functions with a minimum of effort.

There are other ways to examine the results of such test cases to make sure they are sensible. Since we know the radii of all three rods, we know in advance what

```
1    double threeRodX (double xLeft, double yLeft,
2                      double xRight, double yRight,
3                      double rLeft, double rRight, double rTop)
4    {
5      double a, b, c, d, e;
6      double cosA, sinA, cosB, sinB;
7
8      a = rLeft + rTop;
9      b = rRight + rTop;
10     c = distance(xLeft, yLeft, xRight, yRight);
11     d = xRight - xLeft;
12     e = yRight - yLeft;
13
14     cosA = cosine(a, c, b);
15     sinA = sqrt(1 - sq(cosA));
16     cosB = d/c;
17     sinB = e/c;
18
19     return(xLeft + a * (cosA*cosB - sinA*sinB));
20   }
```

Figure 13.12 Implementation of `threeRodX` (`rod5.c`)

the distances between the left and top rods, and the right and top rods, should be. We can easily verify that the results of each test case meet this constraint.

Returning to the model in Section 13.1 will provide further inspiration. The model was derived using a geometry in which the center of the left supporting rod was lower than the center of the right supporting rod. Does the model hold when the left rod is even with the right rod? When it is higher? It turns out that the model holds in both of these cases, as properly chosen test cases will reveal.

It is more difficult to test whether the results are exactly correct, as opposed to approximately correct. One approach is to create a simple arrangement of rods for which the exact answer is easy to obtain. For example, if all three rods are of the same radius and the supporting rods are tangent and have the same *y*-coordinates, the centers of the three rods will form an equilateral triangle. In this special case it is not hard to work out by hand the exact coordinates that we expect as results.

Assuming that our implementation has survived these preliminary tests, we can finally derive a few test cases using a calculator or *Mathematica* and compare our results with those produced by the driver. We must bear in mind, however, that any discrepancy we find may be the fault of the derivation process rather than the program, and that an illusory consistency could easily result from basing both the derivation and the program on the same faulty model.

13.3.6 Implementing `main`

We are now ready to implement the `main` function. Along the way to this point, we have implemented and discarded five different `main` functions that served as drivers. We created them to test the individual functions as they were implemented, and we then put them aside.

The final version of `main`, which ties everything together, appears in Figure 13.13. The variables are assigned values according to Table 13.1 on lines 7–10 and the resulting output appears in Figure 13.14. The `main` function does not require a driver to test, of course, because it is the top-level function of the program. With the exception of the `printf` statements on lines 27–32, its implementation is straightforward.

The `printf` statements are contrived to display the results in aligned columns. The floating-point conversion specifications that we have used are all of the form `%7.2f`. This means that a floating-point number is to be displayed with two digits after the decimal point using a minimum of seven characters. If the number turns out to require fewer than seven characters to print, it is padded on the left with spaces.

Similar formatting information can also be included with the `%g`, `%e`, and `%d` format specifications. For example, `%5.3g` means that a floating-point number is to be displayed with a three-digit mantissa using a minimum of five characters; `%9.2e` means that a floating-point number is to be displayed with two digits after the decimal point using a minimum of nine characters; and `%6d` means that an integer is to be displayed using a minimum of six characters.

We used *two* conversion specifications and *three* parameters in each of the last five `printf` statements of `rod6.c`. The first conversion specification in the format

```
1    void main (void)
2    {
3      double x1, x2, x3, x4, x5, x6, x7, x8;
4      double y1, y2, y3, y4, y5, y6, y7, y8;
5      double r1, r2, r3, r4, r5, r6, r7, r8;
6
7      x1 = 3; y1 = 3; r1 = 3;
8      x2 = 20; y2 = 4; r2 = 4;
9      x3 = 38; y3 = 2; r3 = 2;
10     r4 = 9; r5 = 8; r6 = 3; r7 = 5; r8 = 4;
11
12     x4 = threeRodX(x1, y1, x2, y2, r1, r2, r4);
13     y4 = threeRodY(x1, y1, x2, y2, r1, r2, r4);
14
15     x5 = threeRodX(x2, y2, x3, y3, r2, r3, r5);
16     y5 = threeRodY(x2, y2, x3, y3, r2, r3, r5);
17
18     x6 = threeRodX(x4, y4, x5, y5, r4, r5, r6);
19     y6 = threeRodY(x4, y4, x5, y5, r4, r5, r6);
20
21     x7 = threeRodX(x6, y6, x5, y5, r6, r5, r7);
22     y7 = threeRodY(x6, y6, x5, y5, r6, r5, r7);
23
24     x8 = threeRodX(x4, y4, x7, y7, r4, r7, r8);
25     y8 = threeRodY(x4, y4, x7, y7, r4, r7, r8);
26
27     printf("Rod     X        Y\n");
28     printf(" 4 %7.2f %7.2f\n", x4, y4);
29     printf(" 5 %7.2f %7.2f\n", x5, y5);
30     printf(" 6 %7.2f %7.2f\n", x6, y6);
31     printf(" 7 %7.2f %7.2f\n", x7, y7);
32     printf(" 8 %7.2f %7.2f\n", x8, y8);
33   }
```

Figure 13.13 Implementation of main (rod6.c)

```
Rod    X        Y
  4   10.23   12.58
  5   30.89    9.03
  6   21.90   15.37
  7   27.08   21.46
  8   18.19   22.85
```

Figure 13.14 Output produced by rod6.c

string is replaced with the value of the second parameter, and the second conversion specification is replaced with the value of the third parameter.

Testing `main` poses even more of a problem than testing `threeRodY` and `threeRodX`. We wrote the program to solve a specific rod-stacking problem, presumably because we didn't know the answer. It would be asking a bit much for us to calculate independently what the answer should be in order to verify the results printed by `main`.

We can certainly verify that the results shown in Figure 13.14 are roughly correct by sketching the rod stack on graph paper. Beyond that, though, there is not much we can do in the way of testing. Fortunately, if we are convinced that `threeRodY` and `threeRodX` are correct, if we verify that we have correctly entered all of the constants into our program, and if we check that we have passed the right parameters in the right order to `threeRodY` and `threeRodX`, we are justified in trusting the results of the program.

13.4 Assessment

The model we have developed in this chapter is useful only if we know the radii and relative positions of a stack of rods, know the center coordinates of all of the rods supported by the ground, and wish to calculate the center coordinates of all of the supported rods. This limitation is due to the fact that we must know which two rods support a third rod before we can calculate the third rod's center. A more sophisticated model might also determine which rods end up supporting which other rods when the rods are placed in turn at the center of the stack and allowed to settle into position.

Although our implementation works, it is more complicated than it needs to be. The functions `threeRodX` and `threeRodY` are almost identical. They take the same parameters, declare the same local variables, and contain exactly the same sequence of assignment statements. The only difference other than the function name occurs in the last statement, where the return value is calculated.

It would be appealing to combine `threeRodX` and `threeRodY` into a single `threeRod` function that calculated both coordinates, but we are limited by the fact that a C function can return only one result. The `threeRod` function that we are envisioning would have to return two coordinates.

If C provided a built-in type called `point`, we would not have this problem. We could write a single `threeRod` function that returned a value of type `point`; this value would be a compound value containing the two coordinates. Although C does not provide such a type, it is possible for a programmer to define and use one. In Chapter 16 we will see how to do this and then exploit this capability to streamline our implementation of the rod-stacking problem.

13.5 Key Concepts

Functional decomposition. A program must be designed before it can be implemented. This becomes increasingly important as programs become more complicated. Functional decomposition is a simple but effective design strategy.

Beginning from a statement of the problem to be solved, we create prototypes for functions that would make the implementation of `main` straightforward. We repeat this process on the newly designed functions until we work our way down to C library functions.

Incremental implementation.   Functional decomposition produces a design consisting of a collection of function prototypes. An effective strategy for implementing these prototypes is to work in the opposite order from which they were designed. This way, each function that we implement can rely on the completed implementations of the functions on which it depends. As we implement each function we create a driver, which is a temporary `main` function whose purpose is to give us a way to test the function we are implementing.

13.6 Exercises

13.1   Modify `rod3.c` so that it issues separate prompts for each of the four coordinates.

13.2   Modify the output produced by `rod6.c` so that each number is displayed with a minimum of eight characters, with three digits after the decimal point.

13.3   Modify the output produced by `rod6.c` so that each number is displayed with a minimum of eight characters with exactly three digits in the mantissa.

13.4   Modify the output produced by `block5.c` in Figure 12.10 so that the position of the block is reported using a four-digit mantissa.

13.5   Modify the output produced by `rod6.c` so that the coordinates of rods 1–3, and the radii of all eight rods, are displayed.

13.6   Modify `rod6.c` so that rod 8 has a radius of 6.00.

13.7   Modify `rod6.c` so that it also finds the position of a ninth rod of radius 6.00 supported by rods 7 and 8.

13.8   There are five pairs of assignment statements on lines 12–25 of Figure 13.13. In what other ways could the pairs be ordered without affecting the results produced by the program?

13.9   Reverse the order of the six functions in the complete implementation of `rod6.c` and attempt to compile it. The error messages you see are typical of those produced when a program uses a function before it has been declared.

13.10 Replace the sq function from rod6.c with its prototype and attempt to compile it. The error message you see is typical of those produced when a program uses a function that is declared but not implemented.

13.11 Modify the implementation of sq in rod6.c so that it uses the library function pow, instead of the multiplication operator, to square its parameter. Verify that the program continues to work properly after this change.

13.12 Equation 13.3 defines the length of c using the Cartesian distance formula. It could also have defined the length of c in terms of the lengths of d and e using the Pythagorean theorem. Design a function called hypot that calculates the length of the hypotenuse of a right triangle given the lengths of the other two sides. Replace the distance function in rod6.c with an implementation of hypot, and modify threeRodX and threeRodY to use hypot instead of distance. Verify that the program continues to work properly after this change.

13.13 The model that we created in this chapter calculates the sine and cosine of the angles α and β, but never explicitly calculates the values of the angles themselves. Devise a different model that determines the values of α and β (using inverse trigonometric functions) and then uses the sine and cosine of $\alpha + \beta$ to calculate X_{top} and Y_{top}.

13.14 Modify rod6.c so that it uses your model from Exercise 13.13. How do the results produced by the modified rod6.c compare with the original? Which implementation is better?

13.15 We designed and implemented rod6.c to solve the specific rod-stacking problem from Figure 13.1 and Table 13.1. Use graph paper to sketch a different stack of rods and modify rod6.c to find the coordinates of each of that stack's supported rods.

13.16 If we had insisted on implementing each function in rod6.c immediately after we designed it, how might we have gone about testing?

13.17 Suppose you were to modify rod6.c by eliminating sq, cosine, distance, threeRodX, and threeRodY and moving all of the calculations that they do into main. Would the modified program be longer or shorter than the original? By approximately what factor?

13.18 In light of Exercise 13.17, does dividing a program into functions make it longer or shorter? Harder or easier to understand? More or less difficult to test?

13.19   The model of the rod-stacking problem was based in part on Figure 13.3, in which the center of the right supporting rod happened to be higher than the center of the left supporting rod. Verify that the model still holds when this orientation is reversed.

13.20   If C had a `point` type, as envisioned in Section 13.4, what might the prototype of a `threeRod` function be?

14

Newton's Beam: Repetition

A cubical box, each of whose sides is 1 meter long, is placed on the floor against a wall. A beam 4 meters long is leaned against the edge of the box so that one end touches the wall and the other end touches the floor. There are two different ways to position the beam so that these constraints are satisfied, one of which is illustrated in Figure 14.1. What is the distance along the floor from the base of the wall to the bottom of the beam in each of these two configurations?

In Figure 14.1 we denote by y the distance in meters along the beam from the floor to the edge of the box, and by x the distance in meters from the bottom of the

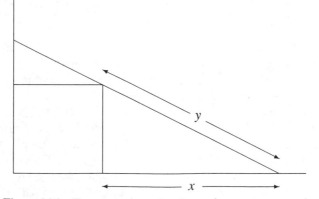

Figure 14.1 Four-meter beam leaning against a one-meter cube

box to the bottom of the beam. Thus, $x + 1$ is the distance in meters from the wall to the bottom of the beam, which is a solution to the problem.

The lower triangle (formed by the floor, the box, and the beam) and the upper triangle (formed by the box, the wall, and the beam) of Figure 14.1 are similar. Thus, the ratio of the base to the hypotenuse of the bottom triangle is the same as the ratio of the base to the hypotenuse of the upper triangle.

$$\frac{x}{y} = \frac{1}{4 - y} \tag{14.1}$$

If we rearrange this equation to solve for y, we obtain

$$y = \frac{4x}{x + 1} \tag{14.2}$$

Both triangles, and in particular the lower one, are right triangles. According to the Pythagorean Theorem

$$x^2 + 1^2 = y^2 \tag{14.3}$$

If we substitute the value of y from Equation 14.2 into Equation 14.3, we obtain

$$x^2 + 1 = \frac{16x^2}{(x + 1)^2} \tag{14.4}$$

This equation can be rearranged and simplified to obtain the equation

$$x^4 + 2x^3 - 14x^2 + 2x + 1 = 0 \tag{14.5}$$

This fourth-order polynomial has two positive and two negative roots. The positive roots are the two solutions to the beam problem. We could, use *Mathematica*'s `FindRoot` function to find the roots. Instead, we will use Equation 14.5 as a motivating example as we pursue the topic of numerical equation solving in more depth. We introduced this topic in Chapter 9, where we explored the bisection method. In this chapter we will explore a more powerful technique called *Newton's method*.

14.1 Newton's Method

Newton's method is a numerical technique for approximating a root of a function f, beginning from an initial guess g_0. The method is illustrated graphically in Figure 14.2. Both the function and the guess are labeled.

We have constructed a straight line that is tangent to f at the point $(g_0, f(g_0))$. This line crosses the x-axis at a point that we have labeled g_1. The principle behind Newton's method is that g_1 will very often be closer to an actual root than g_0.

The intuition behind this principle should be clear from Figure 14.2, in which the tangent line "points" toward the root. This will not always be true, especially if g_0 is far from the actual root or if f is a higher-order function with lots of ups and

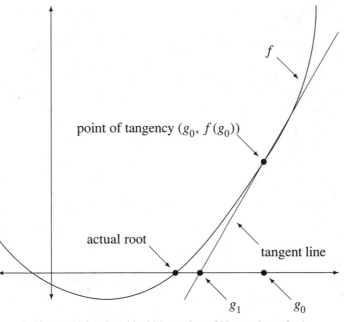

Figure 14.2   Graphical illustration of Newton's method

downs. Fortunately, the principle holds often enough for the kinds of functions that arise in practice to make Newton's method viable.

It is not difficult to determine the value of the unknown g_1 in terms of the knowns f and g_0. Recall that one form of the equation for a straight line is

$$\frac{\Delta y}{\Delta x} = m \tag{14.6}$$

where m is the slope of the line and Δx and Δy are the differences between the x- and y-coordinates, respectively, of two points on the line.

The slope of a line that is tangent to a curve is, by definition, the slope of the curve at the point of tangency. In our case, that slope is $f'(g_0)$, where f' is the first derivative of the function f.

Two of the points on the tangent line are the point of tangency, $(g_0, f(g_0))$, and the x-intercept, $(g_1, 0)$. Thus, Δx is $g_0 - g_1$ and Δy is $f(g_0) - 0$. Plugging everything into Equation 14.6, we obtain

$$\frac{f(g_0) - 0}{g_0 - g_1} = f'(g_0) \tag{14.7}$$

$$g_1 = g_0 - \frac{f(g_0)}{f'(g_0)} \tag{14.8}$$

We can now treat g_1 as an improved guess to the root of f. We can repeat this process as many times as we wish, producing g_2 from g_1, g_3 from g_2, and so on. The process converges to a root of f under a broad set of circumstances.

To help you visualize how Newton's method works, we have created a *Mathematica* package called `Newton` that provides the function `AnimateNewton`. Figure 14.3 shows the last four of the five frames of the animation that results from loading the `Newton` package into *Mathematica*,

$$\boxed{\texttt{In[9]:= Needs["ISP`Newton`"]}} \tag{14.9}$$

defining the following function,

$$\boxed{\texttt{In[10]:= example[x\_] := x\^3 - 9*x}} \tag{14.10}$$

and then evaluating this function call.

$$\boxed{\begin{array}{l}\texttt{In[11]:= AnimateNewton[example, 1.3, 4, \{-3.5, 3.5\}]} \\ \\ \texttt{Out[11]= (See Figure 14.3)}\end{array}} \tag{14.11}$$

The four parameters to `AnimateNewton` are the function whose root we are finding, an initial guess for the root, the number of repetitions of the root-finding process to animate, and the range of the x-axis to display.

Each of the frames in Figure 14.3 shows the graph of $x^3 - 9x$, which doesn't change from frame to frame. Each frame also shows a straight line that is tangent to the graph of the function. Figure 14.3a shows the tangent line corresponding to the initial guess. Each successive frame shows the tangent line corresponding to an improved guess. Notice how the x-intercept of the tangent line jumps around a bit but quickly settles down and approaches one of the three x-intercepts of the function. You should use *Mathematica* and our `AnimateNewton` function to experiment with functions and guesses of your own choosing.

We will investigate convergence criteria, and get an idea of how things can go wrong, in Section 14.4. For the time being, however, we will assume that the method always works and proceed with producing a C implementation of it.

14.2 Implementation of Newton's Method

Our underlying goal as we develop an implementation of Newton's method will be to explore the use of `while` statements in C. We briefly encountered `While` expressions (also known as `While` *loops*) in the context of *Mathematica* when we studied the bisection method in Chapter 9. From that encounter, you should

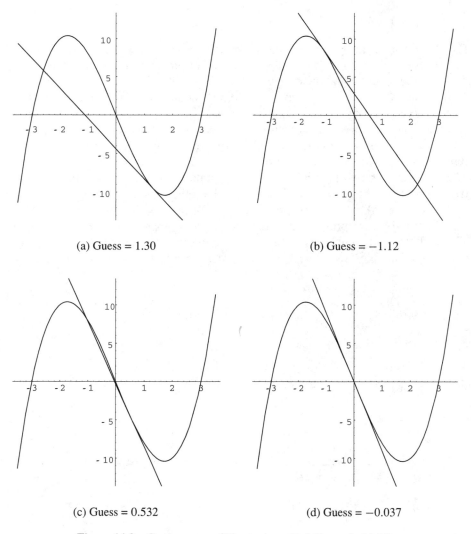

(a) Guess = 1.30

(b) Guess = −1.12

(c) Guess = 0.532

(d) Guess = −0.037

Figure 14.3   Convergence of Newton's method (Example 14.11)

appreciate the fact that `while` loops afford a way to arrange for a program to repeatedly execute a sequence of statements.

We will begin by designing our implementation as a collection of four separate functions. We will be able to implement three of these functions immediately, leaving only the function that actually realizes Newton's method.

We will develop an implementation for this fourth function via a series of three successively more refined versions. Because the first two versions will serve primarily to motivate the need for repetition, the third version will be the only one that contains a `while` loop.

14.2.1 Design

Figure 14.4 gives the prototypes of the four functions into which we have divided our program. Before we deal with the issue of implementing the prototypes, let's consider the ideas behind the design.

Because Newton's method involves doing calculations with both the function whose root is sought (f) and its first derivative (f'), it is only natural to implement f and fprime explicitly. This organization makes it easy to modify the program to find the roots of different functions. Whenever we modify f, however, we must also remember to modify fprime appropriately.

The remaining two functions serve to separate the details of Newton's method (newton) from the user interactions required to obtain an initial guess and display the result (main). The design principle at work here is that newton should do exactly one thing: improve a guess until it is a good approximation to a root of f. This will make it much easier to use the implementation of newton in other contexts, where the user interaction requirements are likely to be much different.

```
/* Computes f(x), the function whose root we seek. */

double f (double x);


/* Computes f'(x), the first derivative of f(x). */

double fprime (double x);


/* Uses Newton's method to approximate a root of f(x),
 beginning with the guess g0. */

double newton (double g0);


/* Prompts the user for an initial guess to a root of f(x)
 and then displays the result of calling newton. */

void main (void);
```

Figure 14.4   Function prototypes for Newton's method program

14.2.2 Preliminary Implementation

Figure 14.5 shows an implementation of each of the four functions that we designed in Figure 14.4. The implementations of f, fprime, and main are complete. The implementation of newton, which will be our focus of attention, is not.

```
1    #include <stdio.h>
2    #include <math.h>
3
4    double f (double x)
5    {
6      return(1 + x*(2 + x*(-14 + x*(2 + x))));
7    }
8
9    double fprime (double x)
10   {
11     return(2 + x*(-28 + x*(6 + x*4)));
12   }
13
14   double newton (double g0)
15   {
16     return(g0);
17   }
18
19   void main (void)
20   {
21     double guess, root;
22     printf("Please enter an initial guess: ");
23     scanf("%lf", &guess);
24     root = newton(guess);
25     printf("An approximate root is %.10g\n", root);
26     printf("f(approximate root) is %.10g\n", f(root));
27   }
```

Figure 14.5 First implementation of Newton's method (newton1.c)

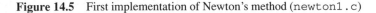

The function f is an implementation of the expression on the left-hand side of Equation 14.5, while fprime is its first derivative. We have implemented these two functions as in Figure 14.5, instead of in the more straightforward fashion suggested by Equation 14.5, to minimize the number of arithmetic operations required. The strategy for evaluating polynomials embodied in f and fprime is called *Horner's rule*.

The main function displays both the approximate root computed by newton and the value of f at that approximate root. This will make it easier for us to judge how well the root-finding method is working.

Although newton1.c will compile and run, it will not produce a very good approximation to the root of f. Its implementation of newton, after all, does nothing but return its parameter. This implementation of newton, which is correct so far as C is concerned but which is incomplete from our point of view, is called a *stub*.

A stub is a syntactically correct but deliberately incomplete implementation of a function. Its purpose is to temporarily stand in, during the early part of the program development process, for the complete implementation that will eventually be written. By employing a stub for newton, we can write and test the other three functions before tackling `newton`. We could not otherwise test `main`, because it calls `newton`.

As we continue to develop our implementation of Newton's method, we will need only to modify the `newton` function. Accordingly, we will show only the changing implementation of `newton` in subsequent figures.

14.2.3 Two Straight-Line Implementations

Equation 14.8 embodies the key computational idea behind Newton's method. It shows how, under broad circumstances, we can improve an approximation to the root of a function. Figure 14.6 shows an implementation of the `newton` function that translates this idea directly into C. Beginning with the initial guess `g0`, this implementation improves it five times, producing `g1`, `g2`, `g3`, `g4`, and `g5`, whose value is finally returned.

The flaw in this implementation should be readily apparent: How do we know that improving the initial guess five times will produce a close approximation to a root? Perhaps more repetitions are necessary, or perhaps fewer will suffice. As we will discover in Section 14.4, in fact, it is not possible to decide in advance how many repetitions to perform. The required number of repetitions depends on the quality of the initial guess `g0`, the nature of the function `f`, and the degree of accuracy that we would like in our final answer.

The solution is to arrange for the program to repeat the guess-improvement process for as long as is necessary to closely approximate a root. Employing a `while` loop in the program will make this possible. A `while` loop will repeatedly execute the same sequence of statements so long as a condition, supplied by the programmer, remains true.

Unfortunately, the five assignment statements that we employ in Figure 14.6, although similar, are *not* identical. If we are going to employ a `while` loop, we must find a way to solve our problem by repeatedly executing exactly the same statement (or sequence of statements). Figure 14.7 shows just such an implementation.

```
1    double newton (double g0)
2    {
3        double g1, g2, g3, g4, g5;
4
5        g1 = g0 - f(g0)/fprime(g0);
6        g2 = g1 - f(g1)/fprime(g1);
7        g3 = g2 - f(g2)/fprime(g2);
8        g4 = g3 - f(g3)/fprime(g3);
9        g5 = g4 - f(g4)/fprime(g4);
10
11       return(g5);
12   }
```

Figure 14.6   Second implementation of Newton's method (`newton2.c`)

```
1    double newton (double g0)
2    {
3        g0 = g0 - f(g0)/fprime(g0);
4        g0 = g0 - f(g0)/fprime(g0);
5        g0 = g0 - f(g0)/fprime(g0);
6        g0 = g0 - f(g0)/fprime(g0);
7        g0 = g0 - f(g0)/fprime(g0);
8
9        return(g0);
10   }
```

Figure 14.7 Third implementation of Newton's method (newton3.c)

This implementation takes advantage of the fact that once g0 is used on the right-hand side on the assignment on line 5 of Figure 14.6, it is no longer used. Thus, we are free to assign the result of that computation to g0 itself, which is what we do in Figure 14.7. By reusing g0 in this fashion, all five assignments can be rendered identical. This will lead directly to an implementation based on a while loop.

Figure 14.8 shows the results of running newton3.c three times with different initial guesses. With an initial guess of 1.0, five repetitions of the guess-improvement process are sufficient to come up with an extremely close approximation to a root near 0.36. With an initial guess of −1.0, we get fairly close to a root near −0.20. With an initial guess of 2.0, we remain far from any root after five repetitions. You can get a better understanding of *why* Newton's method behaves this way by animating these three examples in *Mathematica* with AnimateNewton.

```
Please enter an initial guess: 1.0
An approximate root is 0.3621999927
f(approximate root) is -1.940447802e-12

Please enter an initial guess: -1.0
An approximate root is -0.2032583902
f(approximate root) is -3.842556149e-07

Please enter an initial guess: 2.0
An approximate root is 3.979082758
f(approximate root) is 163.9842411
```

Figure 14.8 Three runs of newton3.c

14.2.4 While Statements

The results in Figure 14.8 reveal that the number of repetitions required to find a good approximation to a root is sensitive to the initial guess. By using a while statement, we can leave it to the program to decide how many repetitions are appropriate.

The general form of a C `while` statement is shown below.

```
while (CONDITION) {
    STATEMENTS
}
```

It consists of the keyword `while`, a Boolean expression enclosed in parentheses, and a sequence of statements enclosed in braces.

The Boolean expressions that are used to control `while` loops are exactly the same as those used to control `if` statements, as described in Chapter 12. The statements that make up the body of a `while` loop can be any statements whatsoever, including assignments, conditionals, and other loops.

When C encounters a `while` loop:

1. It evaluates the condition.
2. If the condition is false, it terminates the loop and skips ahead to the statement immediately following the `while` loop.
3. If the condition is true, it evaluates the statements that make up the body of the `while` loop and then returns to step 1.

This means that the statements that make up the body of a `while` loop may be executed zero or more times. If the condition is initially false, the body will not be executed at all. If the condition never becomes false, the body will execute until the program is interrupted by the user.

Figure 14.9 shows a version of the `newton` function that exploits a `while` loop that improves on an initial guess until the absolute value of the function applied to the guess is no greater than 10^{-10}. The body consists of the single assignment statement that we employed five consecutive times in Figure 14.7.

Figure 14.10 shows the results of running `newton4.c` using the same three initial guesses with which we experimented in Figure 14.8. In all three cases we have succeeded in getting within 10^{-11} of a root. In the first case, we are close to a root at 0.36; in the second case, to a root at -0.20; and in the third case, to a root at 2.76. (There is a fourth root that we have not yet found.)

Table 14.1 shows what happens when the `newton` function in Figure 14.9 is called with an actual parameter of 1.0. The values that `g0` and `f(g0)` have before the loop executes, as well as after each of its five repetitions, are given.

```
1    double newton (double g0)
2    {
3        while (fabs(f(g0)) > 1e-10) {
4            g0 = g0 - f(g0)/fprime(g0);
5        }
6        return(g0);
7    }
```

Figure 14.9   Fourth implementation of Newton's method (`newton4.c`)

```
Please enter an initial guess: 1.0
An approximate root is 0.3621999927
f(approximate root) is -1.940447802e-12

Please enter an initial guess: -1.0
An approximate root is -0.2032583416
f(approximate root) is -3.508304758e-14

Please enter an initial guess: 2.0
An approximate root is 2.760905633
f(approximate root) is 1.331157407e-13
```

Figure 14.10   Three runs of `newton4.c`

Table 14.1   Calling `newton` with a parameter of 1.0

Execution Point	g0	f(g0)
Before loop	1.0	−8.0
After one repetition	0.5	−1.1875
After two repetitions	0.38125	−0.1404642319
After three repetitions	0.362722104	−0.003743623325
After four repetitions	0.3622004119	$-3.00360377 \times 10^{-6}$
After five repetitions	0.3621999927	$-1.940447802 \times 10^{-12}$

When the loop begins executing, the absolute value of `f(g0)` is 8.0. Since this is much larger than 10^{-10}, the body of the loop is executed. This changes `g0` to 0.5 and `f(g0)` to −1.1875. Since the absolute value of `f(g0)` is still larger than 10^{-10}, the loop body is executed once again. This continues for five repetitions until the absolute value of `f(g0)` shrinks to almost 10^{-12}. At this point the loop terminates. The `return` statement that follows is executed, and the current value of `g0` (0.3621999927) is returned as the approximate root.

Figure 14.11 shows a version of `newton` into which we have inserted `printf` statements. These statements display the kind of information that appears in

```
1    double newton (double g0)
2    {
3        while (fabs(f(g0)) > 1e-10) {
4            printf("g0 = %.10g, f(g0) = %.10g\n", g0, f(g0));
5            g0 = g0 - f(g0)/fprime(g0);
6        }
7        return(g0);
8    }
```

Figure 14.11   Producing diagnostic output as `newton` runs (`newton5.c`)

Table 14.1 as `newton` is running. This can be a useful diagnostic technique, as it is otherwise difficult to tell exactly what is going on inside a `while` loop that is under development. If you insert `printf` statements into your loops, however, be sure to remove them when you have completed the program. The users of a function will rarely be interested in diagnostic output.

The conversion specification `%.10g` that is used on line 4 of Figure 14.11 means that a floating-point number is to be displayed using a ten-digit mantissa but with no minimum number of characters. Similarly, `%.8f` and `%.8e` both mean to display a floating-point number using eight digits after the decimal point but with no minimum number of characters. You should compare these variants of the floating-point format specifications with those discussed in Section 13.3.6.

The `while` loop in the augmented `newton` function of Figure 14.11 now contains two statements, a `printf` and an assignment. Each time the loop body is executed, these two statements are executed in order. Figure 14.12 shows the output of `newton5.c` for two different initial guesses. It reveals that an initial guess of -1.0 leads to six repetitions, and that an initial guess of 2.0 leads to 11 repetitions.

```
Please enter an initial guess: -1.0
g0 = -1, f(g0) = -16
g0 = -0.5, f(g0) = -3.6875
g0 = -0.2830882353, f(g0) = -0.7270722922
g0 = -0.2126119929, f(g0) = -0.07525639551
g0 = -0.2034185928, f(g0) = -0.00126725472
g0 = -0.2032583902, f(g0) = -3.842556149e-07
An approximate root is -0.2032583416
f(approximate root) is -3.508304758e-14

Please enter an initial guess: 2.0
g0 = 2, f(g0) = -19
g0 = 11.5, f(g0) = 18704.3125
g0 = 8.647428321, f(g0) = 5856.429463
g0 = 6.552172383, f(g0) = 1818.720582
g0 = 5.038198449, f(g0) = 555.8010481
g0 = 3.979082758, f(g0) = 163.9842411
g0 = 3.28888183, f(g0) = 44.29527967
g0 = 2.910650917, f(g0) = 9.305179926
g0 = 2.777658884, f(g0) = 0.9285359062
g0 = 2.761148866, f(g0) = 0.01328615536
g0 = 2.760905685, f(g0) = 2.8567518e-06
An approximate root is 2.760905633
f(approximate root) is 1.331157407e-13
```

Figure 14.12   Two runs of `newton5.c`

14.3 Bisection Method Implementation

Now that we have a working version of Newton's method, we will reexamine the C implementation of the bisection method that we presented without much comment in Chapter 10. Like our implementation of Newton's method, this implementation has a `while` loop at its heart.

14.3.1 Do Statements

Figures 14.13–14.14 show a slightly different implementation of the bisection method from the one that we originally gave in Figures 10.1–10.2.

- We have modified f to compute $x^4 + 2x^3 - 14x^2 + 2x + 1$.
- We have changed `bisection` so that it uses the same criterion to terminate its `while` loop as is used in `newton`.
- We have added a pair of loops to `main` to enforce restrictions on the two initial guesses provided by the user.

```
1    #include <stdio.h>
2    #include <math.h>
3
4    double f (double x)
5    {
6      return(1 + x*(2 + x*(-14 + x*(2 + x))));
7    }
8
9
10   double bisection (double pos, double neg)
11   {
12     double avg;
13     avg = (pos + neg) / 2.0;
14
15     while (fabs(f(avg)) > 1e-10) {
16       avg = (pos + neg) / 2.0;
17       if (f(avg) >= 0) {
18         pos = avg;
19       }
20       else {
21         neg = avg;
22       }
23     }
24
25     return(avg);
26   }
```

Figure 14.13   Implementation of the bisection method (continued in Figure 14.14)

```
27    void main ()
28    {
29      double pos, neg, root;
30
31      do {
32        printf("Enter positive guess: ");
33        scanf("%lf", &pos);
34      }
35      while (f(pos) <= 0);
36
37      do {
38        printf("Enter negative guess: ");
39        scanf("%lf", &neg);
40      }
41      while (f(neg) >= 0);
42
43      root = bisection(pos, neg);
44      printf("An approximate root is %.10g\n", root);
45      printf("f(approximate root) is %.10g\n", f(root));
46    }
```

Figure 14.14   Implementation of the bisection method (bisect2.c)

The bisection function contains a more complicated while loop than does newton. Although the condition that controls the bisection loop is essentially identical to the one that controls the newton loop, the bisection loop contains two statements. One is an assignment that updates the value of avg, and the other is a two-way conditional that updates the value of either pos or neg, as appropriate.

The bisection method will not work unless the two guesses (pos and neg) provided by the user satisfy the inequalities f(pos)>0 and f(neg)<0. None of our previous implementations of the bisection method have enforced these restrictions. Figure 14.14 shows how the main function can be written so as to guarantee these essential properties of pos and neg.

The main function employs two do loops, which are distinct from, but related to, while loops. The general form of a do statement is

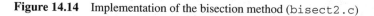

```
do {
    STATEMENTS
}
while (CONDITION);
```

A do statement consists of the keyword do, a sequence of statements enclosed in braces, the keyword while, and a condition enclosed in parentheses. The key difference between a do loop and a while loop is that the statements of a do loop are executed *before* the condition is evaluated. As a result, the body of a do loop will always be executed at least once.

A do loop is often employed, as in Figure 14.14, for input validation. The first loop in the main function prompts for and reads in a value for pos. If it turns out that f(pos) is not positive, the loop continues and the user is prompted again. This continues until an acceptable value for pos is entered. Because the user must be given at least one opportunity to enter a value, it is important that the loop body be executed at least once.

14.4 Assessment

In this section we will do two different forms of assessment. We will begin by assessing the convergence criterion that we used in our implementation of Newton's method, which will lead to an improved criterion. We will then compare Newton's method with the bisection method.

14.4.1 Improving Convergence

We used a simple and straightforward convergence criterion in our implementation of Newton's method. When looking for the root of a function f, we stop as soon as we obtain a guess x such that $|f(x)| < 10^{-10}$. (This tolerance is arbitrary, and could be adjusted up or down several orders of magnitude.)

Although this approach appears reasonable at first glance, it is in fact seriously flawed. It has worked well to this point only because the function we have been solving has moderately sized roots. With functions that have either extremely large or extremely small roots, however, our convergence criterion fails completely.

Consider the function $f_1(x) = x^2 - 10^{-50}$, which has the two extremely small roots $\pm 10^{-25}$. Beginning with a guess of 1, our implementation stops after 17 repetitions and reports $1.52587890625 \times 10^{-5}$ as the root. Although this is over 20 orders of magnitude distant from an actual root, our implementation is satisfied because the value of f_1 at this point is less than 10^{-10}.

Next consider the function $f_2(x) = x^2 - 10^{50}$, which has the two extremely large roots $\pm 10^{25}$. Beginning with a guess of 1, our implementation goes into an infinite loop and never reports an answer. Diagnostic output reveals that it converges to a guess that is different only in the last digit of its mantissa from the true root of 10^{25}. (Roundoff error prevents convergence to the exact root.) Unfortunately, even this tiny error in the last digit of the mantissa causes the value of f_2 at this point to be greater than 10^{34}.

A far superior approach to judging convergence is to stop as soon as the relative difference between two successive guesses becomes small. In symbols, if g_n and g_{n+1} are two successive guesses and

$$\left| \frac{g_{n+1} - g_n}{g_{n+1}} \right| < \epsilon \tag{14.12}$$

where ϵ is a small tolerance that we choose, we stop and report g_{n+1} as our approximate root. Because we are dealing with *relative* differences, this has the effect of requiring a small *absolute* difference for small roots while allowing a large absolute difference for large roots.

Figure 14.15 shows the implementation of a `newton` function based on this idea. We have replaced the `while` loop with a do loop, and employ the variable `oldg0` to track the previous value of `g0`.

When we use `newton6.c` to solve the function f_1 defined above, we obtain a root of 10^{-25} (when rounded to ten digits) at which f_1 has a value of approximately 1.2×10^{-66}. When we solve the function f_2, we obtain a root of 10^{25} (again rounded to ten digits) at which f_2 has a value of approximately 2.1×10^{34}. Although there is still a small amount of error in the low-order digits of the mantissa, the error is insignificant in a relative sense.

Exactly the same convergence criterion can be exploited in an implementation of the bisection method, as illustrated in Figure 14.16.

We will close our discussion of the convergence of Newton's method by pointing out that our new approach to convergence introduces a new problem. If any of the guesses produced on the way to a root are zero, our convergence test will attempt to perform a division by zero.

```
1    double newton (double g0)
2    {
3      double oldg0;
4
5      do {
6        oldg0 = g0;
7        g0 = g0 - f(g0)/fprime(g0);
8      }
9      while (fabs((g0-oldg0)/g0) > 1e-10);
10
11     return(g0);
12   }
```

Figure 14.15   An improved convergence criterion (`newton6.c`)

14.4.2 Newton's Method versus the Bisection Method

The bisection method has two major drawbacks. It requires the user to supply two guesses that are guaranteed to bracket a root, and it takes much longer on average to converge to a root than does Newton's method. For example, with a positive guess of zero and a negative guess of 1, `bisect3.c` requires 33 repetitions to converge to an approximate root. In contrast, `newton6.c` requires only four repetitions to

```
1      double bisection (double pos, double neg)
2      {
3        double avg, oldavg;
4        avg = (pos+neg)/2.0;
5
6        do {
7          oldavg = avg;
8          if (f(avg) >= 0) {
9            pos = avg;
10         }
11         else {
12           neg = avg;
13         }
14         avg = (pos + neg) / 2.0;
15       }
16       while (fabs((oldavg-avg)/oldavg) > 1e-10);
17
18       return(avg);
19     }
```

Figure 14.16   Improved convergence for bisection (bisect3.c)

converge to an approximate root from an initial guess of 1. The two approximations, incidentally, are identical in their first ten digits.

Newton's method has several subtle drawbacks, two of which we will discuss here. If one of the guesses produced during the repetitions of Newton's method occurs near a local minimum (or maximum), problems can arise. If the guess is *exactly* at such a point, division by zero will occur because the first derivative at that point will be zero. If the guess is close to but not exactly at a minimum or maximum, the number of repetitions required to converge to a root can skyrocket.

For example, suppose we are using Newton's method to find the root of $(x - 2)(x - 4)$ (Figure 14.17a), which has roots at 2 and 4 and a minimum at 3. If our initial guess is 3, the first repetition of the improvement process will try to divide by zero. If the initial guess is 3.0000001, the next guess produced is 5000003 and 28 repetitions are required to converge back to the root at 4.

A related problem occurs when a function tends toward but never reaches a finite limit at infinity. In this case, Newton's method can chase off toward infinity in search of a nonexistent root. For example, consider the function xe^{-x} (Figure 14.17b), which has a root at zero, reaches a maximum at 1, and then tends toward but never quite reaches the x-axis. If the initial guess is less than 1, Newton's method converges quickly to the root at zero. But if the initial guess is greater than 1, Newton's method diverges in search of a root in the direction of positive infinity.

Newton's method has been studied extensively by numerical analysts, and solutions to all of these problems have been identified. We will touch on some of them in the exercises.

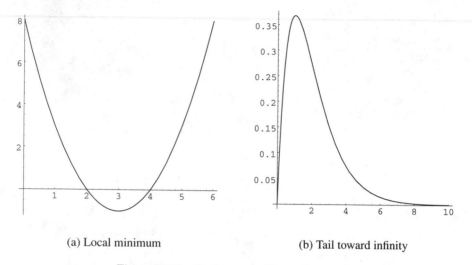

(a) Local minimum   (b) Tail toward infinity

Figure 14.17   Challenges for Newton's method

14.4.3 Solution to the Beam Problem

We have so far found three close approximations to the roots of Equation 14.5. One is approximately 0.36, one is approximately 2.76, and one is approximately −0.20. Since the equation is a fourth-order polynomial, it has a fourth root. This root is negative, and you will locate it as an exercise.

We were solving Equation 14.5 to find possible values for x from Figure 14.1, in which x represented the distance along the floor from the box to the bottom of the beam. The two negative roots are clearly physical impossibilities, so we will discard them. The remaining roots tell us that the beam will satisfy the constraints of the problem when it is either approximately 1.36 or 3.76 meters from the wall.

We must ask ourselves at this point whether it makes sense that we have two different solutions to the problem. In fact, it does, because the problem is symmetrical. When the distance from the wall to the bottom of the ladder is approximately 1.36 meters, the distance from the floor to the top of the ladder will be approximately 3.76 meters. (This is easily verified with the Pythagorean Theorem.) If we rotate Figure 14.1 90 degrees so that the wall becomes the floor and the floor becomes the wall, the two measurements are similarly reversed.

14.5 Key Concepts

Newton's method.   Newton's method is a numerical method for finding a root of a function. It converges much faster than the bisection method and requires only one initial guess, but it fails to find a root more often than does the bisection method. Newton's method improves on a guess g_0 for the root of a function f by using Equation 14.8.

While loops in C.   A C while loop consists of a Boolean expression and a sequence of statements. When C encounters a while loop, it begins by evaluating the Boolean expression. It then repeatedly evaluates the statements and the expression until the expression becomes false.

Do loops in C.   A C do loop consists of a sequence of statements and a Boolean expression. When C encounters a do loop, it repeatedly evaluates the statements and the expression until the expression becomes false. Whereas the statements of a while loop are executed *zero* or more times, the statements of a do loop are executed *one* or more times. This is the only important difference between do loops and while loops.

Programming with stubs.   When implementing a complicated function, a good strategy is to begin with a stub. A stub has the same name, parameters, and return type as the function being implemented, but contains only a return statement. Once the rest of the functions of the program have been tested along with the stub, the stub can be incrementally fleshed out into the desired final form.

14.6 Exercises

14.1   Use newton6.c to find the fourth root of Equation 14.5.

14.2   Use *Mathematica* to plot the left-hand side of Equation 14.5 between -5.5 and 3.5. By consulting the plot, predict which roots Newton's method will find when starting with initial guesses of $-5, -4, -3, -2, -1, 0, 1, 2$, and 3. Use newton6.c to verify your predictions.

14.3   Use your plot from Exercise 14.2 to predict initial guesses from which Newton's method will converge extremely slowly, if at all, to a root. Use newton5.c to verify your predictions.

14.4   Modify bisect3.c so that it displays the same sort of convergence information as newton5.c. Compare the number of repetitions required by the two programs to find roots.

14.5   Modify newton5.c so that it finds the root of the equation $x - 10^{-n} = 0$, where n is a positive integer. Your program should use 1 as the initial guess in all cases, and should prompt the user for the value of n. (The first derivative of $x - 10^{-n}$ is 1.) Does the rate of convergence depend on n? Compare your results with those you obtained in Exercise 9.19.

14.6   Modify newton5.c to find the roots of $1 + xe^{-x}$, the first derivative of which is $e^{-x} - xe^{-x}$. What happens when you supply an initial guess of 2? Use a plot of the function to explain the behavior you observe.

14.7 Horner's rule observes that the fifth-order polynomial

$$p(x) = ax^5 + bx^4 + cx^3 + dx^2 + ex + f \qquad (14.13)$$

is equivalent to

$$p(x) = f + x(e + x(d + x(c + x(b + xa)))) \qquad (14.14)$$

Implement in C a program that prompts the user for x and the six coefficients a through f and evaluates $p(x)$ using both the straightforward approach of Equation 14.13 and the clever approach of Equation 14.14.

14.8 How many fewer multiplications are required to evaluate the left-hand side of Equation 14.5 using Horner's rule than are required using the straightforward approach? (One multiplication is required to compute x^2, and two multiplications are required to compute each of x^3 and x^4.)

14.9 Implement in C a program for solving the block-stacking problem of Chapter 4. Your program should take as input the number of blocks to stack, and should report the extension obtained. (This is the same calculation carried out by the `blockFloat` function from Chapter 4.) Your program should work by using a loop to sum the terms of Series 4.1 beginning with the largest term.

14.10 Compare the times required to sum 10,000 terms of Series 4.1 by your C implementation from Exercise 14.9 and by our `BlockFloat` *Mathematica* function from Chapter 4.

14.11 Repeat Exercise 14.9, but this time sum the terms of Series 4.1 beginning with the smallest term. Explain why the results you obtain are different from those you obtained for Exercise 14.9. (You will need to use a floating-point output format that displays the entire mantissa to see any difference.)

14.12 Develop a model for the leaning beam problem from this chapter in which the beam is a meters long and the box is b meters on a side. What relationship must hold between a and b so that the problem will have a physical solution?

14.13 Use Newton's method as the basis for a C implementation of your model from Exercise 14.12. It should prompt the user for the length of the beam and the size of the box. If the problem does not have a physical solution, your program should notify the user; otherwise, it should calculate and report *both* of the solutions. Your program should supply its own initial guess for use by Newton's method.

14.14 Repeat Exercise 14.13 using the bisection method as the basis for the implementation.

14.15   Use Newton's method as the basis for a C implementation of Old MacDonald's problem from Chapter 9.

14.16   Modify newton5.c to cope with the problem caused when fprime(g0) is zero. If fprime(g0) is either zero or close to zero, your program should perturb the value of g0 slightly before attempting to improve it in the usual way.

14.17   Compare the number of loop repetitions required by your solution to Exercise 14.16 with those required by newton5.c. You should use initial guesses, such as 2.0, that lead to a large number of loop repetitions by newton5.c.

14.18   The value of $f'(x)$ is approximated by

$$f'(x) \approx \frac{f(x + \Delta x) - f(x)}{\Delta x} \qquad (14.15)$$

when Δx is small. Write a C program that prompts the user for values of x and Δx and then displays the exact and approximate values for $f'(x)$, where

$$f(x) = x^4 + 2x^3 - 14x^2 + 2x + 1 \qquad (14.16)$$
$$f'(x) = 4x^3 + 6x^2 - 28x + 2 \qquad (14.17)$$

14.19   Newton's method must have access both to the function f whose root is sought and its first derivative f'. This is a disadvantage relative to the bisection method, in which only f is needed. One solution to this problem is to compute the first derivative numerically using the approach described in Exercise 14.18. Modify newton6.c by replacing the implementation of fprime with one that computes the values of the first derivative of f numerically.

14.20   Compare the quality of the roots produced by your modified implementation of Newton's method from Exercise 14.19 with those produced by newton6.c.

15

Corrugated Sheets: Multiple-File Programs

A factory receives an order for corrugated steel sheets. Each sheet is to be 5 meters wide by 10 meters long and is to be corrugated in a sinusoidal pattern. When viewed in cross section along either of its long ends, each sheet should appear as illustrated in Figure 15.1. There are to be 100 ridges along the length of each sheet. Every ridge is to have a height and width of 10 centimeters, and every pair of ridges is to be the square of a complete cycle of a sine wave.

Each corrugated sheet is formed by bending a 5-meter-wide flat sheet. How long should each flat sheet be so that it will be 10 meters long after it has been corrugated?

If the corrugated sheet is to be 10 meters long, each flat sheet must have a length that is 100 times greater than the length of a single ridge of the curve from Figure 15.1. The length of a curve is the distance a particle travels when

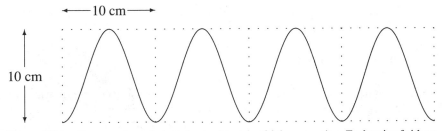

Figure 15.1 Cross section of steel sheet with sinusoidal corrugation. Each pair of ridges is the square of a complete cycle of a sine wave. Only four of the 100 ridges are shown.

it moves along the curve from one end to the other, and can be calculated using the formula

$$\int_a^b \sqrt{1 + f'(x)^2}\, dx \tag{15.1}$$

where f is the function describing the curve, f' is the first derivative of f, a is the x-coordinate of the left endpoint of the curve, and b is the x-coordinate of the right endpoint of the curve.

The sinusoidal cross section from Figure 15.1 is given by

$$10 \sin^2 \left(\frac{\pi x}{10} \right) \tag{15.2}$$

By defining this expression in *Mathematica*

```
In[3]:= f = 10 * Sin[Pi*x/10]^2

Out[3]= 10 Sin[πx/10]²
```
$$\tag{15.3}$$

and finding its first derivative

```
In[4]:= fprime = D[f, x]

Out[4]= 2π Cos[πx/10] Sin[πx/10]
```
$$\tag{15.4}$$

we can apply Equation 15.1 to find that the distance along one ridge is

$$\int_0^{10} \sqrt{1 + 4\pi^2 \sin^2 \left(\frac{\pi x}{10} \right) \cos^2 \left(\frac{\pi x}{10} \right)}\, dx \tag{15.5}$$

The total length of the sheet must be 100 times this, or

$$100 \int_0^{10} \sqrt{1 + 4\pi^2 \sin^2 \left(\frac{\pi x}{10} \right) \cos^2 \left(\frac{\pi x}{10} \right)}\, dx \tag{15.6}$$

If we can evaluate this expression, we will have solved our problem.

15.1 Numerical Integration

You are probably most familiar with symbolic methods for doing integration. Consider, for example, the problem of evaluating the definite integral

$$\int_0^8 (\sin(x) + 2)\, dx \tag{15.7}$$

Since the indefinite integral can be found symbolically, it can be evaluated at the limits of integration to determine the definite integral.

$$\int_0^8 (\sin(x) + 2)\, dx = (2x - \cos(x))\big|_0^8 \tag{15.8}$$

$$= (16 - \cos(8)) - (0 - \cos(0)) \tag{15.9}$$

$$\approx 17.1455 \tag{15.10}$$

Mathematica is very capable doing symbolic integration. For example, we can ask *Mathematica* to evaluate the indefinite form of the integral from Formula 15.7 via

```
In[11]:= Integrate[Sin[x]+2, x]

Out[11]= 2x - Cos[x]
```
(15.11)

and the definite form of the integral from Formula 15.7 via

```
In[12]:= Integrate[Sin[x]+2, {x, 0.0, 8.0}]

Out[12]= 17.1455
```
(15.12)

Unfortunately, not every integral can be evaluated symbolically. For example, even *Mathematica* is stumped by the integral from Formula 15.5:

```
In[13]:= Integrate[Sqrt[1 + fprime^2], {x, 0, 10}]
```
$$\text{Out[13]}= \int_0^{10} \sqrt{1 + 4\pi^2 \cos\left[\frac{\pi x}{10}\right]^2 \sin\left[\frac{\pi x}{10}\right]^2}\, dx$$
(15.13)

By displaying the unevaluated integral, *Mathematica* is admitting failure.

Just as there are numerical methods, such as the bisection method and Newton's method, for solving equations that defy symbolic solution, there are numerical methods for evaluating definite integrals that cannot be evaluated symbolically. We can ask *Mathematica* to use a numerical method to evaluate the integral from Example 15.13 via

```
In[14]:= NIntegrate[Sqrt[1 + fprime^2], {x, 0, 10}]

Out[14]= 23.0489 cm
```
(15.14)

We use `NIntegrate` instead of `Integrate`. This gives us the answer to our problem. If one ridge of the corrugation is approximately 23.05 centimeters long, then 100 ridges will be approximately 23.05 meters long, which is the required length of the flat sheet.

If we were interested strictly in the solution to our problem, we would be done at this point. In this chapter, however, we are also interested in investigating how numerical methods for solving definite integrals work. We will study two particularly simple techniques, the rectangular and trapezoidal methods, and will implement them in C.

Both the rectangular and trapezoidal methods hinge on the fact that the problem of definite integration can be understood graphically. For example, finding the integral of $\sin(x) + 2$ (the integrand) between zero and 8 (the limits of integration) is equivalent to determining the area that lies between the curve $\sin(x) + 2$ and the x-axis and is bounded on the sides at $x = 0$ and $x = 8$. This is illustrated in Figure 15.2.

Because of this correspondence between definite integration and the problem of determining the area of a region bounded by curves, the development of techniques for numerical integration predates the seventeenth century invention of calculus by almost 2000 years. The trapezoidal method that we will study in Section 15.4, for example, was known to the ancient Greeks and perhaps even to the Babylonians.

Bryson, one of the followers of Pythagoras, developed a technique for approximating the area bounded by a closed curve in around 450 B.C. His approach, known as the method of exhaustion, was to inscribe a polygon inside of the curve. The area of the polygon, which could be readily determined, approximated the area surrounded by the curve. By increasing the number of sides of the polygon, he could reduce the amount of error. A little more than 200 years later, Archimedes used the method of exhaustion in three dimensions to find the volume of the sphere, which allowed him to derive bounds for π of $\frac{223}{71} < \pi < \frac{22}{7}$.

We will use the integral in Formula 15.7 as an example throughout most of this chapter, even though it can be evaluated symbolically. This will work to our advantage, as it is easy to visualize and will give us a standard of comparison to help us understand how well the rectangular and trapezoidal methods work. When we have completed our implementations of the two methods, we will apply them to the integral in Formula 15.5 in the exercises.

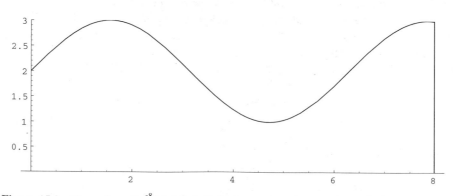

Figure 15.2   The value of $\int_0^8 (\sin(x) + 2)\, dx$ is equal to the area beneath the curve $\sin(x) + 2$.

15.2 Rectangular Method

The idea behind the rectangular method is most easily understood graphically. Figure 15.3 shows two different ways in which the area beneath the graph of the integrand given symbolically in Formula 15.7 and graphically in Figure 15.2 can be divided into a collection of vertical rectangles. The base of each rectangle lies on the x-axis, and the top of each rectangle intersects the integrand. The areas of the rectangles can be summed to obtain an approximation to the integral.

The total area of the rectangles will give only an approximation to the integral because the tops of the rectangles do not match up exactly with the graph of the integrand. Each of the four rectangles in Figure 15.3a, for example, either includes a region that is above the graph of the integrand or excludes a region that is below the graph of the integrand, or both. If we are extremely lucky, these errors might cancel each other out, but that is rarely the case.

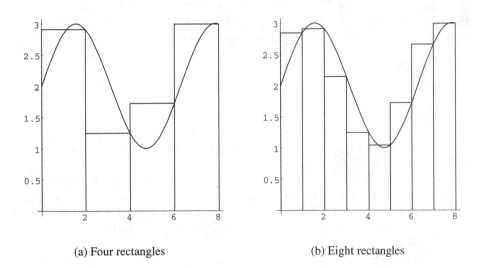

(a) Four rectangles   (b) Eight rectangles

Figure 15.3   Two frames from Example 15.17 showing how $\int_0^8 (\sin(x) + 2)\, dx$ can be approximated with rectangles.

The amount of error can generally be decreased by increasing the number of rectangles. Figure 15.3b, for example, uses eight rectangles to approximate the integral. For the kinds of functions that occur in practice, the amount of error can be made arbitrarily small by using enough rectangles. In the limit, with an infinite number of infinitesimally thin rectangles, the approximation becomes exact.

The plots in Figure 15.3 are two frames from a *Mathematica* animation of the rectangular method. This particular animation was produced by using the `Integ` package, which is on the diskette included with this book.

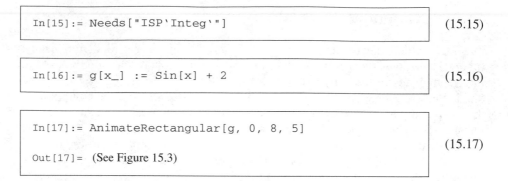

```
In[15]:= Needs["ISP`Integ`"]
```
(15.15)

```
In[16]:= g[x_] := Sin[x] + 2
```
(15.16)

```
In[17]:= AnimateRectangular[g, 0, 8, 5]

Out[17]= (See Figure 15.3)
```
(15.17)

The first parameter to `AnimateRectangular` is the integrand, the next two parameters are the low and high bounds of integration, and the fourth parameter is the number of frames to produce. The first frame in the animation shows the integral approximated by two rectangles, and in each subsequent frame the number of rectangles is doubled. Figure 15.3a is thus the second frame in the animation, and Figure 15.3b is the third. You should experiment with `AnimateRectangular` to better appreciate how the rectangular method works.

15.2.1 Examples

The rectangles in Figure 15.3 are not drawn arbitrarily. The four rectangles in Figure 15.3a are of uniform width. Because the integral is being computed between zero and 8, each of the rectangles has a width of 2. Similarly, each of the eight rectangles in Figure 15.3b has a width of 1.

Furthermore, each of the rectangles is drawn so that its upper-right corner lies exactly on the graph of the integrand. Systematically drawing the rectangles in this way makes it easy to calculate their respective heights. Beginning from the left, the heights of the four rectangles in Figure 15.3a are $\sin(2) + 2$, $\sin(4) + 2$, $\sin(6) + 2$, and $\sin(8) + 2$.

From the height and width of each rectangle we can calculate the individual areas, which we can sum to obtain the overall area that serves as an approximation to the integral.

Table 15.1a summarizes the area calculation for the four rectangles in Figure 15.3a, and Table 15.1b summarizes the area calculation for the eight rectangles in Figure 15.3b. For each rectangle, we show

- the x-coordinate of the right side of the rectangle;
- the height of the rectangle, which is $\sin(x) + 2$; and
- the area of the rectangle, which is the product of its height and width.

The true value of the integral, rounded to six digits, is 17.1455. The four-rectangle approximation has an error of less than 3.4%, while the eight-rectangle approximation has an error of less than 2.4%.

Table 15.1 Upper-right corner and area of each rectangle in Figure 15.3

Upper-right corner		
x	$\sin(x) + 2$	Area
2.0000	2.9093	5.8186
4.0000	1.2432	2.4864
6.0000	1.7206	3.4412
8.0000	2.9894	5.9778
Total		17.7240

Upper-right corner		
x	$\sin(x) + 2$	Area
1.0000	2.8415	2.8415
2.0000	2.9093	2.9093
3.0000	2.1411	2.1411
4.0000	1.2432	1.2432
5.0000	1.0411	1.0411
6.0000	1.7206	1.7206
7.0000	2.6570	2.6570
8.0000	2.9894	2.9894
Total		17.5432

(a) Four rectangles   (b) Eight rectangles

15.2.2 General Case

We will now turn from the two specific examples that we just considered to the problem of developing the rectangular method for the general case.

Suppose that we wish to approximate the definite integral

$$\int_a^b f(x)\, dx \tag{15.18}$$

using the rectangular method with four rectangles. Each of the four rectangles will be of uniform width $h = \frac{b-a}{4}$. The x-coordinates of the right sides of the rectangles will be $a + h, a + 2h, a + 3h$, and $a + 4h$, which means that the heights of the four rectangles will be $f(a + h)$, $f(a + 2h)$, $f(a + 3h)$, and $f(a + 4h)$. The total area of the four rectangles, which approximate the integral, will be

$$\int_a^b f(x)\, dx \approx hf(a + h) + hf(a + 2h) + hf(a + 3h) + hf(a + 4h) \tag{15.19}$$

$$= h\left(f(a + h) + f(a + 2h) + f(a + 3h) + f(a + 4h)\right) \tag{15.20}$$

If we use n rectangles, this generalizes to

$$\int_a^b f(x)\, dx \approx hf(a + h) + hf(a + 2h) + \cdots + hf(a + nh) \tag{15.21}$$

$$= h\sum_{i=1}^{n} f(a + ih) \tag{15.22}$$

where $h = \frac{b-a}{n}$. Implementing the rectangular method involves writing a function to compute this sum.

15.3 Rectangular Method Implementation

Each of the programs that we will present in this section consists of three functions:

- `integrand` is an implementation of the function whose integral we are computing.
- `rectangular` is an implementation of the rectangular method.
- `main` is a driver that will prompt for the bounds of integration and the desired number of rectangles, then use `rectangular` to approximate the definite integral of `integrand`, and then display the results.

In all of the programs that we consider, `integrand` and `main` will not change. Because neither raises any interesting implementation issues, we will not show them in the figures. Instead, we will focus exclusively on our successive implementations of `rectangular`.

The `rectangular` function will take three parameters: the low bound of the integral (a), the high bound of the integral (b), and the number of rectangles (n). The value returned by `rectangular` will be the approximation to the integral of `integrand` between a and b obtained using the rectangular method with n rectangles.

We will consider four different versions of `rectangular`. In the first two we will experiment with using straight-line code. Although neither experiment will succeed, the two attempts will point the way to a third implementation that employs a `while` loop. Finally, we will discuss the C `for` loop and see how it can be used to produce a fourth, more streamlined, implementation.

15.3.1 Two Straight-Line Implementations

Figure 15.4 shows an attempt to implement `rectangular` without using a loop. It is not a correct implementation, however, because it ignores the parameter n by always using four rectangles to approximate the integral. The four assignment statements on lines 8–11 sum the heights of the four rectangles, and this sum is multiplied by the rectangles' width and the product is returned as the result.

No implementation that uses a separate statement to calculate the area of each rectangle can correctly realize `rectangular`. Because the number of rectangles that the function must employ is a parameter, some sort of loop is required. The implementation in Figure 15.4, however, does not directly lead to a loop-based implementation. Because every statement is different, it would serve no purpose to repeat any of them.

Figure 15.5 shows a second straight-line implementation that is more amenable to conversion into one that uses a loop. This version uses the variable i to count up from 1. As a result, the four assignments to `sum` on lines 10, 13, 16, and 19 are identical. After each of these assignments to `sum`, the counting variable i is incremented. Thus, the sum of the heights of the four rectangles is computed by executing the pair of statements

```
1    double rectangular (double a, double b, int n)
2    {
3      double width, sum;
4
5      width = (b - a) / 4;
6      sum = 0;
7
8      sum = sum + integrand(a + 1*width);
9      sum = sum + integrand(a + 2*width);
10     sum = sum + integrand(a + 3*width);
11     sum = sum + integrand(a + 4*width);
12
13     return(width * sum);
14   }
```

Figure 15.4   Straight-line rectangular method implementation (`rect1.c`)

```
1    double rectangular (double a, double b, int n)
2    {
3      double width, sum;
4      int i;
5
6      width = (b - a) / 4;
7      sum = 0;
8      i = 1;
9
10     sum = sum + integrand(a + i*width);
11     i = i+1;
12
13     sum = sum + integrand(a + i*width);
14     i = i+1;
15
16     sum = sum + integrand(a + i*width);
17     i = i+1;
18
19     sum = sum + integrand(a + i*width);
20     i = i+1;
21
22     return(width * sum);
23   }
```

Figure 15.5   Modified straight-line rectangular method implementation (`rect2.c`)

```
sum = sum + integrand(a + i*width);
i = i+1;
```

four times in succession.

Although this implementation still always employs four rectangles regardless of the value of the parameter n, it points the way to a correct implementation that exploits a `while` loop.

15.3.2 Counting Loops

Figure 15.6 shows our first correct implementation of the `rectangular` function. We have taken the two assignment statements that were repeated four times in Figure 15.5 and made them the body of the `while` loop on lines 10–13. By setting i to 1 before the loop begins and continuing until i exceeds n, we arrange for the loop body to be repeated n times.

A loop such as this one, which is controlled by a variable that counts from some initial value to some limit, is called a *counting* loop. All counting loops have three things in common.

1. Before the loop begins, a count variable (i in Figure 15.6) is set to an initial value (1).

2. The test that determines whether the loop will continue involves the count variable (i <= n).

3. The count variable is modified at the end of each repetition of the loop (i = i+1).

```
1    double rectangular (double a, double b, int n)
2    {
3      double width, sum;
4      int i;
5
6      width = (b - a) / n;
7      sum = 0;
8
9      i = 1;
10     while (i <= n) {
11       sum = sum + integrand(a + i*width);
12       i = i+1;
13     }
14
15     return(width * sum);
16   }
```

Figure 15.6   Rectangular method implementation with `while` loop (`rect3.c`)

15.3.3 **For** Statements

Because counting loops are common, C provides a third form of loop—the `for` statement—that makes them easier to write. As we study `for` loops, bear in mind that anything that can be done with a `for` loop can also be done with a `while` loop. C provides `for` loops as a convenience to the programmer.

Figure 15.7 shows an implementation of `rectangular` that uses a `for` loop instead of a `while` loop. The general form of a `for` loop is shown below:

```
for (INITIALIZE; CONDITION; INCREMENT) {
    STATEMENTS
}
```

INITIALIZE is typically an assignment to the counting variable; *CONDITION* is typically a Boolean expression involving the counting variable; and *INCREMENT* is typically an assignment to the counting variable. Notice that *INITIALIZE*, *CONDITION*, and *INCREMENT* are *separated* by, not *terminated* by, semicolons. These three components of the `for` loop correspond to the three characteristics of counting loops that we detailed in the previous section.

```
1    double rectangular (double a, double b, int n)
2    {
3      double width, sum;
4      int i;
5
6      width = (b - a) / n;
7      sum = 0;
8
9      for (i = 1; i <= n; i = i+1) {
10         sum = sum + integrand(a + i*width);
11     }
12
13     return(width * sum);
14   }
```

Figure 15.7   Rectangular method implementation with `for` loop (`rect4.c`)

The `for` loop given above is equivalent to the following `while` loop:

```
INITIALIZE;
while (CONDITION) do {
    STATEMENTS;
    INCREMENT;
}
```

The advantage of using a `for` loop to implement a counting loop is that it puts all of the elements that control the loop together in one place. If you compare lines 9–11 of Figure 15.7 to lines 9–13 of Figure 15.6, you will find that they compose exactly the same statements and expressions in different ways.

15.4 Trapezoidal Method

The trapezoidal method is similar to the rectangular method, except that we divide the area beneath the graph of the integrand into trapezoids instead of rectangles. Figure 15.8 shows two different ways in which the area beneath the graph of the integrand given symbolically in Formula 15.7 and graphically in Figure 15.2 can be divided into a collection of vertical trapezoids. The base of each trapezoid lies on the *x*-axis, and the top corners of each trapezoid both touch the graph of the integrand. The areas of the trapezoids can be summed to obtain an approximation to the integral.

Because a trapezoid can more closely follow the contour of a curve than can a rectangle of the same width, a trapezoidal approximation to an integral that uses *n* trapezoids is usually more accurate than a rectangular approximation to the same integral that uses *n* rectangles. You should compare Figure 15.8 with Figure 15.3 to appreciate the difference between the trapezoidal and rectangular methods. As with the rectangular method, the amount of error in a trapezoidal approximation can be decreased by increasing the number of trapezoids.

The two plots in Figure 15.8 are frames from a *Mathematica* animation of the trapezoidal method produced using our `Integ` package. This particular animation was produced by

```
In[23]:= AnimateTrapezoidal[g, 0, 8, 5]

Out[23]= (See Figure 15.8)
```

(15.23)

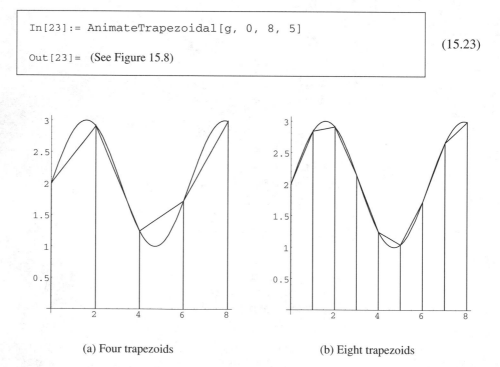

(a) Four trapezoids   (b) Eight trapezoids

Figure 15.8   Two frames from Example 15.23 showing how $\int_0^8 \sin(x) + 2\,dx$ can be approximated with trapezoids.

The `AnimateTrapezoidal` function takes the same four parameters as does the `AnimateRectangular` function discussed in Section 15.2. You should experiment with `AnimateTrapezoidal` to better appreciate how the trapezoidal method works.

15.4.1 Examples

The trapezoids in Figure 15.8 are of uniform width, and the top corners of each trapezoid lie on the graph of the integrand. This makes it easy to calculate the lengths of the vertical sides of the trapezoids. Beginning from the left, the sides of the four trapezoids in Figure 15.8a are $\sin(0) + 2$ and $\sin(2) + 2$, $\sin(2) + 2$ and $\sin(4) + 2$, $\sin(4) + 2$ and $\sin(6) + 2$, and $\sin(6) + 2$ and $\sin(8) + 2$. Recall that if a trapezoid has a base width h and parallel sides of length l_1 and l_2, its area is $\frac{h}{2}(l_1 + l_2)$.

From the side lengths and width of each trapezoid we can calculate the individual areas, which we can sum to obtain the overall area that serves as an approximation to the integral.

Table 15.2a summarizes the area calculation for the four trapezoids in Figure 15.8a, and Table 15.2b summaries the area calculation for the eight trapezoids in Figure 15.8b. For each trapezoid we show

- the x-coordinates (x_1 and x_2) of the left and right sides of the trapezoid;
- the lengths of both vertical sides of the trapezoid, which are $\sin(x_1) + 2$ and $\sin(x_2) + 2$; and
- the area of the trapezoid.

Recall that the true value of the integral, rounded to six digits, is 17.1455. The four-trapezoid approximation has an error of less than 2.4%, while the eight-trapezoid approximation has an error of less than 0.6%.

15.4.2 General Case

Suppose that we wish to approximate the definite integral

$$\int_a^b f(x)\,dx \tag{15.24}$$

using the trapezoidal method with four trapezoids. Each of the four trapezoids will be of uniform width $h = \frac{b-a}{4}$. The x-coordinates of the sides of the trapezoids will be a and $a + h$, $a + h$ and $a + 2h$, $a + 2h$ and $a + 3h$, and $a + 3h$ and $a + 4h$. This means that the lengths of the sides of the four trapezoids will be, respectively, $f(a)$ and $f(a + h)$, $f(a + h)$ and $f(a + 2h)$, $f(a + 2h)$ and $f(a + 3h)$, and $f(a + 3h)$ and $f(a + 4h)$. The total area of the four trapezoids, which will approximate the

Table 15.2 Upper corners and area of each trapezoid in Figure 15.8

Upper-left corner		Upper-right corner		
x_1	$\sin(x_1) + 2$	x_2	$\sin(x_2) + 2$	Area
0.0000	2.0000	2.0000	2.9093	4.9093
2.0000	2.9093	4.0000	1.2432	4.1525
4.0000	1.2432	6.0000	1.7206	2.9638
6.0000	1.7206	8.0000	2.9894	4.7100
		Total		16.7356

(a) Four trapezoids

Upper-left corner		Upper-right corner		
x_1	$\sin(x_1) + 2$	x_2	$\sin(x_2) + 2$	Area
0.0000	2.0000	1.0000	2.8415	2.4208
1.0000	2.8415	2.0000	2.9093	2.8754
2.0000	2.9093	3.0000	2.1411	2.5252
3.0000	2.1411	4.0000	1.2432	1.6922
4.0000	1.2432	5.0000	1.0411	1.1422
5.0000	1.0411	6.0000	1.7206	1.3809
6.0000	1.7206	7.0000	2.6570	2.1888
7.0000	2.6570	8.0000	2.9894	2.8232
		Total		17.0487

(b) Eight trapezoids

integral, will be

$$\int_a^b f(x)\, dx \approx \frac{h}{2}\Big(f(a) + f(a+h)\Big) + \frac{h}{2}\Big(f(a+h) + f(a+2h)\Big) +$$

$$\frac{h}{2}\Big(f(a+2h) + f(a+3h)\Big) + \frac{h}{2}\Big(f(a+3h) + f(a+4h)\Big) \qquad (15.25)$$

$$= \frac{h}{2}\Big(f(a) + 2f(a+h) + 2f(a+2h) + 2f(a+3h) + f(a+4h)\Big) \qquad (15.26)$$

$$= h\left(\frac{1}{2}f(a) + f(a+h) + f(a+2h) + f(a+3h) + \frac{1}{2}f(b)\right) \qquad (15.27)$$

If we use *n* trapezoids, this generalizes to

$$\int_a^b f(x)\,dx \approx h\left(\frac{1}{2}f(a) + \frac{1}{2}f(b) + \sum_{i=1}^{n-1} f(a+ih)\right) \qquad (15.28)$$

where $h = \frac{b-a}{n}$. Implementing the trapezoidal method involves writing a function to compute this sum.

15.5 Trapezoidal Method Implementation

A `trapezoidal` function with the same interface as the `rectangular` function from Section 15.3 appears in Figure 15.9. As Figure 15.9 is a complete program, it also contains implementations of `integrand` and `main`. The `trapezoidal` function makes use of a `for` loop on lines 18–20 to compute the summation from Equation 15.28. This loop counts from 1 up to $n - 1$.

In its current form, `trapezoidal` can only be used to integrate functions named `integrand`. While this might not seem like much of a limitation, imagine that we wish to write a program in which `trapezoidal` is used to integrate *two* different functions, neither named `integrand`. We could probably rename one of the functions to be `integrand`, but we could certainly not rename *both* of them to have the same name.

Multiple functions aside, why should we (or anyone else who wishes to use `trapezoidal`) be forced to give a particular name to the function we wish to integrate? The name of the function being integrated had no significance when we developed the trapezoidal method, so why should it have any significance now that we are implementing that method?

The `trapezoidal` function takes as its three parameters the two limits of integration and the number of trapezoids to use. It would be more general if it took the function to integrate as a fourth parameter. This is not the first time we have encountered the idea of passing a function as a parameter. We experimented with exactly this notion in Chapter 9, where we developed a *Mathematica* implementation of the bisection method.

Figure 15.10 is a complete program containing a version of `trapezoidal` that takes the integrand as its fourth parameter. The header of `trapezoidal` is now

```
double trapezoidal (double a, double b, int n,
                    double integrand (double))
```

This specifies that `trapezoidal` takes four parameters:

- two numbers of type `double` (the bounds of integration a and b),
- one number of type `int` (the number n of trapezoids to use), and
- one function that takes a `double` as its parameter and returns a `double` as its result (the `integrand`).

```
1    #include <stdio.h>
2    #include <math.h>
3
4    double integrand (double x)
5    {
6      return(sin(x) + 2);
7    }
8
9
10   double trapezoidal (double a, double b, int n)
11   {
12     double width, sum;
13     int i;
14
15     width = (b - a) / n;
16     sum = (integrand(a) + integrand(b)) / 2;
17
18     for (i = 1; i <= n-1; i = i+1) {
19       sum = sum + integrand(a+i*width);
20     }
21
22     return(width * sum);
23   }
24
25
26   void main (void)
27   {
28     int ntraps;
29     double low, high;
30
31     printf("Enter number of trapezoids to use: ");
32     scanf("%d", &ntraps);
33
34     printf("Enter low bound of integration: ");
35     scanf("%lf", &low);
36
37     printf("Enter high bound of integration: ");
38     scanf("%lf", &high);
39
40     printf("%g\n", trapezoidal(low, high, ntraps));
41   }
```

Figure 15.9 Trapezoidal method implementation (`trap1.c`)

Just as `trapezoidal` must specify the type of each of its numerical parameters, it must also specify the type of its function parameter. The type of a function parameter is given by specifying (in the form of a header without parameter

```
1    #include <stdio.h>
2    #include <math.h>
3
4    double trapezoidal (double a, double b, int n,
5                        double integrand (double))
6    {
7      double width, sum;
8      int i;
9
10     width = (b - a) / n;
11     sum = (integrand(a) + integrand(b)) / 2;
12
13     for (i = 1; i <= n-1; i = i+1) {
14       sum = sum + integrand(a+i*width);
15     }
16
17     return(width * sum);
18   }
19
20
21   double f (double x)
22   {
23     return(sin(x) + 2);
24   }
25
26
27   void main (void)
28   {
29     int ntraps;
30     double low, high;
31
32     printf("Enter number of trapezoids to use: ");
33     scanf("%d", &ntraps);
34
35     printf("Enter low bound of integration: ");
36     scanf("%lf", &low);
37
38     printf("Enter high bound of integration: ");
39     scanf("%lf", &high);
40
41     printf("%g\n", trapezoidal(low, high, ntraps, f));
42   }
```

Figure 15.10   Trapezoidal method implementation with function parameter (`trap2.c`)

names) the parameter types and return type that the function parameter must have.
Much as

```
double a
```

in the header of `trapezoidal` specifies that the first parameter to `trapezoidal` must be a number of type `double`,

```
double integrand (double)
```

specifies that the fourth parameter to `trapezoidal` must be a function of one `double` parameter that returns a `double`.

The remainder of `trap2.c` consists of the implementations of two functions, `f` and `main`. In the call to `trapezoidal` on line 41, `f` is passed as the fourth parameter to `trapezoidal`. This is consistent with the header of `trapezoidal` since `f` takes a `double` parameter and returns a `double` result.

We elected to name the function that is being integrated `f` instead of `integrand` to emphasize the point that the names of formal and actual function parameters need not be the same, any more than the names of formal and actual `double` or `int` parameters need to be the same.

15.6 Multiple-File Programs

All of the C programs that we have examined to this point in the book have been contained entirely within a single file. This is typical only for small programs. Even moderately large C programs are usually divided up into more than one file. No single file of such a multiple-file program can stand alone: All of the files that make it up are combined by the compiler to produce a single executable.

Large programs are divided into more than one file to make it easier to develop, test, and reuse the program's components. We will focus on the issue of *reuse*. Suppose, for example, that we wanted to write a comparison program to compare the performance of the rectangular and trapezoidal methods when integrating the function $f(x) = \sin(x) + 2$. The program would consist of a single file containing implementations of `rectangular`, `trapezoidal`, `f`, and `main`.

If we were to ever write another program that needed to do numerical integration, we would likely want to reuse the implementations of `rectangular` and `trapezoidal` from the comparison program discussed above. If the comparison program were implemented as a single-file program, we would have to use a text editor to copy and paste the two implementations into the new program. By structuring the comparison program as a multiple-file program, we can make it easier to reuse `rectangular` and `trapezoidal`.

Under the multiple-file approach, the comparison program would consist of one file containing the implementations of `f` and `main` and another file containing the implementations of `rectangular` and `trapezoidal`. If we later wrote another program that needed to do numerical integration, we could use the second file unaltered as a component of that program. For that matter, if we had located a file containing implementations of the rectangular and trapezoidal methods, we could have incorporated that file into the comparison program in the first place and saved ourselves some programming effort.

15.6.1 Two Implementation Files

We will consider two versions of our multiple-file numerical integration method comparison program. The first version consists of two implementation files: compare1.c, which appears in Figure 15.11, and integ1.c, which appears in Figure 15.12.

The file compare1.c contains implementations of f and main, while the file integ1.c contains implementations of rectangular and trapezoidal. As we know from our experience with single-file programs, every function that is used in a file must be declared in that file. Because rectangular is used on line 33

```
1    #include <stdio.h>
2    #include <math.h>
3
4    double rectangular (double a, double b, int n,
5                        double integrand (double));
6
7
8    double trapezoidal (double a, double b, int n,
9                        double integrand (double));
10
11
12   double f (double x)
13   {
14      return(sin(x) + 2);
15   }
16
17
18   void main (void)
19   {
20      int ndivisions;
21      double low, high;
22
23      printf("Enter number of divisions: ");
24      scanf("%d", &ndivisions);
25
26      printf("Enter low bound of integration: ");
27      scanf("%lf", &low);
28
29      printf("Enter high bound of integration: ");
30      scanf("%lf", &high);
31
32      printf("Rectangular = %g, trapezoidal = %g\n",
33              rectangular(low, high, ndivisions, f),
34              trapezoidal(low, high, ndivisions, f));
35   }
```

Figure 15.11 Comparison of rectangular and trapezoidal methods (compare1.c)

```
1    double rectangular (double a, double b, int n,
2                               double integrand (double))
3    {
4      double width, sum;
5      int i;
6
7      width = (b - a) / n;
8      sum = 0;
9
10     for (i = 1; i <= n; i = i+1) {
11        sum = sum + integrand(a + i*width);
12     }
13
14     return(width * sum);
15   }
16
17
18   double trapezoidal (double a, double b, int n,
19                               double integrand (double))
20   {
21     double sum, width;
22     int i;
23
24     width = (b - a) / n;
25     sum = (integrand(a) + integrand(b)) / 2;
26
27     for (i = 1; i <= n-1; i = i+1) {
28        sum = sum + integrand(a+i*width);
29     }
30
31     return(width * sum);
32   }
```

Figure 15.12   Implementations of rectangular and trapezoidal methods (`integ1.c`)

and `trapezoidal` on line 34 of `compare1.c`, the prototypes on lines 4–9 of `compare1.c` are mandatory. On the other hand, prototypes for `f` and `main` need not appear in `integ1.c` because neither `f` nor `main` is used there.

Various versions of C take different approaches to creating an executable from a multiple-file program. All require, however, that each file be self-contained in the following sense. Each of the functions that is used but not implemented in a file must be declared with a prototype (as with `rectangular` and `trapezoidal` in `compare1.c`) or with the inclusion of the appropriate header file (as with `sin`, `scanf`, and `printf` in `compare1.c`).

The use of prototypes in multiple-file programs has one major drawback. Suppose we are writing a new program that needs to make use of the two functions contained in `integ1.c`. By reusing `integ1.c`, we need only create a second implementation file containing the remaining functions of the program. Unfortunately, we must also copy the headers for `rectangular` and `trapezoidal` from `integ1.c` and insert them into the new implementation file. This is a tedious and potentially error-prone task.

15.6.2 Header Files

We can solve this problem by creating a programmer-defined header file. We have already made extensive use in our programs of the C library header files `stdio.h` and `math.h`, which contain prototypes for some of the functions supplied by the C library. Similarly, we can create and use a file `integ.h` (Figure 15.13) containing the prototypes for the numerical integration functions.

```
1    double rectangular (double a, double b, int n,
2                              double integrand (double));
3
4    double trapezoidal (double a, double b, int n,
5                              double integrand (double));
```

Figure 15.13   Programmer-defined header file (`integ.h`)

An improved version of `compare1.c` appears as `compare2.c` in Figure 15.14. We have replaced the two prototypes that originally appeared in `compare1.c` with the inclusion directive

```
#include "integ.h"
```

on line 3. This directs the C compiler to include the prototypes from `integ.h` into `compare2.c`. Notice that this inclusion directive uses double quotes instead of angle brackets around the name of the header file. Double quotes are used to denote programmer-created header files, while angle brackets are used to denote C library header files.

A slightly modified version of `integ1.c` appears as `integ2.c` in Figure 15.15. We have added the inclusion directive

```
#include "integ.h"
```

as the first line of the file. We do this so that the C compiler can verify for us that the prototypes contained in `integ.h` are consistent with the implementations in `integ2.c`. Our complete program now consists of *three* files: `integ.h` and `integ2.c`, which together implement `rectangular` and `trapezoidal`, and `compare2.c`.

Under this approach, if we wish to reuse the implementations of the integration functions we will need to reuse both `integ2.c` and `integ.h`. By including `integ.h` into any implementation file that uses `rectangular` or `trapezoidal`, we can avoid having to explicitly copy either prototype into that

```
1    #include <stdio.h>
2    #include <math.h>
3    #include "integ.h"
4
5    double f (double x)
6    {
7      return(sin(x) + 2);
8    }
9
10
11   void main (void)
12   {
13     int ndivisions;
14     double low, high;
15
16     printf("Enter number of divisions: ");
17     scanf("%d", &ndivisions);
18
19     printf("Enter low bound of integration: ");
20     scanf("%lf", &low);
21
22     printf("Enter high bound of integration: ");
23     scanf("%lf", &high);
24
25     printf("Rectangular = %g, trapezoidal = %g\n",
26             rectangular(low, high, ndivisions, f),
27             trapezoidal(low, high, ndivisions, f));
28   }
```

Figure 15.14   Using a programmer-created header file (`compare2.c`)

file. We will follow this approach for the rest of the book whenever we create multiple-file programs.

The prototypes that appear in a header file should be well commented. Ideally, a programmer who wishes to reuse an implementation/header file pair should be able to learn everything required to make use of them by reading only the header file.

15.7 Comparison of Rectangular and Trapezoidal Methods

When using the rectangular method with n rectangles, the integrand must be evaluated at n evenly spaced points. When using the trapezoidal method with n trapezoids, the integrand must be evaluated at $n + 1$ evenly spaced points. Once n becomes at all large, this difference is negligible. Thus, it is accurate to say that the rectangular

```
1    #include "integ.h"
2
3    double rectangular (double a, double b, int n,
4                            double integrand (double))
5    {
6       double width, sum;
7       int i;
8
9       width = (b - a) / n;
10      sum = 0;
11
12      for (i = 1; i <= n; i = i+1) {
13         sum = sum + integrand(a + i*width);
14      }
15
16      return(width * sum);
17   }
18
19
20   double trapezoidal (double a, double b, int n,
21                           double integrand (double))
22   {
23      double width, sum;
24      int i;
25
26      width = (b - a) / n;
27      sum = (integrand(a) + integrand(b)) / 2;
28
29      for (i = 1; i <= n-1; i = i+1) {
30         sum = sum + integrand(a+i*width);
31      }
32
33      return(width * sum);
34   }
```

Figure 15.15   Slightly modified implementations of rectangular and trapezoidal methods
(integ2.c)

and trapezoidal methods require the same amount of work. This makes possible a
fair comparison of their performance.

   Table 15.3 shows a comparison of the rectangular and trapezoidal methods when
used to compute two different integrals. For each integral, the number of rectangles
or trapezoids used is indicated along with the percentage error.

   Two key facts about the rectangular and trapezoidal methods are revealed by
Table 15.3.

   1. For the same value of n, the trapezoidal method is generally more precise than
the rectangular method.

Table 15.3   Percentage errors introduced by the rectangular and trapezoidal methods, using *n* subintervals

n	Rect	Trap		n	Rect	Trap
4	3.4	2.4		4	150	39
8	2.3	0.57		8	67	10
16	1.3	0.14		16	31	2.6
32	0.69	0.035		32	15	0.66
64	0.35	0.0087		64	7.2	0.16
128	0.18	0.0022		128	3.6	0.041

(a) $\int_0^8 \sin(x) + 2\,dx$ (b) $\int_1^{10} e^x\,dx$

2. Doubling the number of rectangles tends to cut the error approximately in half, whereas doubling the number of trapezoids tends to cut the error approximately in a quarter. As the value of *n* increases, then, the advantage of the trapezoidal method over the rectangular method increases.

These facts can be established by a more detailed analysis of the mathematics behind the rectangular and trapezoidal methods than we are prepared to undertake here.

There is a better version of the rectangular method than the one we have described in this chapter. Instead of arranging for the upper-right corner of each rectangle to touch the graph of the integrand, it is better to let the center of the top of each rectangle touch the graph of the integrand. This *midpoint method* produces a version of the rectangular method that is more competitive with the trapezoidal method.

In practice, neither the rectangular nor trapezoidal methods is widely used to do numerical integration. More advanced adaptive methods are used in systems such as *Mathematica*. While more complicated to implement and understand, these adaptive methods are based on the same general principle as the methods we have studied, and supply even more accuracy for a given amount of computational effort.

15.8 Key Concepts

Numerical integration.   Just as there are both symbolic and numerical methods for solving equations, there are also symbolic and numerical methods for evaluating definite integrals. These numerical methods involve using simple geometric figures such as rectangles and trapezoids to approximate the area underneath a curve. By using more geometric figures to approximate a given area, greater accuracy can be obtained at the cost of increased computational effort.

Counting loops.   The rectangular and trapezoidal methods rely on counting loops in their implementations. A counting loop executes a sequence of statements a fixed number of times. Although it is possible to implement counting loops by using the C `while` statement, the `for` statement is often more convenient for this purpose.

Functions as parameters. A function (say g) can be passed as a parameter to another function (say f). Just as with int and double parameters, the type of the function parameter g must be specified in the header of f. The type of a function parameter is specified by giving the header that is required of it.

Multiple-file programs. Dividing a program into multiple files promotes the reuse of program components. When an implementation file contains functions that can be used in more than one program, a header file consisting of commented prototypes for those functions should be created and used along with the implementation file.

15.9 Exercises

15.1 Sketch an integrand for which the trapezoidal method is exact but the rectangular method is not.

15.2 In both the rectangular and the trapezoidal methods, we divide the area beneath the graph of the integrand vertically. Would it work just as well to divide the area horizontally?

15.3 Modify rect4.c so that it evaluates Formula 15.5.

15.4 Modify trap2.c so that it evaluates Formula 15.5.

15.5 Modify compare2.c so that it evaluates Formula 15.5.

15.6 Use a multiple-file program composed of your modified compare2.c from Exercise 15.5, integ.h, and integ2.c to create a table similar to Tables 15.3a and 15.3b that compares the performance of the rectangular and trapezoidal methods on Formula 15.5. Explain the surprising results.

15.7 In the rectangular method as presented in this chapter, each rectangle is chosen so that its upper-right corner touches the graph of the integrand. An alternative approach, known as the midpoint method, is to choose each rectangle so that the integrand intersects the center of its top end. Develop this idea further by finding an equation similar to Equation 15.22 that embodies the midpoint method.

15.8 Add an implementation of the midpoint method of Exercise 15.7 to the multiple-file program consisting of integ.h, integ2.c, and compare2.c. You will need to put the prototype of a midpoint function into integ.h, put the implementation of that prototype into integ2.c, and call midpoint (along with rectangular and trapezoidal) from compare2.c.

15.9   Use your multiple-file program from Exercise 15.8 to compare the performance of the midpoint, rectangular, and trapezoidal methods on a selection of functions. What conclusions can you draw?

15.10   Repeat Exercise 14.9, this time using a `for` loop instead of a `while` loop.

15.11   Modify `block6.c` from Chapter 12 so that it displays the position of the block every second until the block stops moving. Do this without modifying the implementation of the `position` function.

15.12   Implement a C program based on the ballistic trajectory problem from Chapter 7. It should prompt the user for the direction and magnitude of the projectile's initial velocity. It should then calculate and report the x- and y-coordinates of the projectile at one-second intervals until it hits the ground. Your program should make use of functions that compute the horizontal and vertical positions of the projectile as functions of time.

15.13   Implement a C program that computes the position of the destroyer from Chapter 8 when it follows the trajectory described by Equation 8.37. Your program should report the position of the destroyer at 30-second intervals. Your program should make use of a function that computes the position of the destroyer as a function of time.

15.14   Modify your solution to Exercise 12.20 so that it displays the value of each investment at the end of each year.

15.15   Modify `bisect3.c` from Chapter 14 so that `bisection` takes as a parameter the function whose root it is to find.

15.16   Modify `newton6.c` from Chapter 14 so that `newton` takes as parameters both the function whose root is sought and its first derivative.

15.17   Modify your solution to Exercise 14.19 so that the root-finding function takes as a parameter the function whose root it is to find.

15.18   Create a root-finding library consisting of one file containing the implementations of the root-finding functions from Exercises 15.15–15.17 and another file containing their prototypes.

15.19   Create a multiple-file program consisting of your root-finding library from Exercise 15.18 and a file containing a `main` function that compares the behavior of the three root-finding functions.

15.20   Generalize your solution to Exercise 14.7 by writing a C program that prompts the user for a positive integer n and a floating-point number x, and then evaluates the nth order polynomial

$$a_n x^n + a_{n-1} x^{n-1} + \cdots + a_1 x + a_0 \qquad (15.29)$$

using Horner's rule. After reading n and x, your program should read the $n + 1$ coefficients a_i in order from a_n to a_0.

16

Harmonic Oscillation: Structures and Abstract Datatypes

An object of mass m is attached to opposite walls of a room with two identical springs, as illustrated in Figure 16.1. Neither spring is under tension when the object is equidistant from the walls. If we move the object x_0 meters toward one wall, the opposing spring will resist by exerting a force of κx_0 newtons. We say that κ is the *spring constant* of the spring.

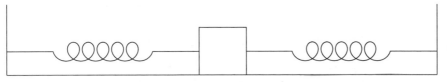

Figure 16.1   An object attached with a pair of springs to opposing walls

If we now release the object, it will begin moving back and forth. If no forces other than those generated by the springs work to oppose the object's motion, it will oscillate indefinitely by shuttling between two points x_0 meters on either side of the center of the room. In the real world, of course, friction would cause the amplitude of the oscillation to grow gradually smaller until the object came to rest.

Now let's transport our object to the bottom of a swimming pool and attach it to opposite walls with the same springs. Even if our object has negligible friction with the bottom of the pool, the drag exerted by the water in the pool is too great to ignore. As with the destroyer in Chapter 8, the drag exerted on the object when it is moving at v m/sec will be βv newtons, where β is the drag coefficient of the object in water.

Suppose that the mass of the object is $m = 1$ kilogram, the spring constant of each spring is $\kappa = 4.25$ kg/sec^2, and the drag coefficient of the object is $\beta = 4$ kg/sec. If we move the object $x_0 = 1$ meter to the right and release it at time zero, what will be its position $x(t)$ as a function of time in seconds?

This problem is an example of *damped harmonic oscillation*. It is well beyond the scope of this book to derive the model we are about to present, so you will have to take it on faith. If the solutions to the quadratic equation

$$ms^2 + \beta s + \kappa = 0 \qquad (16.1)$$

are the two complex numbers $-a + bi$ and $-a - bi$, the position of the object relative to the center of the room as a function of time is

$$x(t) = \frac{x_0}{a} e^{-bt} (a \cos(at) + b \sin(at)) \qquad (16.2)$$

(If the solutions to Equation 16.1 are not complex, which will be the case when the damping coefficient is much larger than the spring constant, a solution other than Equation 16.2 applies.)

In our problem, Equation 16.1 reduces to

$$s^2 + 4s + 4.25 = 0 \qquad (16.3)$$

We can find both roots of this equation by using *Mathematica*'s `Solve` function.

```
In[4]:= Solve[s^2 + 4*s + 4.25 == 0, s]

Out[4]= {{s → -2. - 0.5I}, {s → -2. + 0.5I}}
```
(16.4)

(Notice that *Mathematica* uses I to stand for i, the square root of -1.)

Plugging in 2 for a and $\frac{1}{2}$ for b, Equation 16.2 reduces to

$$x(t) = e^{-t/2} \left(2 \cos(2t) + \frac{1}{4} \sin(2t) \right) \qquad (16.5)$$

If we plot this equation with t along the horizontal axis and $x(t)$ along the vertical axis, we can visualize the damped harmonic oscillation of the object.

```
In[6]:= Plot[Exp[-t/2] * (Cos[2*t] + 1/4*Sin[2*t]),
            {t,0,10}, AxesLabel->{"sec","m"}]

Out[6]= (See Figure 16.2)
```
(16.6)

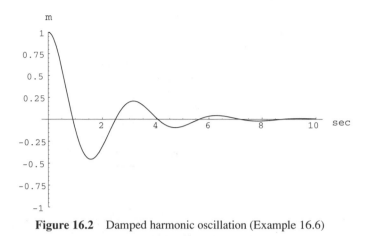

Figure 16.2   Damped harmonic oscillation (Example 16.6)

16.1 Newton's Method with Complex Roots

The problem of damped harmonic oscillation is just one example of how complex numbers are used in science and engineering. Because measurements of the natural world are made using real numbers, many students mistakenly believe that complex numbers are merely mathematical curiosities. As we have just seen, they can be useful tools for deriving real-valued results.

To apply our model for damped harmonic oscillation to specific situations, we only need to solve for the complex roots of Equation 16.1. While Equation 16.1 is quadratic and can be solved symbolically, equations with complex roots that require numerical solutions can easily arise.

We have already studied two methods for approximating the roots of equations: the bisection method and Newton's method. Although we have used these two methods only to find real roots, Newton's method also works for complex roots. In the remainder of this chapter we will modify the implementation of Newton's method from Chapter 14 so that it can find complex roots.

Suppose we want to use Newton's method to find the roots of $f(x) = x^2 - 6x + 13$, which are $3 \pm 2i$ as we can determine algebraically. The first derivative of $f(x)$ is $f'(x) = 2x - 6$, so we can improve an approximation g by calculating

$$g - \frac{f(g)}{f'(g)} = g - \frac{g^2 - 6g + 13}{2g - 6} \tag{16.7}$$

By beginning with a complex guess, we can approximate a complex root. Table 16.1 shows how, beginning with a guess of $1 + 1i$, only six repetitions are required to obtain the first ten digits of both the real and imaginary parts of a root of $f(x)$.

As we modify our C implementation of Newton's method so that it can find complex roots, we will have to overcome the fact that C does not provide built-in

Table 16.1   Newton's method with complex numbers

Repetition	Approximation
0	$1.000000000 + 1.000000000i$
1	$2.800000000 + 0.900000000i$
2	$3.370588235 + 2.567647059i$
3	$3.075166289 + 2.046852045i$
4	$3.001749176 + 1.999220292i$
5	$2.999999317 + 1.999999385i$
6	$3.000000000 + 2.000000000i$

complex numbers. To do this, we will exploit the mechanism that C provides for programmers to define new types of compound data.

To this point in our study of C we have worked exclusively with two types of *scalar data*: integers and floating-point numbers. *Compound data* are formed by combining two or more pieces of data, such as two floating-point numbers, into a single unit. C supports several kinds of compound data, including structures and arrays. We will study structures in this chapter and arrays in Chapter 17.

Before returning to the problem of implementing a version of Newton's method that can find complex roots, we will examine a simpler problem as a means of learning the basics of structures. We will use structures to improve the design and modify the implementation of our solution to the rod-stacking problem from Chapter 13. A simpler design can be obtained by using structures to create compound data.

16.2 Rod Stacking Revisited

Figure 16.3 shows the prototypes for the five functions (excluding `main`) contained in our original design for the rod-stacking implementation. Notice that every parameter and return value is a `double`. In fact, every variable used in the implementation of the program is also a `double`. This shouldn't be particularly surprising, of course, because the only other type of data that we had to choose from at the time was `int`, and it wasn't appropriate to the problem.

When we were creating the original design, it was apparent that the rod-stacking program dealt extensively with points in the plane. The entire purpose of the program, after all, was to calculate the center points of five rods from eight radii and three other center points. Just as we decided during the design process that programmer-defined functions such as `sq` and `cosine` would be useful, we should also have decided that a programmer-defined type called `point` would be useful.

For the time being, let's pretend that C provides a type called `point` such that

- a value of type `point` is a compound value containing two `doubles` that represent the *x-y* coordinates of a point;

```
double sq (double x);

double cosine (double adj1, double adj2, double opp);

double distance (double x1, double y1, double x2, double y2);

double threeRodX (double xLeft, double yLeft,
                  double xRight, double yRight,
                  double rLeft, double rRight, double rTop);

double threeRodY (double xLeft, double yLeft,
                  double xRight, double yRight,
                  double rLeft, double rRight, double rTop);
```

Figure 16.3   Original prototypes for the rod-stacking implementation

- given a point, it is possible to extract and modify the two coordinates that it contains; and
- it is possible to pass a point as a parameter, return a point as a result, and declare a variable of type point.

16.2.1 Exploiting a point Type in Design

It is possible for a programmer to create a point type that behaves exactly as described above. Before we learn how to create compound data, however, we will consider how our hypothetical point type would simplify the design of the rod-stacking implementation.

Figure 16.4 shows how the function prototypes for the rod-stacking design can be simplified by exploiting a point type. While the prototypes for sq and cosine (which don't deal with points) are unchanged, we have simplified distance, threeRodX, and threeRodY.

```
double sq (double x);

double cosine (double adj1, double adj2, double opp);

double distance (point p1, point p2);

double threeRodX (point pLeft, point pRight,
                  double rLeft, double rRight, double rTop);

double threeRodY (point pLeft, point pRight,
                  double rLeft, double rRight, double rTop);
```

Figure 16.4   Better prototypes for the rod-stacking implementation

The original version of distance took four doubles, representing two pairs of *x-y* coordinates, as parameters. The new version takes two points as parameters. The same information is being passed in both cases—the point parameters p1 and p2 will contain *x-y* coordinates that can be extracted as needed—but the packaging is more natural in the second version. The function distance, after all, is more easily understood as one that operates on a pair of points than as one that operates on four coordinates.

Each of the original versions of threeRodX and threeRodY took the same seven parameters: two pairs of *x-y* coordinates and three radii. Now, each takes only five parameters: two points and three radii. The same information, packaged into a more convenient and more easily understood form, is still being passed to these two functions.

By postulating the existence of a point type we were able to streamline the prototypes of three functions. The implementation will be similarly streamlined. But we can make an even more significant improvement. The implementations of threeRodX and threeRodY are virtually identical. If you compare Figures 13.11 and 13.12, you will find that, except for their names, the implementations of the two functions differ only in their final return statements.

Even though they do almost the same thing, we designed threeRodX and threeRodY separately because a function can return only one value via its return statement. Although we could have easily written one function to compute both coordinates, until now we did not have a way to return two coordinates as results.

The solution to this dilemma appears in Figure 16.5, which shows our third and final design for the rod-stacking implementation. We have replaced threeRodX and threeRodY with a single function called threeRod. This function takes the same five parameters as the functions it replaces, but returns a point as its result. Thus, one call to threeRod will yield the same information, with only half the computational effort, as a pair of calls to threeRodX and threeRodY.

```
double sq (double x);

double cosine (double adj1, double adj2, double opp);

double distance (point p1, point p2);

point threeRod (point pLeft, point pRight,
                double rLeft, double rRight, double rTop);
```

Figure 16.5   Even better prototypes for the rod-stacking implementation

16.2.2 Creating a point Type

Figure 16.6a shows C code that defines a new point type whose values will behave as we require. A type definition such as this consists of

- the keywords `typedef struct`;
- what looks like a sequence of variable declarations enclosed in braces; and
- a new type name followed by a semicolon.

The particular definition in Figure 16.6a means that

- there will be a new type named `point` that can be used in function headers and variable declarations;
- every value of type `point` will be a structure containing two fields named `x` and `y`; and
- both of these fields will contain a `double`.

A structure is a compound value consisting of one or more named *fields*, with each field containing a value of a particular type. Two typical structures of type `point` are illustrated in Figure 16.6b. Thus, the `x` and `y` fields of p1 contain 3.0 and 3.0, while the `x` and `y` fields of p2 contain 20.0 and 4.0. A structure can be treated as a single unit, or its fields can be manipulated individually, as it suits the programmer's purposes. We will see examples of this below.

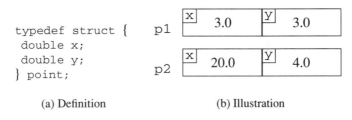

| (a) Definition | (b) Illustration |

Figure 16.6   Definition and illustration of type `point`

16.2.3 Implementing the New Design

Figures 16.7–16.8 show an implementation of the rod-stacking problem based on the prototypes from Figure 16.5 and the definition of the `point` type from Figure 16.6a. The first thing to notice is the structure of the program. In addition to the five functions, the program also contains the definition of the `point` type on lines 4–7 of Figure 16.7. A new type name must be defined before it can be used. While we might have elected to move the `point` definition below the implementation of the `cosine` function, the `point` definition *must* occur before the `distance` function, which is the first to use it.

Next look at the `main` function in Figure 16.8. On line 47 we declare eight variables of type `point`, exactly as we declare eight variables of type `double` on the line that follows. On line 55 we pass the two `points` p1 and p2 as the first two parameters to `threeRod`, exactly as we pass the three `doubles` r1, r2, and r4

```
1    #include <stdio.h>
2    #include <math.h>
3
4    typedef struct {
5      double x;
6      double y;
7    } point;
8
9
10   double sq (double x) {
11     return(x*x);
12   }
13
14   double cosine (double adj1, double adj2, double opp)
15   {
16     return((sq(adj1) + sq(adj2) - sq(opp)) / (2*adj1*adj2));
17   }
18
19   double distance (point pt1, point pt2) {
20     return(sqrt(sq(pt1.x-pt2.x) + sq(pt1.y-pt2.y)));
21   }
22
23   point threeRod (point pLeft, point pRight,
24                   double rLeft, double rRight, double rTop)
25   {
26     double a, b, c, d, e;
27     double cosA, sinA, cosB, sinB;
28     point pTop;
29
30     a = rLeft + rTop;
31     b = rRight + rTop;
32     c = distance(pLeft, pRight);
33     d = pRight.x - pLeft.x;
34     e = pRight.y - pLeft.y;
35
36     cosA = cosine(a, c, b);
37     sinA = sqrt(1 - sq(cosA));
38     cosB = d/c;
39     sinB = e/c;
40
41     pTop.x = pLeft.x + a * (cosA*cosB - sinA*sinB);
42     pTop.y = pLeft.y + a * (sinA*cosB + sinB*cosA);
43     return(pTop);
44   }
```

Figure 16.7   Implementation of rod-stacking problem using programmer-defined `point` type
(continued in Figure 16.8)

```
45   void main (void)
46   {
47      point p1, p2, p3, p4, p5, p6, p7, p8;
48      double r1, r2, r3, r4, r5, r6, r7, r8;
49
50      p1.x = 3; p1.y = 3; r1 = 3;
51      p2.x = 20; p2.y = 4; r2 = 4;
52      p3.x = 38; p3.y = 2; r3 = 2;
53      r4 = 9; r5 = 8; r6 = 3; r7 = 5; r8 = 4;
54
55      p4 = threeRod(p1, p2, r1, r2, r4);
56      p5 = threeRod(p2, p3, r2, r3, r5);
57      p6 = threeRod(p4, p5, r4, r5, r6);
58      p7 = threeRod(p6, p5, r6, r5, r7);
59      p8 = threeRod(p4, p7, r4, r7, r8);
60
61      printf("Rod    X        Y\n");
62      printf(" 4 %7.2f %7.2f\n", p4.x, p4.y);
63      printf(" 5 %7.2f %7.2f\n", p5.x, p5.y);
64      printf(" 6 %7.2f %7.2f\n", p6.x, p6.y);
65      printf(" 7 %7.2f %7.2f\n", p7.x, p7.y);
66      printf(" 8 %7.2f %7.2f\n", p8.x, p8.y);
67   }
```

Figure 16.8   Implementation of rod-stacking problem using programmer-defined `point` type
(`rod7.c`)

as the final three parameters to `threeRod`. The value returned by `threeRod` is a
`point`, and we assign it to the `point` variable p4 in the usual way.

The individual fields that make up a particular structure can be specified with the
dot operator. Thus, `p1.x` is the x field, and `p1.y` the y field, of the `point` p1
declared on line 47. A field that is identified in this way can be used as an ordinary
variable. Thus, on line 50 we initialize the two fields of p1, and on line 51 we
initialize the two fields of p2. After this is done, the values of p1 and p2 will be
exactly as illustrated in Figure 16.6b.

Now consider the `distance` function on lines 19–21 of Figure 16.7. It receives
the two `points` pt1 and pt2 as its parameters. On line 20 it uses the dot operator
to access and perform calculations on the fields of pt1 and pt2. If the actual
parameters are p1 and p2 from Figure 16.6b, the return result will be 17.0294.

Finally, consider the `threeRod` function in Figure 16.7. It provides a good
illustration of how the fields of a structure can be treated as individual values, or as
components of a single compound value, according to our purposes. The function
accesses the individual field values of pLeft and pRight on lines 33–34, but
passes the two structures as single values to the `distance` function on line 32.
On lines 41–42, it initializes the two fields of pTop via assignments, but on line 43
returns pTop as a single value.

16.3 Newton's Method Revisited

We are now prepared to revisit the problem that motivated our study of compound data, which was to modify our implementation of Newton's method so that it could find complex roots. The difficulty we face is that C does not provide a complex number type. In this section we will use what we have just learned about structures to create and exploit a complex number type.

16.3.1 An Incorrect Implementation

A complex number $a + bi$ can be treated as an ordered pair of real numbers, where a is called the *real part* and b the *imaginary part*. This suggests that we can represent a complex number in C as a structure consisting of a pair of floating-point numbers, as defined on lines 4–7 of Figure 16.9.

Figure 16.9 shows the result of *incorrectly* modifying the program for Newton's method from Figure 14.15 so that it uses `complex` values in place of `doubles`. It is written in an attempt to solve the function $f(x) = x^2 - 6x + 13$ that we discussed at the beginning of this chapter. The definition of the `complex` type is followed by implementations of `f`, `fprime`, `newton`, and `main`. We have written these four functions so that all of the variables, function parameters, and return values are of type `complex` instead of `double`.

Although the high-level structure of the program is correct, it will not compile—let alone run—because it is riddled with errors. All of the errors are connected with misuse of the `complex` values that the program manipulates. On line 11, for example, we attempt to subtract `6*x` from `x*x` and then add `13`. These operations will work when applied to `doubles`, but they will not work with values of type `complex`. C does not know how to add, subtract, or multiply structures; it only knows how to apply the dot operator.

There are similar errors in each function:

- On lines 16, 24, and 26 we attempt to subtract, multiply, divide, and find the absolute value of `complex` values.
- On line 34 we attempt to scan a value into a `complex` using a `double` format.
- On line 36 we attempt to print a `complex` using a `double` format.

16.3.2 A Correct but Obscure Implementation

The solution is to rewrite the offending expressions and statements to make proper use of the *fields* of the `complex` values, which *are* floating-point numbers; we have done this in Figures 16.10 and 16.11. In the corrected implementation of `fprime`, for example, we declare a `complex` variable, set its `real` and `imag` fields individually, and return it.

Let's consider the mathematical reasoning that led us to the implementation of `fprime` in Figure 16.10. The purpose of the function is to compute $2x - 6$ (the

```
1    #include <stdio.h>
2    #include <math.h>
3
4    typedef struct {
5      double real;
6      double imag;
7    } complex;
8
9    complex f (complex x)
10   {
11     return(x*x - 6*x + 13);
12   }
13
14   complex fprime (complex x)
15   {
16     return(2*x - 6);
17   }
18
19   complex newton (complex g0)
20   {
21     complex oldg0;
22     do {
23       oldg0 = g0;
24       g0 = g0 - f(g0)/fprime(g0);
25     }
26     while (fabs((g0-oldg0) / g0) > 1e-10);
27     return(g0);
28   }
29
30   void main (void)
31   {
32     complex guess, root;
33     printf("Please enter an initial guess: ");
34     scanf("%lf", &guess);
35     root = newton(guess);
36     printf("An approximate root is %g\n", guess);
37   }
```

Figure 16.9   Incorrect implementation of Newton's method (`newton7.c`)

first derivative of f), where x is a complex number. Because x is of the form $a + bi$, where a and b are real numbers,

$$2x - 6 = 2(a + bi) - 6 \tag{16.8}$$
$$= 2a + 2bi - 6 \tag{16.9}$$
$$= (2a - 6) + (2b)i \tag{16.10}$$

```
1    #include <stdio.h>
2    #include <math.h>
3
4    typedef struct {
5      double real;
6      double imag;
7    } complex;
8
9    complex f (complex x)
10   {
11     complex result;
12     result.real = x.real*x.real - x.imag*x.imag -
13                        6*x.real + 13;
14     result.imag = 2*x.real*x.imag - 6*x.imag;
15     return(result);
16   }
17
18   complex fprime (complex x)
19   {
20     complex result;
21     result.real = 2*x.real - 6;
22     result.imag = 2*x.imag;
23     return(result);
24   }
```

Figure 16.10   Correct but obscure version of Newton's method (continued in Figure 16.11)

Thus, the real part of the result is twice the real part of x minus 6, and the imaginary part of the result is twice the imaginary part of x. These derivations are reflected by the assignment statements on lines 21–22 of Figure 16.10. Because these statements operate on the fields of x and not on x itself, they require only floating-point arithmetic.

By following this approach of breaking calculations involving complex numbers into separate calculations involving their real and imaginary parts, we have produced correct implementations for each of the four functions. Unfortunately, the resulting code is not always easy to read or understand. As an extreme example, consider the implementation of the newton function in Figure 16.11.

The version of newton that worked with floating-point numbers was short, straightforward, and easy to understand. This new version is almost frighteningly complex. This complexity results from the fact that finding quotients and absolute values of complex numbers requires some intricate calculations. A first-time reader of the newton function would have an extremely hard time figuring out what it does.

16.3.3 A Better Implementation

The floating-point implementation of Newton's method in Figure 14.15 is simple because it relies on existing C operators (+, -, *, /), library functions (printf,

```
25   complex newton (complex g0)
26   {
27     complex oldg0, fg0, fpg0;
28
29     do {
30       oldg0 = g0;
31       fg0 = f(g0);
32       fpg0 = fprime(g0);
33
34       g0.real = g0.real -
35         (fg0.real*fpg0.real + fg0.imag*fpg0.imag) /
36           (fg0.real*fg0.real + fpg0.imag*fpg0.imag);
37
38       g0.imag = g0.imag -
39         (fg0.imag*fpg0.real - fg0.real*fpg0.imag) /
40           (fg0.real*fg0.real + fpg0.imag*fpg0.imag);
41     }
42     while
43       (sqrt(pow(1-(oldg0.real*g0.real + oldg0.imag*g0.imag) /
44                 (g0.real*g0.real + g0.imag*g0.imag), 2) +
45           pow((oldg0.imag*g0.real - oldg0.real*g0.imag) /
46                 (g0.real*g0.real + g0.imag*g0.imag), 2)) >
47       1e-10);
48
49     return(g0);
50   }
51
52
53   void main (void)
54   {
55     complex guess, root;
56     printf("Please enter an initial guess: ");
57     scanf("%lf%lfi", &guess.real, &guess.imag);
58     root = newton(guess);
59     printf("An approximate root is %.10g%+.10gi\n",
60             root.real, root.imag);
61   }
```

Figure 16.11   Correct but obscure version of Newton's method (newton8.c)

scanf, fabs), and constants (2, 6, 13) to create and manipulate its floating-point numbers. This suggests that we can simplify newton8.c by first creating functions to do the complex number manipulations that we require, and then using these functions to implement Newton's method.

Suppose we have available the seven functions whose prototypes appear in Figure 16.12, and that they behave as follows:

- make creates a complex number from its real and imaginary parts.

```
1    typedef struct {
2      double real;
3      double imag;
4    } complex;
5
6    complex make (double real, double imaginary);
7
8    complex read (void);
9
10   void write (complex c);
11
12   complex add (complex c1, complex c2);
13
14   complex sub (complex c1, complex c2);
15
16   complex mul (complex c1, complex c2);
17
18   complex div (complex c1, complex c2);
19
20   double absval (complex c);
```

Figure 16.12   Prototypes for complex number functions (`complex.h`)

- `read` returns the `complex` number that it reads from the keyboard, and `write` writes c to the display.
- `add`, `sub`, `mul`, and `div` return the result of performing the appropriate operation on c1 and c2.
- `absval` returns the absolute value of c.

Using these functions we can produce the implementation of Newton's method that appears in Figure 16.13. It is not quite as straightforward as the floating-point version because it must use functions such as `add` instead of operators such as +, and because it must create `complex` constants using `make` instead of using `double` constants such as 13. Nevertheless, it represents a big improvement in clarity over the version in Figures 16.10–16.11.

To obtain a complete program, of course, we must also implement the seven functions from Figure 16.12. We have done this in Figures 16.14–16.15. Taken individually, each of these functions is also straightforward and easy to understand so long as you know something about complex number arithmetic.

Our experience in this section underscores the fact that the complexity of a program is not directly related to its length. Although the multiple-file solution spread across `complex.h`, `complex.c`, and `newton9.c` (with 132 lines) is over twice as long as the single-file solution contained in `newton8.c` (61 lines), it is much simpler to explain, understand, modify, test, and maintain. The advantage enjoyed by the multiple-file solution is that we have implemented it using an *abstract datatype*, as we will explain in Section 16.4.

```
1    #include <stdio.h>
2    #include <math.h>
3    #include "complex.h"
4
5    complex f (complex x)
6    {
7      complex c6, c13;
8
9      c6 = make(6,0);
10     c13 = make(13, 0);
11
12     return(add(sub(mul(x,x), mul(c6,x)), c13));
13   }
14
15   complex fprime (complex x)
16   {
17     complex c2, c6;
18
19     c2 = make(2,0);
20     c6 = make(6,0);
21
22     return(sub(mul(c2,x), c6));
23   }
24
25   complex newton (complex g0)
26   {
27     complex oldg0;
28
29     do {
30       oldg0 = g0;
31       g0 = sub(g0, div(f(g0), fprime(g0)));
32     }
33     while (absval(div(sub(g0, oldg0), g0)) > 1e-10);
34
35     return(g0);
36   }
37
38   void main (void)
39   {
40     complex guess, root;
41
42     printf("Please enter an initial guess: ");
43     guess = read();
44     root = newton(guess);
45     printf("An approximate root is ");
46     write(root);
47     printf("\n");
48   }
```

Figure 16.13 Better implementation of Newton's method (newton9.c)

```
1    #include <stdio.h>
2    #include <math.h>
3    #include "complex.h"
4
5    complex make (double real, double imaginary)
6    {
7      complex c;
8      c.real = real;
9      c.imag = imaginary;
10     return(c);
11   }
12
13   complex read (void)
14   {
15     complex c;
16     scanf("%lf%lfi", &c.real, &c.imag);
17     return(c);
18   }
19
20   void write (complex c)
21   {
22     printf("%.10g%+.10gi", c.real, c.imag);
23   }
24
25   complex add (complex c1, complex c2)
26   {
27     complex result;
28     result.real = c1.real + c2.real;
29     result.imag = c1.imag + c2.imag;
30     return(result);
31   }
32
33   complex sub (complex c1, complex c2)
34   {
35     complex result;
36     result.real = c1.real - c2.real;
37     result.imag = c1.imag - c2.imag;
38     return(result);
39   }
```

Figure 16.14   Implementations of `complex` functions (continued in Figure 16.15)

16.3.4 Input/Output Fine Points

The implementations of `write` and `read` in Figure 16.14 exploit aspects of `printf` and `scanf` that we have not previously seen.

The `printf` statement in `write`

```
printf("%.10g%+.10gi", c.real, c.imag);
```

```
40    complex mul (complex c1, complex c2)
41    {
42      complex result;
43      result.real = c1.real*c2.real - c1.imag*c2.imag;
44      result.imag = c1.real*c2.imag + c1.imag*c2.real;
45      return(result);
46    }
47
48    complex div (complex c1, complex c2)
49    {
50      complex result;
51      double denom;
52
53      denom = c2.real*c2.real + c2.imag*c2.imag;
54      result.real = (c1.real*c2.real + c1.imag*c2.imag) /
55                        denom;
56      result.imag = (c1.imag*c2.real - c1.real*c2.imag) /
57                        denom;
58
59      return(result);
60    }
61
62    double absval (complex c) {
63      return(sqrt(c.real*c.real + c.imag*c.imag));
64    }
```

Figure 16.15   Implementations of `complex` functions (`complex.c`).

contains two conversion specifications, the second of which is `%+.10g`. The `+` symbol that follows the percent sign directs `printf` to explicitly display the sign of `c.imag` even if it is positive. This ensures that the real and imaginary parts of the displayed complex number will be separated by either + or –.

The `scanf` statement in `read`

```
scanf("%lf%lfi", &c.real, &c.imag);
```

contains two conversion specifications followed by the character `i`. This directs `scanf` to look for two doubles followed by an `i`. The doubles will be stored in `c.real` and `c.imag`, while the `i` will be read and discarded. This allows the user to enter complex numbers in their standard form, for example `2.5+1.2i`.

16.4 Assessment

Before we created our design for solving the rod-stacking problem in Section 13.2, we observed that the essence of the design process is to answer two questions.

1. How should the program's data be represented?

2. What programmer-defined functions are needed?

Although we asserted that the first question is the more important and fundamental of the two, we gave it no real consideration. Instead, we decided that every piece of data in our program would be represented as a `double` and then began designing the programmer-defined functions that would make up our eventual implementation.

The natural progression for a beginning programmer is to learn how to exploit programmer-defined functions before learning how to exploit programmer-defined types. This is partly because programmer-defined functions are easier to understand than programmer-defined types, and partly because programmer-defined types cannot be fully exploited in the absence of programmer-defined functions.

In this chapter, we have seen two examples of how, by carefully choosing the data to be manipulated by a program, we can significantly simplify the resulting design and implementation. In this section we will step back from the two implementations that we produced and focus on the bigger picture.

16.4.1 Built-In Types

Before we began working with programmer-defined functions, we first gained a thorough understanding of built-in functions. Similarly, it is pointless to discuss programmer-defined types unless we are first clear on exactly what a built-in type is. You are most familiar with the built-in types `int` and `double`. Each of these types provides

- a type name;
- a set of values of that type; and
- a means of manipulating values of that type.

We can illustrate these three points by considering the `int` type, with which we are already familiar.

- The type name (`int`) gives us a way to declare variables that can take on integer values. It also gives us a way to create headers and prototypes for functions that can take integers as parameters and return them as results.
- The values provided by a type are what our programs store in variables, pass as parameters, and return as results. The values provided by the `int` type, for example, are all of the mathematical integers between a smallest and a largest permissible value.
- Having a collection of values that we can store in variables, pass as parameters, and return as results is not particularly interesting unless we can *do* something with them. Thus, a type must provide a means of manipulating its values. The `int` type, for example, provides operators for doing addition, subtraction, multiplication, and division.

16.4.2 Programmer-Defined Types

Although we were able to write our original implementation of the rod-stacking problem using only `doubles`, we discovered in Section 16.2 that we could make our implementation shorter, simpler, and more efficient by using the `typedef struct` construct to define a new type of compound data.

The new type that we created, `point`, provided the three ingredients that we identified above.

- The type name was `point`.
- The values were pairs of `doubles`, which we took to represent the *x*-*y* coordinates of a point.
- The dot operator gave us a way to extract the individual coordinates from a `point` as well as a way to change the coordinates contained by a particular point.

Do not be misled about how the use of structures simplified the rod-stacking implementation. The advantage of the `point` type was *not* simply that it reduced the number of parameters to some of the functions. If that had been of primary interest, we could have easily defined a structure for each function with enough fields to hold *all* of its parameters. This wouldn't have simplified (or even shortened) the program, however. We would have had to devote a lot of code to assembling the structures before making function calls and disassembling them afterwards.

The real value of the `point` type was that it was naturally matched to the type of program we were writing. Our program manipulated points and radii, which we represented with `points` and `doubles`. It is important to define appropriate types of data when writing a program.

16.4.3 Abstract Datatypes

When we extended our implementation of Newton's method to use complex numbers, we discovered that it was not sufficient merely to define a new `complex` structure containing `real` and `imag` fields; although doing this provided the new type name that we needed, it didn't provide the appropriate set of operations. Because the only available operations on `complex` values were those provided by the dot operator, our initial implementation was so complicated as to be unreadable.

Sometimes a structure, and the simple operations that come with it, are entirely appropriate, as was the case in the rod-stacking implementation. More often, however, the definition of a new type via `typedef struct` must be augmented with an appropriate set of functions to manipulate the values of the new type. In our implementation of Newton's method, we quickly discovered that we needed functions to perform such operations as addition and multiplication.

Our final implementation of Newton's method breaks naturally into three files.

1. The first file, `complex.h`, contains the definition of the new `complex` structure along with the prototypes of seven functions for manipulating `complex` values.

2. The second file, `complex.c`, contains implementations of the seven functions from `complex.h`. Notice that `complex.c` includes `complex.h` in order to have access to the `complex` structure definition.

3. The third file, `newton9.c`, contains the actual implementation of Newton's method. It includes `complex.h` in order to have access to both the `complex` type definition and the prototypes for the seven `complex` functions.

The files `complex.h` and `complex.c` together define and implement what is called an *abstract datatype* (ADT). The two files are concerned only with providing a type name, a set of values, and a set of operations for complex numbers. The `complex` ADT that the two files provide has no particular connection to Newton's method, which is why it is separated out from `newton9.c`. This way, the ADT can be easily reused in any program requiring complex numbers. Abstract datatypes and programmer-defined functions are the two important building blocks out of which large programs are constructed.

The `complex` ADT is deliberately divided into an interface (`complex.h`) and an implementation (`complex.c`). The interface file supplies all of the information (type definition and function prototypes) that is needed by functions that *use* the ADT. The implementation file supplies implementations for the function prototypes of the interface.

The file `newton9.c` uses the `complex` ADT. If you read the implementation of `newton` carefully, you will discover that it makes absolutely no assumptions about how `complex` numbers are represented. It never accesses a field of a `complex` number directly. When it needs to create, read, write, add, subtract, multiply, divide, or find the absolute value of a complex number, it uses the functions provided for that purpose by the ADT interface.

This separation between the abstract datatype and the code that uses it is as deliberate as it is beneficial. The only assumptions that Newton's method makes about the complex number type are that the type name is `complex` and that functions whose prototypes appear in Figure 16.12 are available. Any other version of `complex.c` could be substituted without affecting `newton9.c`, so long as that version of `complex.c` correctly implemented the interface presented in `complex.h`.

An appreciation of the power of abstract datatypes is arguably the most important factor in becoming a truly proficient programmer. It is the key insight behind *object-oriented programming*, about which much has been written, even in the popular press. Although we have only been able to touch on the topic here, be aware that your growth as a programmer will ultimately be tied to mastering ADTs.

16.5 Key Concepts

Programmer-defined types.   C cannot possibly provide all of the functions that any programmer might want. By the same token, it cannot provide all of the types of data that any programmer might want. Just as C provides a means for the programmer to create new functions, it also provides a means for the programmer to define new types of data. By using the `typedef struct` construct, a programmer can define new types of composite data.

Abstract datatypes. A built-in datatype consists of a set of values along with a set of operations to manipulate them. An abstract datatype consists of a set of values—often defined with the `typedef struct` construct—and a set of programmer-defined functions to manipulate those values. The key to defining an abstract datatype is to make sure that only the functions that belong to the datatype are allowed access to the representation of the values.

16.6 Exercises

16.1 Could the design and implementation of `rod7.c` be simplified if the programmer-defined type

```
typedef struct {
 point center;
 double radius;
} rod;
```

were used to represent rods?

16.2 Assess your answer to Exercise 16.1 by modifying `rod7.c` so that the prototype of `threeRod` is

```
point threeRod (rod left, rod right, double rTop);
```

16.3 Repackage `rod7.c` into a multiple-file program consisting of

- a file (`rod.h`) containing the definition of the `point` type and the prototypes for `sq`, `cosine`, `distance`, and `threeRod`;
- a file (`rod.c`) containing the implementations of `sq`, `cosine`, `distance`, and `threeRod`; and
- a file `rod-main.c` containing the `main` function.

16.4 Of what advantage is it to repackage `rod7.c` as in Exercise 16.3?

16.5 Why can't the bisection method be generalized to find complex roots?

16.6 Modify `newton8.c` so that it finds the roots of Equation 16.3.

16.7 Modify `newton9.c` so that it finds the roots of Equation 16.3.

16.8 Which was easier to do, Exercise 16.6 or Exercise 16.7? Explain the difference.

16.9 Modify `newton9.c` so that `newton` takes as parameters both the function whose roots are sought and its first derivative.

16.10 Repackage the multiple-file program that consists of `complex.h`, `complex.c`, and `newton9.c`. Leaving `complex.h` and `complex.c` unchanged, divide `newton9.c`, as modified in Exercise 16.9, into

- a file (`newton.h`) containing the prototype of `newton`;
- a file (`newton.c`) containing the implementation of `newton`; and
- a file (`newton-main.c`) containing the implementations of `f`, `fprime`, and `main`.

16.11 Use `complex.h`, `complex.c`, `newton.h`, and `newton.c` from Exercise 16.10 as components in a new multiple-file program. Your program should prompt the user for a mass m, a drag coefficient β, and a spring constant κ. All three values should be positive, and $\beta^2 - 4m\kappa$ should be negative. If it has been given valid values, your program should compute and display the two roots of Equation 16.1.

16.12 Extend your solution to Exercise 16.11 so that it also prompts for a positive time t in seconds. It should then calculate and display the value of $x(t)$ as defined by Equation 16.2.

16.13 Modify your solution to Exercise 14.18 so that it finds $f'(x)$ for complex values of x.

16.14 Modify your solution to Exercise 16.10 along the lines of Exercises 14.19 and 16.13 so that it finds first derivatives numerically.

16.15 Modify your solution to Exercise 16.14 so that it copes with the problem identified in Exercise 14.16.

16.16 In this chapter we designed and implemented a complex number abstract datatype so that we could more easily write programs to manipulate complex numbers. Suppose that we wanted to write programs to manipulate exact rational numbers such as $\frac{1}{3}$ and $\frac{5}{7}$. Design the functional interface of a rational number ADT that would make this possible.

16.17 Before implementing the rational ADT from Exercise 16.16, you must first pick a representation. Design a new programmer-defined type called `rational` for representing rational numbers.

16.18 Implement the rational number ADT that you designed in Exercise 16.16 using the representation that you designed in Exercise 16.17. Create a header file (`rational.h`) that contains the type definition and the function prototypes, and an implementation file (`rational.c`) that contains the function implementations.

16.19 Write a `main` function to help you test your implementation of the rational number ADT from Exercise 16.18.

16.20 What applications might your rational number ADT from Exercise 16.18 have? Would it make sense to use it to implement a version of Newton's method that finds rational approximations to roots?

17

Heat Transfer in a Rod: Arrays

Imagine that we have a silver rod 10 centimeters long with a constant cross section. The rod is encased in thermal insulation along its entire length so that no heat can enter or leave except at its ends. It is initially at a temperature of $0°C$. At time $t = 0$, heat sources are placed in contact with its left and right ends. The left heat source has a temperature of $100°C$, and the right heat source has a temperature of $50°C$.

The rod begins warming up, more quickly at its left end than at its right, but the temperatures of the heat sources do not change. If we wait long enough, the temperature of the rod will vary linearly from $100°C$ on the left to $50°C$ on the right and will be $75°C$ in the middle. Before that happens, what is the temperature halfway between the rod's ends as a function of time?

The problem of mathematically characterizing the process of heat transfer in solid bodies, of which the problem we have just described is but a simple example, caught the attention of the French mathematician and physicist Joseph Fourier around 1804, when he was 36. He devoted much of the remaining 26 years of his life to studying the physics and mathematics of heat transfer.

Fourier's mathematical techniques were suspect in the eyes of the French scientific establishment. He was also in a delicate political situation because of his activities during the French Revolution and his subsequent associations with Napoleon. As a result, during most of the period between 1802 and 1816 when he was developing his theory of heat, he lived in a kind of exile as a government official in the French provincial cities of Grenoble and Lyons, isolated from the centers of scientific activity in Paris.

The most controversial aspect of Fourier's approach was his use of infinite series of trigonometric functions. Such series, which we know today as Fourier Series, can converge to discontinuous functions and figured large in the solutions to the heat transfer problems that he studied. Two seminal papers, the first in 1807 and the second in 1811, began to turn the tide of mathematical opinion, and he was elected to the French Academy of Sciences in 1817. Fourier's crowning achievement was the publication of his book *The Analytical Theory of Heat* in 1822; he was elevated to the post of permanent secretary of the Academy the same year.

17.1 Modeling Heat Flow

Although the problem that we stated in the introduction is specific concerning the composition of the rod and the initial temperatures, we will develop a more general model. We will assume only that the rod is made of a homogeneous material, has a constant cross section, is thermally insulated, and has a uniform initial temperature. We will then be able to use the resulting general model to solve specific problems.

Our model will use a large number of symbols to stand for the various physical constants that characterize the rod. For convenient reference, Figure 17.1 contains a glossary of the symbols we will be using.

17.1.1 Three Physical Principles

Most of the problems we have considered in this book have been based on familiar physical principles. You probably don't have a similar grasp of the physics behind

Rod Characteristics

L	length (cm)
A	cross-sectional area (cm^2)
T	initial temperature (degrees Celsius)
κ	thermal conductivity (cal/sec/cm/C°)
ρ	density (gm/cm^3)
σ	specific heat (cal/gm/C°)
c^2	thermal diffusivity ($\kappa/(\rho\sigma)$ cm^2/sec)

Heat Sources

T_L	temperature of left source (C°)
T_R	temperature of right source (C°)

Discretization

n	number of segments into which rod is divided
h	L/n, length of each segment (cm)
Δt	discrete simulated time interval (sec)
T_i	temperature at segment boundary $i = 0 \ldots n$ at time t (C°)
T_i'	temperature at segment boundary $i = 0 \ldots n$ at time $t + \Delta t$ (C°)

Figure 17.1   Glossary of constants used in model

heat transfer and temperature change. As we develop our model, we will draw on the following three basic facts.

1. If we leave the rod in contact with the heat sources long enough, it will eventually reach a steady state. At steady state, the temperature along the rod will vary linearly between the temperatures of the two heat sources. This means that the temperature halfway between the two ends will be

$$\frac{1}{2}(T_L + T_R) \text{ degrees Celsius} \tag{17.1}$$

where T_L is the temperature of the left heat source and T_R is the temperature of the right heat source.

2. Once the rod reaches steady state, heat energy will flow from the hotter heat source, through the rod, to the cooler heat source. The amount of heat that will flow from left to right during an interval of Δt seconds will be

$$\frac{\kappa \Delta t A(T_L - T_R)}{L} \text{ calories} \tag{17.2}$$

where L is the length of the rod, A is the cross-sectional area of the rod, and κ is the thermal conductivity of the material from which the rod is made. (This is Fourier's law of conduction.)

3. When heat energy enters the rod, the rod's temperature will rise. If the energy entering is ΔH calories, the rod's temperature will rise by

$$\frac{\Delta H}{AL\rho\sigma} \text{ degrees Celsius} \tag{17.3}$$

where ρ is the density and σ the specific heat of the material from which the rod is made. (ΔH is a signed quantity, so if heat leaves the rod, its temperature will fall.)

17.1.2 Discretizing the Rod

Equations 17.1 and 17.2 describe what will happen once the rod *reaches* steady state. We, however, must model what happens to the rod as it *approaches* steady state. We will deal with this problem by modeling the rod as consisting of n individual segments, each of length $h = L/n$.

Figure 17.2 shows how the rod can be divided into $n = 6$ segments, each of length $L/6$. The six segments form seven boundaries with one another and the ends of the rod. The temperatures at those seven boundaries are labeled T_0 through T_6. T_0 (at the left end) is the same as T_L, while T_6 (at the right end) is the same as T_R.

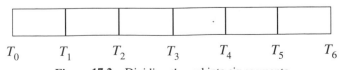

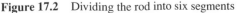

Figure 17.2   Dividing the rod into six segments

We will assume that each of the six segments is at steady state. Thus, we are assuming that the temperature in the first segment varies linearly between T_0 and T_1, that the temperature in the second segment varies linearly between T_1 and T_2, and so forth. Although this is an unrealistic assumption, it is not nearly as unrealistic as assuming that the entire rod is at steady state. In fact, by dividing the rod into a large enough number of segments, we can minimize the errors introduced by this assumption.

Our strategy for developing a model will be as follows. Initially, $T_0 = T_L$, $T_n = T_R$, and all of the other $T_i = T$. With each instant of time, heat will flow among the segments and the interior boundary temperatures will change. Because we are assuming that each of the segments is at steady state, we will be able to use Equations 17.1–17.3 to model how the heat flows between the segments and how the temperatures at the boundaries change.

17.1.3 Discretizing Time

Suppose that the temperatures at the $n + 1$ segment boundaries at time t are T_0 through T_n. Using our assumption that the individual segments are at steady state, what will be the temperatures T_0' through T_n' at the segment boundaries at time $t + \Delta t$, for some small Δt?

Because they are in contact with the heat sources, the temperatures T_0 and T_n at the ends will remain unchanged. To help determine the temperatures at the interior boundaries, we will draw on Figure 17.3, which provides a detailed view of two contiguous segments of the rod. The temperatures at each of the three boundaries are labeled T_{i-1}, T_i, and T_{i+1}. We have divided both segments in half with dashed lines and labeled the resulting half-segments for reference. Each of the segments is of length $h = L/n$, and each of the half-segments is of length $h/2$.

We will determine how much heat flows into the two half-segments b and c during the interval Δt. This will give us a way of determining how the average temperature changes within those two half-segments and consequently at their shared boundary. Heat will flow across a into (or out of) b, and across d into (or out of) c.

From Equation 17.1, we know that the temperature at the boundary between a and b is $\frac{1}{2}(T_{i-1} + T_i)$. Equation 17.2 tells us that the amount of heat that will flow across a into b during the interval Δt will be

$$\Delta H_b = \frac{\kappa \Delta t \, A(T_{i-1} - \frac{1}{2}(T_{i-1} + T_i)}{h/2} \tag{17.4}$$

$$= \frac{\kappa \Delta t \, A(T_{i-1} - T_i)}{h} \tag{17.5}$$

Similarly, the heat flowing across d into c will be

$$\Delta H_c = \frac{\kappa \Delta t \, A(T_{i+1} - T_i)}{h} \tag{17.6}$$

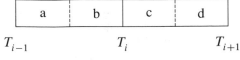

$$T_{i-1} \qquad\qquad T_i \qquad\qquad T_{i+1}$$

Figure 17.3   Two adjacent segments

Putting these two results together, the total heat change in the half-segments b and c will be

$$\Delta H = \Delta H_b + \Delta H_c = \frac{\kappa \Delta t\, A(T_{i-1} + T_{i+1} - 2T_i)}{h} \qquad (17.7)$$

Because the combined length of the two half-segments b and c is h, Equation 17.3 tells us that the change in the average temperature of these two segments will be

$$\Delta T = \frac{\Delta H}{A h \rho \sigma} \qquad (17.8)$$

$$= \frac{\kappa \Delta t\,(T_{i-1} + T_{i+1} - 2T_i)}{\rho \sigma h^2} \qquad (17.9)$$

Finally, T_i' will be $T_i + \Delta T$.

Putting everything together, we now know that the temperatures at the segment boundaries at time $t + \Delta t$ are approximated by

$$T_0' = T_0 \qquad (17.10)$$
$$T_n' = T_n \qquad (17.11)$$
$$T_i' = T_i + \frac{c^2 \Delta t}{h^2}\left(T_{i-1} + T_{i+1} - 2T_i\right) \quad (0 < i < n) \qquad (17.12)$$

where $c^2 = \kappa/\rho\sigma$ is called the *thermal diffusivity* of the rod material.

17.2 A Finite-Element Method

The model that we have just developed is called a *finite-element* model. We developed the model by dividing the rod into a finite number (n) of elements (the segments) and then applying physical analysis to the segments. We facilitated this analysis by making the simplifying assumption that each segment was at steady state. This would have been a grossly inaccurate assumption to make for the entire rod, but it introduces only a small amount of error when the individual segments are small. As the number of segments increases, the amount of error decreases. In the limiting case, when we have an infinite number of segments, we can reach an exact solution.

We will stop short of using an infinite number of segments. Equations 17.10–17.12 give us a way to approximate the temperatures at the segment boundaries at time $t + \Delta t$ so long as we know their approximate temperatures at time t. Because we know the temperatures at each boundary at time zero, we can determine temperatures arbitrarily far into the future by repeatedly applying Equations 17.10–17.12.

To apply these equations, we need to know the length of the rod and the initial temperatures (which are given), the thermal diffusivity of the rod material (which we can look up), the number of segments into which the rod should be divided, and the length of the time step Δt.

We can simplify our computational task if we always choose Δt such that

$$\Delta t = \frac{h^2}{2c^2} \tag{17.13}$$

in which case Equation 17.12 reduces to

$$T_i' = \frac{T_{i-1} + T_{i+1}}{2} \quad (0 < i < n) \tag{17.14}$$

Table 17.1 shows how we can determine the temperatures at the segment boundaries over multiple time steps. We are modeling a 10-centimeter silver bar whose thermal diffusivity is 1.752 cm^2/sec. We have divided it into ten segments and show the results of simulating ten time steps, each of length 0.571 seconds as derived from Equation 17.13. Initially, the bar is at 0°C while the left and right heat sources are at 100°C and 50°C, respectively. After 5.71 seconds, the center of the rod is at 16.4°C.

Table 17.1   Ten repetitions of finite-element method

0	1	2	3	4	5	6	7	8	9	10
100.0	0.0	0.0	0.0	0.0	0.0	0.0	0.0	0.0	0.0	50.0
100.0	50.0	0.0	0.0	0.0	0.0	0.0	0.0	0.0	25.0	50.0
100.0	50.0	25.0	0.0	0.0	0.0	0.0	0.0	12.5	25.0	50.0
100.0	62.5	25.0	12.5	0.0	0.0	0.0	6.3	12.5	31.3	50.0
100.0	62.5	37.5	12.5	6.3	0.0	3.1	6.3	18.8	31.3	50.0
100.0	68.8	37.5	21.9	6.3	4.7	3.12	10.9	18.8	34.4	50.0
100.0	68.8	45.3	21.9	13.3	4.7	7.8	10.9	22.7	34.4	50.0
100.0	72.7	45.3	29.3	13.3	10.6	7.8	15.2	22.7	36.3	50.0
100.0	72.7	51.0	29.3	19.9	10.6	12.9	15.2	25.8	36.3	50.0
100.0	75.5	51.0	35.5	19.9	16.4	12.9	19.3	25.8	37.9	50.0
100.0	75.5	55.5	35.5	25.9	16.4	17.9	19.3	28.6	37.9	50.0

17.3 Implementation

We will develop our solution to the heat transfer problem by crafting a sequence of six increasingly more sophisticated implementations. Each of our programs will consist of the two functions `main` and `simulate`. The `main` function is a driver that prompts for the values of the physical parameters, passes them to `simulate`, and displays the result. An example of the kind of interaction that we envision is in Figure 17.4.

```
Enter thermal diffusivity (cm^2/sec): 1.752
Enter length of rod (cm): 10
Enter temperature at left end (degrees C): 100
Enter temperature of rod (degrees C): 0
Enter temperature at right end (degrees C): 50
Enter length of simulation (sec): 5
Enter number of segments: 5
Temperature at center after 5 seconds is 34.8076 degrees C
```

Figure 17.4   Example interaction with heat transfer program developed in this chapter

The focus of our attention in this section will be the function `simulate`, which takes the following six parameters.

- The thermal `diffusivity` $\frac{\kappa}{\rho\sigma}$, derived from the thermal conductivity κ, the density ρ, and the specific heat σ of the material from which the rod is made.
- The `length` of the rod in centimeters.
- The initial temperatures in degrees Celsius of the `left` heat source, `right` heat source, and `rod`.
- The `duration` of the simulation in seconds.

Given these parameters, `simulate` will return the approximate temperature in the middle of the rod `duration` seconds after the heat sources are applied to its ends.

17.3.1 Initial Implementation

Our first version of `simulate`, which appears in Figure 17.5, uses the same six segments to model the rod that we used in Section 17.1.2. In addition to its parameters, `simulate` uses 13 local variables.

- `delta` is Δt from Equation 17.13, calculated on line 9 from the diffusivity constant and the length of each segment.
- `cur0` through `cur6` keep track of the simulated temperatures at the seven boundaries of the six segments, which are numbered as in Figure 17.2. They are initialized on lines 11–17: `cur0` from the temperature of the left heat source, `cur1` through `cur5` from the initial temperature of the rod, and `cur6` from the temperature of the right heat source.
- `nxt1` through `nxt5` are used to temporarily record the simulated temperatures of the five interior segment boundaries after each tick of simulated time.

The major part of `simulate` is a `while` loop. Each repetition of the loop corresponds to one tick of simulated time, with each tick representing the passage of `delta` seconds.

During each repetition, the new temperatures at the five interior boundaries are calculated using Equation 17.14 and stored in the variables `nxt1` through `nxt5`. After they have been calculated, the temperatures are assigned back into `cur1`

```
 1    double simulate (double diffusivity, double length,
 2                          double left, double right, double rod,
 3                          double duration)
 4    {
 5       double delta;
 6       double cur0, cur1, cur2, cur3, cur4, cur5, cur6;
 7       double nxt1, nxt2, nxt3, nxt4, nxt5;
 8
 9       delta = (length*length/36) / (2*diffusivity);
10
11       cur0 = left;
12       cur1 = rod;
13       cur2 = rod;
14       cur3 = rod;
15       cur4 = rod;
16       cur5 = rod;
17       cur6 = right;
18
19       while (duration > 0) {
20
21          nxt1 = (cur0 + cur2) / 2;
22          nxt2 = (cur1 + cur3) / 2;
23          nxt3 = (cur2 + cur4) / 2;
24          nxt4 = (cur3 + cur5) / 2;
25          nxt5 = (cur4 + cur6) / 2;
26
27          cur1 = nxt1;
28          cur2 = nxt2;
29          cur3 = nxt3;
30          cur4 = nxt4;
31          cur5 = nxt5;
32
33          duration = duration - delta;
34       }
35
36       return(cur3);
37    }
```

Figure 17.5   First implementation of `simulate` (`heat1.c`)

through `cur5`. The values of `cur0` and `cur6` never change, as they represent the temperatures of the heat sources.

The total amount of time covered by the simulation is given by the parameter `duration`. At the end of each repetition, we reduce the value of `duration` by `delta`. When `duration` reaches zero the loop terminates and the value of `cur3`, which represents the simulated temperature at the middle of the rod, is returned.

There are three major problems with `simulate` that we will address in succeeding versions.

1. Because it uses only six segments to model the rod, `simulate` will almost never yield an adequate approximation to the underlying physical process. We could always add more `cur` and `nxt` variables to improve the granularity of the simulation, but to do this for hundreds or thousands of segments would require a huge number of variable declarations.

2. Even if we were willing to declare thousands of local variables, a loop body with thousands of assignment statements would then be required to carry out the simulation.

3. Even if we were willing to create the huge program required to simulate thousands of rod segments, there would be no way for the user to control the number of segments being simulated. We would have to write a new program each time we wanted to change the size of the simulation.

17.3.2 A Structure-Based Implementation

The seven variables `cur0` through `cur6` from Figure 17.5 each contain temperatures from different points in the rod. Our program might be simpler if the seven values they contain were appropriately packaged into a single compound value. Similar reasoning applies to the five variables `nxt1` through `nxt5`.

We could, for example, define a `segments` type as in lines 1–9 of Figure 17.6. Each value of type `segment` is a structure containing seven temperatures in fields named `seg0` through `seg6`. With the help of this new type we could implement `simulate` as in Figure 17.6.

This new version of `simulate` contains only three local variables, two of which (`cur` and `nxt`) are of the new type `segments`. Unfortunately, this does not represent an improvement over the version of `simulate` in Figure 17.5. Even though we will need only three local variables no matter how large the simulation becomes, we have done nothing but shift the programming burden elsewhere. Increasing the size of the simulation will now entail adding more fields to the definition of `segments`.

To make matters worse, the variables names have become longer without contributing any increased clarity. Everywhere we previously used `cur0`, for example, we must now use `cur.seg0`. This would be acceptable only if there were an improvement in some other aspect of the implementation.

The advantage of using a structure is that it makes it possible to manipulate multiple pieces of data as a single unit. There is no place in `simulate` where we can take advantage of this capability, however, so there is no compelling reason to use structures.

17.3.3 An Array-Based Implementation

In addition to structures, C provides *arrays* as a means of creating compound data. Like a structure, an array is a collection of one or more values that, although packaged as a single unit, can be extracted or modified individually. There are, however, some important differences.

```
1    typedef struct {
2      double seg0;
3      double seg1;
4      double seg2;
5      double seg3;
6      double seg4;
7      double seg5;
8      double seg6;
9    } segments;
10
11
12   double simulate (double diffusivity, double length,
13                    double left, double right, double rod,
14                    double duration)
15   {
16     double delta;
17     segments cur;
18     segments nxt;
19
20     delta = (length*length/36) / (2*diffusivity);
21
22     cur.seg0 = left;
23     cur.seg1 = rod;
24     cur.seg2 = rod;
25     cur.seg3 = rod;
26     cur.seg4 = rod;
27     cur.seg5 = rod;
28     cur.seg6 = right;
29
30     while (duration > 0) {
31
32       nxt.seg1 = (cur.seg0 + cur.seg2) / 2;
33       nxt.seg2 = (cur.seg1 + cur.seg3) / 2;
34       nxt.seg3 = (cur.seg2 + cur.seg4) / 2;
35       nxt.seg4 = (cur.seg3 + cur.seg5) / 2;
36       nxt.seg5 = (cur.seg4 + cur.seg6) / 2;
37
38       cur.seg1 = nxt.seg1;
39       cur.seg2 = nxt.seg2;
40       cur.seg3 = nxt.seg3;
41       cur.seg4 = nxt.seg4;
42       cur.seg5 = nxt.seg5;
43
44       duration = duration - delta;
45     }
46
47     return(cur.seg3);
48   }
```

Figure 17.6   Structure-based implementation of simulate (heat2.c)

- A structure is made up of one or more fields. An array is made up of one or more *elements*. (This is only a difference of terminology.)
- A field of a structure is accessed via a symbolic field name. An element of an array is accessed via an integer index.
- A structure can be assigned to a variable, passed as a parameter, and returned as a result. Although it is possible to do all three of these with an array, some subtle issues are involved that we will not discuss in this chapter.

(Arrays are similar to *Mathematica* lists, which are also made up of one or more elements that can be accessed with an integer index.)

Figure 17.7 contains an implementation of simulate that uses arrays to keep track of the temperatures at the segment boundaries. It illustrates how to declare arrays and how to access their elements.

An array declaration appears on line 6 of Figure 17.7. It consists of the type shared by its elements (double), the name of the array (cur), and the number of elements in the array (7) enclosed in square brackets. Similarly, line 7 is the declaration of an array nxt.

Each of the elements of an array has an integer index. The index of the first element is zero, and the indexes of the succeeding elements continue sequentially. Because the first index is zero (*not* 1), the index of the last element is one less than the size of the array (*not* the size of the array). It is a common C programming error to treat the last element as if its index were the same as the size of the array.

An element of an array can be accessed by specifying its index in square brackets following the name of the array. The seven elements of cur, for example, are cur[0], cur[1], cur[2], cur[3], cur[4], cur[5], and cur[6].

Just like a field of a structure, an element of an array behaves as a variable. On lines 11–17 of Figure 17.7, the seven elements of cur are initialized individually. The body of the while loop contains numerous references to the elements of nxt and cur. On line 21, for example, we average the values of cur[0] and cur[2] and assign the result to nxt[1].

Figure 17.8 shows the state of the cur array after it has been initialized as well as after the first three repetitions of the loop. The left and right heat sources are at 100°C, and the rod is initially at 0°C.

If you write a program that tries to access a nonexistent field of a structure, the compiler will notify you of your error. If your program tries to access a nonexistent element of an array, however, the compiler will not usually notice. Thus, it is important to be especially careful when writing code that indexes an array element. The consequences of making an incorrect array reference depend on the program, the compiler, and the computer, but they can range all the way up to a catastrophic program failure.

We have not yet accomplished much by using arrays in simulate. Our current approach to implementing simulate will not scale up to a large number of segments. Although we have cut down on the number of variable declarations required, we have done nothing to simplify the bulk of the function. In the next section, however, we will treat array indexes in a more sophisticated way that will produce a significantly improved version of simulate.

```
1    double simulate (double diffusivity, double length,
2                     double left, double right, double rod,
3                     double duration)
4    {
5      double delta;
6      double cur[7];
7      double nxt[7];
8
9      delta = (length*length/36) / (2*diffusivity);
10
11     cur[0] = left;
12     cur[1] = rod;
13     cur[2] = rod;
14     cur[3] = rod;
15     cur[4] = rod;
16     cur[5] = rod;
17     cur[6] = right;
18
19     while (duration > 0) {
20
21       nxt[1] = (cur[0] + cur[2]) / 2;
22       nxt[2] = (cur[1] + cur[3]) / 2;
23       nxt[3] = (cur[2] + cur[4]) / 2;
24       nxt[4] = (cur[3] + cur[5]) / 2;
25       nxt[5] = (cur[4] + cur[6]) / 2;
26
27       cur[1] = nxt[1];
28       cur[2] = nxt[2];
29       cur[3] = nxt[3];
30       cur[4] = nxt[4];
31       cur[5] = nxt[5];
32
33       duration = duration - delta;
34     }
35
36     return(cur[3]);
37   }
```

Figure 17.7 Array-based implementation of simulate (heat3.c)

17.3.4 An Improved Array-Based Implementation

Array indexes do not have to be constants; they can be arbitrary integer expressions. This means that they can be *computed* as a program runs. Our fourth implementation of simulate in Figure 17.9 takes advantage of this fact by replacing three sequences of assignment statements with three for loops.

- We have replaced the assignments to cur[1] through cur[5] on lines 12–16 of Figure 17.7 with the loop on lines 12–14 of Figure 17.9. The loop uses the

	0	1	2	3	4	5	6	Indexes
	100	0	0	0	0	0	100	After initialization
	100	50	0	0	0	50	100	After first repetition
	100	50	25	0	25	50	100	After second repetition
	100	62.5	25	25	25	62.5	100	After third repetition

Figure 17.8   Four snapshots of the cur array

variable i to step through the integers 1 through 5. Each repetition of the loop thus assigns rod to a different element of cur. Notice that we still deal separately with cur[0] and cur[6].

- We have replaced the assignments to nxt[1] through nxt[5] on lines 21–25 of Figure 17.7 with the loop on lines 19–21 of Figure 17.9. The body of the loop consists of an assignment statement

```
nxt[i] = (cur[i-1] + cur[i+1]) / 2;
```

containing three different array references. All three are made relative to i, which counts from 1 to 5 as the loop repeats. The first time through the loop, when i is 1, the assignment statement is equivalent to

```
nxt[1] = (cur[0] + cur[2]) / 2;
```

The last time through the loop, when i is 5, the assignment statement is equivalent to

```
nxt[5] = (cur[4] + cur[6]) / 2;
```

- We have replaced the assignments to cur[1] through cur[5] on lines 27–31 of Figure 17.7 with the loop on lines 23–25 of Figure 17.9.

Arrays are commonly used in tandem with for loops in C programs. The loop can count through the indexes of the array, making it possible for the loop body to perform an operation on every element of the array. This style of programming doesn't work for structures because there is no way to count through the names of the fields that make up a structure.

Our use of the three loops to manipulate the cur and nxt arrays is a significant breakthrough in our evolving implementation of simulate. It will now be much

```
1    double simulate (double diffusivity, double length,
2                         double left, double right, double rod,
3                         double duration)
4    {
5      double delta;
6      double cur[7], nxt[7];
7      int i;
8
9      delta = (length*length/36) / (2*diffusivity);
10
11     cur[0] = left;
12     for (i = 1; i < 6; i = i+1) {
13       cur[i] = rod;
14     }
15     cur[6] = right;
16
17     while (duration > 0) {
18
19       for (i = 1; i < 6; i = i+1) {
20         nxt[i] = (cur[i-1] + cur[i+1]) / 2;
21       }
22
23       for (i = 1; i < 6; i = i+1) {
24         cur[i] = nxt[i];
25       }
26
27       duration = duration - delta;
28     }
29
30     return(cur[3]);
31   }
```

Figure 17.9   Improved array-based implementation of `simulate` (`heat4.c`)

easier to modify our program so that it works for any number of segments. We will not have to add more statements, as was the case before. Instead, we will only have to change the sizes of the two arrays and modify all of the places where we make assumptions about their sizes. These places include

- The two array declarations on line 6. (Notice that we have modified the program so that it illustrates how to declare two arrays at once.)
- The constant 36 on line 9, which is the square of the number of segments.
- The constant 6 on lines 12, 15, 19, and 23, which is one less than the size of the array.
- The constant 3 on line 30, which is the index of the middle element of `cur`.

17.3.5 Exploiting a Symbolic Constant

Although it would now be much easier to change the number of segments being simulated, it is not as easy at it could be. We have just identified eight separate places in the program where constants appear that depend on the number of segments. If we were to forget to modify even one of them, the resulting program would be inconsistent and incorrect.

The version of `simulate` in Figure 17.10 addresses this problem by defining a symbolic constant to represent the number of segments. Line 1 of Figure 17.10 contains the *definition directive*

```
1    #define segs 6
2
3    double simulate (double diffusivity, double length,
4                     double left, double right, double rod,
5                     double duration)
6    {
7      double delta;
8      double cur[segs+1], nxt[segs+1];
9      int i;
10
11     delta = (length*length/(segs*segs)) / (2*diffusivity);
12
13     cur[0] = left;
14     for (i = 1; i < segs; i = i+1) {
15       cur[i] = rod;
16     }
17     cur[segs] = right;
18
19     while (duration > 0) {
20
21       for (i = 1; i < segs; i = i+1) {
22         nxt[i] = (cur[i-1] + cur[i+1]) / 2;
23       }
24
25       for (i = 1; i < segs; i = i+1) {
26         cur[i] = nxt[i];
27       }
28
29       duration = duration - delta;
30     }
31
32     return(cur[segs/2]);
33   }
```

Figure 17.10 Array-based implementation of `simulate` with symbolic constant (`heat5.c`)

```
#define segs 6
```

This means that the symbol `segs` is to be treated throughout the file as an abbreviation for 6. It is important to realize that `segs` is not a variable, which means that it cannot be the target of an assignment statement. The symbol `segs` is interchangeable with the constant 6.

The size of an array that appears in its declaration must be a constant. It cannot depend on the value of a parameter or variable. We have defined `segs` to be an abbreviation for 6 so that, on line 8, we can declare the sizes of `cur` and `nxt` to be `segs+1`. Because this expression involves only constants, the compiler can determine that its value is 7.

We have used the symbolic constant `segs` throughout the program in place of the numerical constants that previously appeared. We use `segs*segs` in place of 36 on line 11; `segs` in place of 6 on lines 14, 17, 21, and 25; and `segs/2` in place of 3 on line 32. As a result, if we now wish to change the number of segments being simulated, we need only change the definition of `segs` on line 1.

Whenever you write a program that uses an array, you should use a symbolic constant to represent the size of the array, and you should use this constant to control the loops that manipulate the array. This makes changing the size of the array a matter of changing a single constant definition.

17.3.6 The Final Version

Although we now have an implementation of `simulate` that is much improved from the original, we are still one step away from a final version. The number of segments to use in the simulation is coded directly into the program. If we wish to change it, we must modify and recompile the program. It would be better if the user were able to input the number of segments along with all of the physical constants.

Unfortunately, this is at odds with the capabilities of C. We have already noted that the size of an array must be a constant known to the compiler; it cannot be read in as the program is running. Our final version of `simulate`, in Figure 17.11, shows how we can at least partially get around this problem.

We have made three changes to the previous version of `simulate`. Everything else remains unchanged.

- We replaced the definition of `segs` with a definition of `MAXSIZE` (line 1).
- We made `segs` the seventh parameter to `simulate` (line 5), the value of which, like those of the other parameters, will be passed from `main`.
- We changed the declarations of `cur` and `nxt` so that they are of size `MAXSIZE` instead of size `segs+1` (line 8).

While the arrays `cur` and `nxt` both have size `MAXSIZE`, the remainder of the function is still written as if the two arrays had size `segs+1`. There is no problem with this, however, so long as `segs` is less than 1001. If `segs` is 100, for example, the final 900 elements of the two arrays will be unused but the program will still work correctly. If `segs` is 1000, both arrays will be completely used. If `segs` is

```
1    #define MAXSIZE 1001
2
3    double simulate (double diffusivity, double length,
4                         double left, double right, double rod,
5                         double duration, int segs)
6    {
7      double delta;
8      double cur[MAXSIZE], nxt[MAXSIZE];
9      int i;
10
11     delta = (length*length/(segs*segs)) / (2*diffusivity);
12
13     cur[0] = left;
14     for (i = 1; i < segs; i = i+1) {
15       cur[i] = rod;
16     }
17     cur[segs] = right;
18
19     while (duration > 0) {
20
21       for (i = 1; i < segs; i = i+1) {
22         nxt[i] = (cur[i-1] + cur[i+1]) / 2;
23       }
24
25       for (i = 1; i < segs; i = i+1) {
26         cur[i] = nxt[i];
27       }
28
29       duration = duration - delta;
30     }
31
32     return(cur[segs/2]);
33   }
```

Figure 17.11 Implementation of `simulate` with number of segments as a parameter (continued in Figure 17.12)

any larger, however, the capacity of the arrays will be exceeded and the program will fail.

The final version of `simulate` correctly implements our finite-element model of the heat transfer problem so long as 1000 or fewer segments are required. If more segments are desired, the value of `MAXSIZE` must be changed and the program recompiled.

If we had made the value of `MAXSIZE` larger, say 10,001 or 100,001, our implementation would be even more flexible. There is a tradeoff between flexibility and program size at work here, however. Even if most of the `cur` and `nxt` arrays remain unused, the C compiler must still allocate memory to hold them. If the arrays

```
34    #include <stdio.h>
35
36    void main (void)
37    {
38      double C, length, left, right, rod, time, temp;
39      int segs;
40
41      printf("Enter thermal diffusivity (cm^2/sec): ");
42      scanf("%lf", &C);
43
44      printf("Enter length of rod (cm): ");
45      scanf("%lf", &length);
46
47      printf("Enter temperature at left end (degrees C): ");
48      scanf("%lf", &left);
49
50      printf("Enter temperature of rod (degrees C): ");
51      scanf("%lf", &rod);
52
53      printf("Enter temperature at right end (degrees C): ");
54      scanf("%lf", &right);
55
56      printf("Enter length of simulation (sec): ");
57      scanf("%lf", &time);
58
59      printf("Enter number of segments: ");
60      scanf("%d", &segs);
61
62      temp = simulate(C, length, left, right, rod, time, segs);
63      printf("Temperature at center after %g seconds
64              is %g degrees C\n", time, temp);
65    }
```

Figure 17.12   Driver for `simulate` (`heat6.c`)

become too large, the program will suffer degraded performance and will eventually fail as it becomes too large to fit into the computer's memory. The best policy for declaring arrays is to make them as large as will typically be required, but no larger.

17.4 Assessment

The heat transfer problem that we have studied in this chapter can be modeled using a second-order partial differential equation. Using the advanced mathematical techniques pioneered by Fourier, the partial differential equation can also be solved analytically, producing a function u that gives the temperature x centimeters from the left end of the rod t seconds after the heat sources are applied. Notice that u depends on the convergent sum of an infinite series.

$$u(x, t) = T_L + \frac{x(T_R - T_L)}{L} +$$

$$\frac{2}{\pi} \sum_{n=1}^{\infty} \frac{((-1)^n (T_R - T) - T_L + T) \sin\left(\frac{n\pi x}{L}\right) e^{-\frac{c^2 n^2 \pi^2 t}{L^2}}}{n} \quad (17.15)$$

We do not expect that you know what a partial differential equation is, let alone how to solve one analytically. We have presented Equation 17.15 for two reasons: It gives us a way to judge how well our implementation of `simulate` works, and it helps to illustrate the power and utility of finite-element models.

17.4.1 Convergence

Let's return to the problem with which we began this chapter. We have a 10-centimeter silver rod that has a uniform temperature of $0°C$. At time zero we put a $100°C$ heat source at its left end and a $50°C$ heat source at its right end. Given that the thermal diffusivity of silver is approximately 1.752 cm^2/sec, what will be the temperature at different points of the rod after 5 seconds?

Table 17.2a gives the temperatures at the center of the rod calculated by our final version of `simulate` for different numbers of segments. For comparison, Equation 17.15 gives an answer, rounded to six digits, of $34.7893°C$. Even with only ten segments our finite-element approach works well, and with 640 segments it calculates the temperature exactly.

All other things being equal, the use of more segments will lead to more accurate results. In constructing our finite-element model, we assumed that each segment is always at steady state. The error introduced by this assumption is smaller when the segments are smaller.

Table 17.2 Relationship of segments to convergence

Segments	Temperature	Segments	Temperature
10	35.6609	10	43.0611
20	35.2192	20	46.4833
40	34.8666	40	48.1831
80	34.8404	80	49.0842
160	34.8030	160	49.5417
320	34.7897	320	49.7708
640	34.7893	640	49.8854

(a) Center of rod	(b) .001 cm from right end

A second reason why an increased number of segments produces greater accuracy lies in our use of Equation 17.13 to fix the length of each simulated time step at $\Delta t = \frac{h^2}{2c^2}$. Recall that h is the length of each segment and c^2 is the thermal diffusivity of the rod. This simplifies the calculations that we make in `simulate`, but it also introduces errors.

With ten segments, Δt for our silver rod is approximately .29 seconds; with 100 segments, it is .0029 seconds. Our program simulates the passage of time in increments of Δt. With ten segments, 5.22 seconds of simulated time have actually passed by the time the program reports its results. With 100 segments, only 5.0025 simulated seconds have accumulated.

Table 17.2b gives the temperatures .001 centimeter from the right end of the rod calculated by a slightly modified version of `simulate` for different numbers of segments. Equation 17.15 reports a temperature of 49.6986 degrees. In this case, the answer obtained with ten segments is fairly far off, and even with 640 segments the answer is not exact.

There are two reasons for this discrepancy. We modified `simulate` to report the temperature of the interior segment boundary that is closest to the point .001 centimeter from the right end of the rod. When we divide a 10-centimeter rod into ten segments, each segment is 1 centimeter long. As a result, we are actually reporting the simulated temperature of a point 1 centimeter from the end of the rod. Even with 640 segments, we are reporting the simulated temperature of a point .016 centimeter from the end of the rod. We would need to use at least 10,000 segments to have a resolution of .001 centimeter.

Even with many more segments, however, our results will always be more accurate near the center of the rod than near the ends. The finite-element method is most accurate in those portions of the rod where the temperatures are changing slowly. The temperatures in the center of the rod change slowly, while those near the ends change rapidly.

17.4.2 Static Arrays

The generality of our final implementation is somewhat compromised because it contains an arbitrary upper bound on the number of rod segments that can be treated. The sizes of the arrays `cur` and `nxt`, which are used to keep track of the temperatures at the segment boundaries, can each contain at most 1001 temperatures. This means that our program cannot model rods that are divided into more than 1000 segments. If we want to be able to deal with more segments, we must modify and recompile the program.

Limitations of this kind are common with programs written in languages such as C. It is possible to write array-manipulation programs so that the only limits on the sizes of arrays are those imposed by the size of the computer's memory. Such programs, however, use pointers and dynamic memory allocation, both of which are beyond the scope of this book.

17.4.3 The Finite-Element Method

We modeled heat transfer in a rod by dividing it into small segments and calculating the temperature changes in the individual segments. We did this by assuming that each segment would be in steady state at any given point in time. This allowed us to apply to each segment the physics of heat transfer that hold at steady state. While this steady-state assumption would have been completely unreasonable had we applied it to the entire rod, it introduced only small errors when applied to each segment. We were able to manage the magnitude of the error by increasing the number of segments.

The finite-element method is broadly applicable to problems throughout science and engineering. To apply the method, we

1. Divide a problem involving a single large region into a large number of interrelated subproblems, each involving a much smaller region.
2. Make simplifying assumptions about the smaller regions that would be unreasonable for the large region.
3. Exploit these assumptions to create a simpler physical model for the smaller regions than would be possible for the large one.
4. Write a computer program to solve the small problems to obtain an approximate solution for the original problem.

Equation 17.15 illustrates that it is possible to find an analytical solution to the heat transfer problem by setting up and solving a partial differential equation. This does not mean, however, that the finite-element method is nothing more than a way for the less mathematically sophisticated to obtain approximate solutions. First, Equation 17.15 depends on the sum of an infinite series that must itself be approximated to obtain a numerical solution. Second, while the heat transfer problem that we studied was simple enough to admit an analytical solution, it is not hard to find problems that do not.

We could complicate the problem that we studied, for example, by letting the radius of the rod vary as a function of distance from the left end. This would greatly complicate the problem of finding an analytical solution, but it would require only a small change to the finite-element model. By assuming that the radius of the rod in the region of each segment boundary is constant—which it very nearly will be if the segments are small—we can apply exactly the same physical arguments we used before. The only difference would be that the program would have to calculate which radius to use in each boundary temperature calculation. We will explore similar variations on our problem in the exercises.

While we have applied the finite-element method only in a single dimension, it extends naturally to two or more dimensions. We could, for example, use it to study heat transfer in an arbitrary three-dimensional solid. Combined with a ready source of computing power, the finite-element method makes it possible to solve problems that will not yield to traditional mathematical techniques.

17.5 Key Concepts

Finite-element method.   The finite-element method is a powerful and widely used technique for solving certain kinds of problems that cannot be solved analytically. It entails dividing a large physical problem into a collection of much smaller problems whose solutions can be approximated by making simplifying assumptions that would be invalid for the original problem.

Arrays.   C arrays provide a way of packaging together an arbitrary number of variables of the same type. Each element of an array behaves as an ordinary variable, and is specified by indexing the array with a non-negative integer. Because of this, arrays are often manipulated inside of loops that count through the valid indexes.

Symbolic constants.   Symbolic constants are often used to declare the sizes of arrays and to control the loops that manipulate those arrays. In this way, the size of an array and everything that depends on it can be changed by modifying the declaration of the constant.

17.6 Exercises

17.1   Explain why, in Table 17.1, the temperature at each boundary changes only once every two time intervals. Will this behavior hold no matter how the rod is segmented?

17.2   Run the version of `simulate` from Figure 17.11 using values for `seg` greater than 1000. The range of behaviors that you will observe is typical of programs that attempt to access elements beyond the end of an array.

17.3   Do timing experiments to determine how the amount of time required for `simulate` to run depends on the value of its `segs` parameter.

17.4   Modify `heat6.c` so that the user can specify for what point on the rod the final temperature should be reported.

17.5   Modify `heat6.c` so that it reports the final temperatures at ten evenly separated points along the rod.

17.6   Modify `heat6.c` so that it reports the amount of time it takes for the temperature at the center of the rod to come within $1°C$ of the average of the temperatures of the two ends.

17.7   Modify `heat6.c` so that it reports the amount of time it takes for the temperature at each segment boundary of the rod to come within $1°C$ of the temperature it would have at steady state.

17.8 The size of `delta` in `heat6.c` is determined using Equation 17.13. This leads to large values for `delta` for smaller numbers of segments, which can contribute error to the simulation, as explained in Section 17.4.1. Modify the method from Section 17.2 along with `heat6.c` so that the user can specify the value of `delta`.

17.9 Assess your modified version of the heat transfer program from Exercise 17.8. How does it compare with `heat6.c` for ten segments when the user specifies a value for `delta` that evenly divides the interval of time being simulated? What happens when the user specifies a value for `delta` that is larger than the one used by `heat6.c`?

17.10 The implementation of `simulate` in Figure 17.11 uses the loop on lines 21–23 to store the temperatures of the rod segments for the next time interval into `nxt`, and the loop on lines 25–27 to move the temperatures from `nxt` back into `cur`. Modify `simulate` so that the newly calculated temperatures are stored immediately back into `cur`.

17.11 Does your modified version of `simulate` from Exercise 17.10 calculate different results than the original? Explain your observations.

17.12 The finite-element model for the heat transfer problem summarized by Equations 17.10–17.12 assumes that the thermal diffusivity of the rod is the same at each point of the rod. Develop a model in which the diffusivity varies linearly from $\frac{1}{2}c^2$ on the left end to c^2 in the middle to $2c^2$ on the right end. The resulting model will barely differ from the original. The key idea is to treat the diffusivity in the region of any particular rod boundary as a constant.

17.13 Modify `heat6.c` so that it implements your model from Exercise 17.12. The user should supply the thermal diffusivity at the center of the rod.

17.14 The finite-element model for the heat transfer problem summarized by Equations 17.10–17.12 assumes that the thermal diffusivity of the rod is the same at each instant in time. Develop a model in which the diffusivity at time t is $c^2 e^{-t/2}$, where t is measured in seconds. The resulting model will barely differ from the original. The key idea is to treat the diffusivity during any small interval of time as a constant.

17.15 Modify `heat6.c` so that it implements your model from Exercise 17.14. The user should supply the thermal diffusivity at time zero.

17.16 Write a program that reads 20 integers from the user and then prints them out in reverse order. To do this, your program will need to store the integers in an array as it reads them.

17.17   Write a program that reads integers that are between 1 and 100, inclusive, from the keyboard. When the user enters an integer outside of that range, your program should report which integer was entered most often and how many times it was entered. (You will need to use an array to keep track of the frequency of each number. Because the numbers are integers, you can use them as array indexes.) Be sure to use a symbolic constant so that the top end of the range can be easily changed from 100 if desired.

17.18   Repeat Exercise 17.17, but begin by prompting the user for an integer *n* between 1 and 1000, inclusive. This integer will serve in place of 100 as the upper bound of the range.

17.19   Write a program that prompts the user for two positive 25-digit integers and then displays their sum. Because 25-digit integers are too large to manipulate in C, you will need to use the following array-based strategy.

• Read the integers one digit at a time. The scanf format specification %1d will read a single digit from the input.
• Represent a 25-digit integer with an array of length 25, where each element contains an integer between 0 and 9, inclusive.
• Add the 25-digit integers one digit at a time from right to left, just as you learned in elementary school. Store the sum in an array, one digit per element. Keep in mind that the sum may be 26 digits long.
• If a two-digit integer arises as you are adding the integers, it must be decomposed into its ones digit and its tens digit. If n is two-digit integer, then n/10 will yield the tens digit and n%10 will yield the ones digit. (The % sign is the C *modulus operator*.) For example, 15/10 is 1 and 15%10 is 5.
• Display the digits of the sum one after another with no intervening spaces or line breaks.

17.20   Modify your solution to Exercise 17.19 so that it will work for all positive integers up to 25 digits long. The user should enter, in order, the number of digits in the first addend, the digits in the first addend, the number of digits in the second addend, and the digits in the second addend. Be sure to define the 25-digit limit as a symbolic constant so that it can be easily modified.

18

Visualizing Heat Transfer: Arrays as Parameters

In Chapter 17 we studied the problem of approximating the temperature at the center of a 10-centimeter silver rod at a specific interval of time after heat sources were put into contact with its ends. All of our work led to a program that produced as its output a single temperature. In this chapter we will study the same heat transfer problem as in Chapter 17, but will focus on characterizing how the temperatures throughout the rod change as time passes.

What we will do is a form of *scientific visualization*, which is the use of visual aids to render numerical data more understandable. Scientific visualization is nothing new. Legend holds that the Greek mathematician Archimedes was drawing geometric figures in the sand when he was killed. Johannes Kepler's notebooks, preserved now for almost 400 years, contain sketches of planetary orbits. Scientific visualization is not something confined to the laboratory. The colorful three-dimensional weather radar images that show up on the nightly news, and the magnetic resonance imaging (MRI) scans that are becoming an ever more commonly used medical diagnostic tool, are but two examples.

There are, of course, forms of scientific visualization that require so much computing power to produce that they *are* confined to the laboratory. Professor Chris Johnson's research group at the University of Utah, for example, uses supercomputers to produce simulations and visualizations of the electric fields produced inside the chest by a beating heart. The simulations are based on a finite-element model of the patient. The data for the model are derived from a combination of MRI scans and measurements of voltages on the body surface made with a "jacket" of sensors worn by the patient. The visualizations of the electrical activity of the heart could be a valuable diagnostic tool for physicians if the images could be generated in real time. Currently, however, the computation of a simulation and the generation of the corresponding visualizations require many hours or days even with the fastest available computers.

The nature of computational research, however, is that what is slow and expensive today will be fast and practically free tomorrow. Over the last 20 years, the speed of computer hardware per unit cost has doubled every 18 months, and this trend is continuing. A generation ago, the computer technology required to convert radar data into three-dimensional representations of weather was prohibitively expensive, if it existed at all. A generation from now, the laboratory-based visualization efforts of today will be fast, inexpensive, and commonplace.

As we develop our visualization of the heat transfer problem, we will use exactly the model that we developed in Chapter 17, which is summarized by Equations 17.10–17.12. These equations give us a way to approximate the temperature at any point in the rod at any point in time. Our implementation from Chapter 17, in fact, calculated many such temperatures along the way to computing the one temperature that it ultimately reported.

We will extend our heat transfer program so that it produces an animation showing how the temperatures throughout the rod change every second. Besides giving us a way to visualize the dynamics of heat transfer in an insulated rod, this extension will allow us to assess more thoroughly our original solution. Possible problems in our model, method, and implementation that might not be obvious from a single-temperature result will be laid bare by an animation.

C isn't particularly well equipped to directly produce animations. Rather than write a C program to produce its own animations—which would be by far the most ambitious program in this book—we will modify our heat transfer implementation so that it creates a file containing *Mathematica* commands that, when read and executed by *Mathematica*, produce the desired animation.

Figure 18.1 shows two frames from the kind of animation we want to create. Figure 18.1a shows the initial temperature distribution in the 10-centimeter silver rod that we considered in Chapter 17. Positions along the rod are graphed relative to the horizontal axis, while temperatures are graphed relative to the vertical axis. Figure 18.1b shows the temperature distribution in the same rod after 5 seconds.

Our approach to producing this animation output will be to insert function calls at strategic points in `simulate`. These calls will be to programmer-defined functions that will write the animation commands to a file. Each such function will need to take the array of simulated temperatures as one of its parameters.

In preparation for this, we will first simplify `simulate` by modifying it so that it exploits two new programmer-defined functions to do most of its array manipulations. This will give us the opportunity to explore—independently of the problem of file-directed input and output—the issues raised by functions that take array parameters.

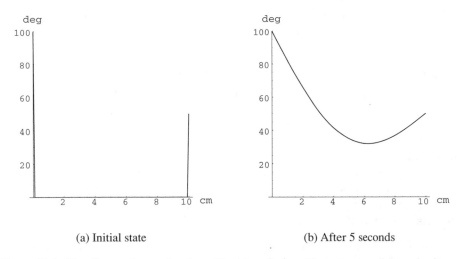

(a) Initial state   (b) After 5 seconds

Figure 18.1   Two frames from animation of heat transfer in a 10-centimeter silver rod using a 100-element model

We will next modify the `main` function of our program so that it reads its input values from a file instead of from the keyboard. This will give us the opportunity to experiment with file-directed input in a relatively simple setting.

Finally, we will modify `simulate` and create programmer-defined functions to actually write the appropriate *Mathematica* commands to a file. When we are done, we will have a program that reads input parameters describing an instance of the heat transfer problem from a file and writes to another file *Mathematica* commands that will produce an animation of that particular heat transfer problem.

18.1 Arrays as Parameters

The conceptual structure of `simulate`, which appears in Figure 17.11, is straight-forward. It

- initializes the `cur` array to represent the temperatures at the boundaries of the rod segments at time zero;
- determines the length of each simulated time step; and
- simulates the passage of time by repeatedly
  - updating the simulated temperatures stored in `cur`, and
  - subtracting one time step from the amount of time remaining in the simulation.

The actual structure of `simulate` is quite a bit more involved. It employs one loop to initialize and two more to update the `cur` array. If we were to replace these loops with calls to appropriately defined functions, the actual and conceptual structures of simulate would be equally straightforward.

18.1.1 Restrictions on Array Manipulation Functions

Let's imagine that we have two functions, initRod and nextRod, that behave as follows.

- initRod takes four parameters (the number of segments to simulate and the initial temperatures of the left end, right end, and middle of the rod) and returns an array of temperatures representing the initial state of the rod.
- nextRod takes two parameters (an array representing the current state of the rod and the number of segments being simulated) and returns an array of temperatures representing the next state of the rod.

If we had these two functions, we might try to write simulate as in Figure 18.2.

Although a version of simulate very similar to the one in Figure 18.2 could be written if we were to use the advanced programming technique of dynamic memory, the version as written is not legal C. In the subset of C that we are using, there are two limitations on the use of arrays.

1. An array cannot be returned as the result of a function. Both initRod and nextRod are designed to do this.

2. An array cannot be assigned to a locally declared array variable. We attempt to do this on lines 10 and 14 of Figure 18.2.

```
1    #define MAXSIZE 1001
2
3    double simulate (double diffusivity, double length,
4                     double left, double right, double rod,
5                     double duration, int segs)
6    {
7      double delta;
8      double cur[MAXSIZE];
9
10     cur = initRod(segs, left, right, rod);
11     delta = (length*length/(segs*segs)) / (2*diffusivity);
12
13     while (duration > 0) {
14       cur = nextRod(cur, segs);
15       duration = duration - delta;
16     }
17
18     return(cur[segs/2]);
19   }
```

Figure 18.2   Incorrect version of simulate (badheat.c)

18.1.2 Designing with Array Manipulation Functions

Let's try again to modify `simulate`, this time using slightly different interfaces for `initRod` and `nextRod`.

- `initRod` takes five parameters (an uninitialized array, the number of segments to simulate, and the initial temperatures of the left end, right end, and middle of the rod) and returns nothing. It *modifies* its array parameter so that it represents the initial state of the rod.
- `nextRod` takes two parameters (an array representing the current state of the rod and the number of segments being simulated) and returns nothing. It *modifies* its array parameter so that it represents the next state of the rod.

Implementations of these two functions appear in Figure 18.3, but we will defer our discussion of them to Section 18.1.3. With these versions of `initRod` and `nextRod`, we can write `simulate` as in Figure 18.4.

```
1    #include <stdio.h>
2
3    #define MAXSIZE 1001
4
5    void initRod (double cur[], int segs,
6                      double left, double right, double rod)
7    {
8      int i;
9
10     cur[0] = left;
11     for (i = 1; i < segs; i = i+1) {
12       cur[i] = rod;
13     }
14     cur[segs] = right;
15   }
16
17   void nextRod (double cur[], int segs)
18   {
19     double nxt[MAXSIZE];
20     int i;
21
22     for (i = 1; i < segs; i = i+1) {
23       nxt[i] = (cur[i-1] + cur[i+1]) / 2;
24     }
25
26     for (i = 1; i < segs; i = i+1) {
27       cur[i] = nxt[i];
28     }
29   }
```

Figure 18.3 Array-manipulation functions that support `simulate` (continued in Figure 18.4)

```
30    double simulate (double diffusivity, double length,
31                     double left, double right, double rod,
32                     double duration, int segs)
33    {
34      double delta;
35      double cur[MAXSIZE];
36
37      initRod(cur, segs, left, right, rod);
38      delta = (length*length/(segs*segs)) / (2*diffusivity);
39
40      while (duration > 0) {
41        nextRod(cur, segs);
42        duration = duration - delta;
43      }
44
45      return(cur[segs/2]);
46
47    }
```

Figure 18.4 Modularizing `simulate` with array-manipulation functions (continued in Figure 18.7)

You should compare the version of simulate in Figure 18.4 with the one in Figure 17.11. We have streamlined the original in three ways.

1. We replaced the statements on lines 13–17 of Figure 17.11, which intitialize the cur array, with the function call

```
initRod(cur, segs, left, right, rod);
```

on line 37 of Figure 18.4.

2. We replaced the two loops on lines 21–27 of Figure 17.11 with the function call

```
nextRod(cur, segs);
```

on line 41 of Figure 18.4.

3. The first two changes allowed us to eliminate several of the local variables that were required in the original version.

18.1.3 Implementing Array Manipulation Functions

So far we have looked at array parameters only from the caller's point of view. We will now consider array parameters from the point of view of the function being called. The implementations of initRod and nextRod in Figure 18.3 show how functions that expect arrays as parameters are implemented.

Let's look first at the headers of these two functions. The first parameter specified in each header is an array of `doubles`. Notice that an array parameter in a function header looks much like an array declaration in the body of a function, with one exception: the size of the array is not specified. Thus, the first parameter of both `initRod` and `nextRod` is written as

```
double cur[]
```

The array that is passed as a parameter to either of these functions *must* be an array of `doubles`, but it can be of any size. This adds an important aspect of flexibility to functions that manipulate arrays. We would not want to have to change the implementations of `initRod` and `nextRod`, for example, every time we needed to deal with a different-sized array of `doubles`.

The body of a function, however, cannot manipulate an array it receives as a parameter unless it knows how big it is. This means that when an array is passed as a parameter, the size of the array must be communicated to the called function. One common way to do this is to include the size of the array as a parameter to the function.

Each of `initRod` and `nextRod` takes as its second parameter the number of rod segments represented by the array. Recall that the number of temperatures in the array is one more than the number of segments. Both `initRod` and `nextRod` use `segs` to control the loops that manipulate the arrays. In fact, the bodies of `initRod` and `nextRod` contain exactly the code that we removed from `simulate` when we inserted the two function calls.

18.1.4 How Array Parameters Work

Both `initRod` and `nextRod` are `void` functions: neither returns a value. This was a deliberate design decision, necessitated by the fact that arrays cannot be returned as results in the subset of C that we are studying. Instead, in our design and implementation, each function indirectly communicates results back to the point of call by modifying its array parameter. This is entirely typical of array manipulation functions in C.

You may be wondering how this can possibly work; if so, you are asking a good question. When an `int` or a `double` or a `struct` is passed as an actual parameter, the function being called receives a *copy* as its parameter. As a result, any change made by the function to the formal parameter has no effect on the actual parameter. This aspect of C is exploited in Figure 18.4, where the formal parameter `duration` is repeatedly modified on line 42. Despite this, the value of the actual parameter that is passed to `simulate` from `main` remains unchanged.

Array parameters behave differently. When an array is used as an actual parameter, it is not copied. The formal parameter inside the function being called refers to the same array as does the actual parameter. As a result, any changes made to the formal array will be visible at the point of call. This is a property of C on which many array manipulation functions—including `initRod` and `nextRod`—depend.

Figure 18.5 shows the formal parameters and local variables of `simulate` and `initRod` about halfway through the execution of `initRod` when called from `simulate`. The four numerical parameters of `initRod`—`segs`, `left`, `right`, and `rod`—are completely independent of the actual parameters that were used to initialize them. The parameter `cur`, however, refers back to the array `cur` that is a local variable of `simulate`.

diffusivity	1.752				
length	10.				
left	100.				
right	50.				
rod	0.				
duration	5.				
segs	5				
delta	0.002854				
cur	100	0	0	0	

cur	
segs	5
left	100.0
right	50.0
rod	0.0
i	3

Figure 18.5   Variables in `simulate` (top) and `initRod` (bottom) midway through call to `initRod`. The cur array is shared by the two functions.

18.2 File Input

The next step in the evolution of our program will be to modify the `main` function so that it reads its input from a file instead of from the keyboard. This can be a useful feature for a program that requires a lot of input, since it allows the user to create the necessary input—and save it for reuse—before running the program.

An example of the type of input file format that we want our program to accept appears in Figure 18.6. It consists of seven numbers separated by white space in this order: the thermal diffusivity; the length; the left heat source, rod, and right heat source temperatures; the number of seconds to simulate; and the number of segments to use. (The numbers in Figure 18.6 are the same numbers, in the same order, that we used as sample input in Figure 17.4.) We will assume throughout that this input has been placed into a file called `heat-input.txt`.

```
1.752
10
100
0
50
5
5
```

Figure 18.6 Sample input to program

18.2.1 File-Oriented Input with `fscanf`

You already know how to use the `scanf` library function to read input from the keyboard. The more general `fscanf` library function will read input either from the keyboard or from a file, as specified by its first parameter. (This is reflected in its name; you should think of `fscanf` as being a *file* scanf.) Figure 18.7 completes the implementation of `heat7.c` by showing the `main` function, which we have written to use `fscanf` instead of `scanf`.

```
48    void main (void)
49    {
50      double C, length, left, right, rod, time, temp;
51      int segs;
52      FILE *inFile;
53
54      inFile = stdin;
55
56      printf("Enter thermal diffusivity: ");
57      fscanf(inFile, "%lf", &C);
58
59      printf("Enter length of rod: ");
60      fscanf(inFile, "%lf", &length);
61
62      printf("Enter left, middle, right rod temperatures: ");
63      fscanf(inFile, "%lf%lf%lf", &left, &rod, &right);
64
65      printf("Enter number of seconds: ");
66      fscanf(inFile, "%lf", &time);
67
68      printf("Enter number of segments: ");
69      fscanf(inFile, "%d", &segs);
70
71      temp = simulate(C, length, left, right, rod, time, segs);
72      printf("Temperature at center is %g\n", temp);
73    }
```

Figure 18.7 Modifying `main` to read input from the keyboard with `fscanf` (`heat7.c`)

We have deliberately written this new version of `main` to read from the keyboard. If you were to run this new version you would find it indistinguishable from the version from `heat6.c` in Figure 17.12. Our plan is to show you how to use `fscanf` to do a familiar thing (keyboard reading) before showing you how to use it to do a new thing (file reading).

We have declared a variable called `inFile` on line 52 of Figure 18.7. This declaration specifies that `inFile` will contain a *pointer to a file*. (The type name `FILE` must be in uppercase. The asterisk in the declaration is what specifies that `inFile` will contain a pointer.) Because this will be our only use of a pointer type in this book, we will not attempt to explain exactly what a pointer is. It is enough to know that this is the proper way to declare variables that are to refer to files.

As is the case with all variables, we must give `inFile` a value before we can use it. We do this on line 54 by assigning it the constant `stdin`. You should think of the keyboard as being a special kind of file, and you should think of `stdin` as being a pointer to that file. The constant `stdin` is defined, along with `fscanf`, in the library file `stdio.h`.

Once the variable `inFile` contains a pointer to a file, we can use it to control the behavior of `fscanf`. The function `fscanf` takes one more parameter than does `scanf`. This parameter, which appears as the first one in the parameter list, must be a pointer to the file from which `fscanf` should read. We use the variable `inFile` as the first parameter to each of the calls to `fscanf` in Figure 18.7. Because `inFile` is initialized to `stdin`, each of these calls will read from the keyboard. The remaining parameters to `fscanf` serve the same purpose as do the parameters to `scanf`.

We now have a revised version (`heat7.c`) of our final implementation of the heat transfer problem from Chapter 17 (`heat6.c`). From the user's point of view, however, the two versions are identical. By restructuring our program with programmer-defined array-manipulation functions and by using `fscanf` in place of `scanf`, we have changed the form but not the function of the program. We have done all this in preparation for the more substantial changes that we will now explore.

18.2.2 Reading from an Input File

The next version of our program, `heat8.c`, behaves just like `heat7.c` except that it takes its input from a file instead of the keyboard. Figure 18.8 contains a `main` function that reads from a file called `heat-input.txt` that is formatted as in Figure 18.6. As before, we declare a file pointer variable `inFile` on line 5. When we initialize it on line 7, however, use the library function `fopen` (declared in `stdio.h`) to obtain a pointer to `heat-input.txt`. The function `fopen` takes two character strings as its parameters. The first parameter gives the name of the file to which we want a pointer, and the second parameter specifies that we want to read from that file. If we wanted to write to the file, we would use `"w"` instead of `"r"`.

```
1     void main (void)
2     {
3       double C, length, left, right, rod, time, temp;
4       int segs;
5       FILE *inFile;
6
7       inFile = fopen("heat-input.txt", "r");
8       if (!inFile) {
9           printf("Could not open heat-input.txt\n");
10          exit(1);
11      }
12
13      fscanf(inFile, "%lf", &C);
14      fscanf(inFile, "%lf", &length);
15      fscanf(inFile, "%lf%lf%lf", &left, &rod, &right);
16      fscanf(inFile, "%lf", &time);
17      fscanf(inFile, "%d", &segs);
18
19      fclose(inFile);
20
21      temp = simulate(C, length, left, right, rod, time, segs);
22      printf("Temperature at center is %g\n", temp);
23    }
```

Figure 18.8   Modifying `main` to read input from a file (`heat8.c`)

If the file that we are trying to open does not exist, or if it does exist but cannot be read for some reason, the call to `fopen` will return an invalid file pointer. Before we use the new value of `inFile`, we must check to be sure that the value returned by `fopen` is valid. We do this on lines 8–11 of Figure 18.8. The condition `!inFile` will be *true* if `inFile` contains an *invalid* pointer. If this is the case, we must not try to read from it. Instead, we print out an explanatory message and then exit the program by calling the C library function `exit`. This function, which is declared in `stdlib.h`, terminates the execution of the program. We take this drastic step because we can make no further progress without a valid file pointer. (The `exit` function takes as its parameter a completion code that is returned to the operating system. By convention, a non-zero completion code indicates abnormal termination.)

Once we have verified that `inFile` contains a valid pointer to the file `heat-input.txt`, we can pass it as a parameter to `fscanf` just as we did in Figure 18.7. This time, of course, the input will be taken from `heat-input.txt` instead of from the keyboard. Because of this, there is no longer any need for the prompts that preceded each call to `fscanf` in Figure 18.7; we have removed them in Figure 18.8.

When we are through with a file pointer obtained via a call to `fopen`, it is our responsibility to notify C of this fact via a call to `fclose`. We take care of this detail on line 19 of Figure 18.8, after we have made all of the necessary calls to `fscanf`. In a well-written program, every file that is opened should eventually be closed.

18.3 File Output

Now that we have modified the `simulate` function to use array-manipulation functions and modified `main` to read its input from a file, the third and final step in the evolution of our program will be to further modify `simulate` so that it creates a file containing *Mathematica* commands that, when read into *Mathematica*, will produce an animation of the heat transfer process. Our goal is to have the program create a file called `heat-output.txt` containing output of the form illustrated in Figure 18.9.

```
ListPlot[{{0,100},{2,0},{4,0},{6,0},{8,0},{10,50}},
        PlotJoined->True, PlotRange->{0,100}]

ListPlot[{{0,100},{2,50},{4,0},{6,0},{8,25},{10,50}},
        PlotJoined->True, PlotRange->{0,100}]

ListPlot[{{0,100},{2,50},{4,25},{6,12.5},{8,25},{10,50}},
        PlotJoined->True, PlotRange->{0,100}]

ListPlot[{{0,100},{2,62.5},{4,31.25},{6,25},{8,31.25},{10,50}},
        PlotJoined->True, PlotRange->{0,100}]

ListPlot[{{0,100},{2,65.625},{4,43.75},{6,31.25},{8,37.5},
        {10,50}}, PlotJoined->True, PlotRange->{0,100}]

ListPlot[{{0,100},{2,71.875},{4,48.4375},{6,40.625},{8,40.625},
        {10,50}}, PlotJoined->True, PlotRange->{0,100}]
```

Figure 18.9   Sample output of heat transfer program

The output in Figure 18.9 is what should be generated when the program is run with the input file from Figure 18.6. The output consists of six calls to the *Mathematica* `ListPlot` function. `ListPlot` takes as its parameter a list consisting of the *x-y* coordinates of a collection of points and creates a plot of those points. The optional parameter `PlotJoined` specifies that `ListPlot` is to connect the points with line segments, and the optional parameter `PlotRange` specifies the range of the vertical axis as usual.

Each of the six calls to `ListPlot` in Figure 18.9 produces one frame in the animation of the behavior of a 10-centimeter rod that is divided into five segments. The frames are spaced at one-second intervals, with the first frame plotting the temperatures at 0 seconds and the sixth frame plotting the temperatures at 5 seconds.

Thus, the third frame shows that the temperature at the left end is 100°C; at 2 centimeter, 50°C; at 4 centimeters, 25°C; at 6 centimeters, 12.5°C; at 8 centimeters, 25°C; and at the right end, 50°C. If we created a two-dimensional plot of these points, with rod coordinates on the horizontal axis and temperatures on the vertical axis, and then connected the points, we would obtain the example animation frame in Figure 18.15b.

18.3.1 File-Oriented Output with `fprintf`

Our strategy for modifying `simulate` to produce animation output is as follows. When the function is first called, we will produce a `ListPlot` command that plots the initial state of the rod. As each second of simulated time passes during the execution of the loop inside of `simulate`, we will produce a `ListPlot` command that plots the temperatures of the rod at that time. If we are simulating the passage of 10 seconds, for example, by the time the loop terminates we will have produced 11 different `ListPlot` commands.

It will be easier to modify `simulate` to produce this output if we first create a programmer-defined function (`createFrame`) that can be called multiple times to create the individual `ListPlot` commands. A version of this function that sends its output to the display appears in Figure 18.10.

The function `createFrame` takes as parameters the array `rod` of temperatures, the number of `segments` represented by the array, the `length` of the rod, and the `low` and `high` temperature bounds of the vertical axis. Each time the function is called, it generates a `ListPlot` command like those shown in Figure 18.9.

```
1    void createFrame (double rod[], int segments,
2                         double length, double low, double high)
3    {
4      int i;
5      FILE *outFile;
6
7      outFile = stdout;
8      fprintf(outFile, "ListPlot[{");
9
10     for (i = 0; i < segments; i = i+1) {
11       fprintf(outFile, "{%g,%g},",
12               (i*length/segments), rod[i]);
13     }
14
15     fprintf(outFile, "{%g,%g}}, ", length, rod[segments]);
16     fprintf(outFile, "PlotJoined->True,
17             PlotRange->{%g,%g}]\n\n", low, high);
18   }
```

Figure 18.10 A function that writes *Mathematica* `ListPlot` commands to the display with `fprintf` (continued in Figure 18.11)

```
19    double simulate (double diffusivity, double length,
20                       double left, double right, double rod,
21                       double duration, int segs)
22    {
23      double delta, clock, low, high;
24      double cur[MAXSIZE];
25
26      initRod(cur, segs, left, right, rod);
27      delta = (length*length/(segs*segs)) / (2*diffusivity);
28
29      low = min(rod, left, right);
30      high = max(rod, left, right);
31
32      createFrame(cur, segs, length, low, high);
33      clock = 0;
34
35      while (duration > 0) {
36
37        nextRod(cur, segs);
38        duration = duration - delta;
39
40        clock = clock + delta;
41        if (clock >= 1 - delta/2) {
42          createFrame(cur, segs, length, low, high);
43          clock = clock - 1;
44        }
45
46      }
47
48      return(cur[segs/2]);
49
50    }
```

Figure 18.11   Exploiting `createFrame` in `simulate` (`heat9.c`)

In `createFrame` we have used `fprintf` in place of the more familiar `printf` to produce the output. The difference between `fprintf` and `printf` is analogous to the difference between `fscanf` and `scanf`. The function `fprintf` takes as its first parameter a pointer to the file to which its output should be directed.

Each call to `fprintf` in Figure 18.10 uses the variable `outFile` as its first parameter. This variable is declared as a pointer to a file and is initialized to contain the constant `stdout`, which stands for "standard output." As a result, all of the output goes to the display, which you should think of as a special output file.

A version of `simulate` that exploits `createFrame` appears in Figure 18.11. In it, we use three new variables.

• We use `clock` to keep track of how much simulated time has passed since the previous frame of the animation was created. We start it out with a value of zero seconds; once it gets acceptably close to one second we create a frame and subtract one second from `clock`.

- We use `low` and `high` to hold the lowest and highest temperatures that will occur anywhere in the rod at any point of the simulation.

After initializing the rod and determining `delta`, we compute the values of `low` and `high` on lines 29 and 30 of Figure 18.11. No temperature in the rod can ever be lower than the lowest of `rod`, `left`, and `right`; similarly, no temperature in the rod can ever be higher than the highest of `rod`, `left`, and `right`. Thus, we use the programmer-defined functions `min` and `max` (which are not shown in the figure) to determine appropriate values for `low` and `high`.

The next step in `simulate` is to create the first frame of the animation with a call to `createFrame` on line 32. This frame describes the initial state of the rod. We then set `clock` to 0.

After each repetition of the loop on lines 35–46, we add `delta` to `clock`. When `clock` gets to within plus or minus one-half `delta` of 1, we create a new frame and subtract 1 from `clock`. When the loop terminates we have produced all of the required frames.

18.3.2 Writing to an Output File

The next step is to modify `createFrame` so that it sends its output to a file instead of to the display. An appealing, but incorrect, attempt to do this appears in Figure 18.12.

All that we have changed about the function is the way the file-pointer variable `outFile` is initialized. Rather than set it to the constant `stdout`, we now use `fopen` to obtain a pointer to the file `heat-output.txt`. We specify that we are opening the file for *writing* by using `"w"` as the second parameter to `fopen` on line 7 of `createFrame`. As before, we are careful to verify that the file was successfully opened.

If the file `heat-output.txt` does not exist, the call to `fopen` will create it. If such a file does exist, the call to `fopen` will delete its contents. In either case, if the call to `fopen` is successful it will return a pointer to an empty file, which is the problem with the implementation in Figure 18.12. Over the course of the program, there will be multiple calls to `createFrame`. Each time it is called, it will open `heat-output.txt`, write a *Mathematica* command into it, and close it. As a result, each of these function calls will overwrite the command created by the previous call. In the end, only the `ListPlot` command for the last frame will be in `heat-output.txt`.

18.3.3 Receiving a File-Pointer Parameter

The solution to this dilemma is to open `heat-output.txt` before the first call to `createFrame`, leave it open during all of the subsequent calls, and close it only after the final call to `createFrame`. To do this we must remove the

```
1    void createFrame (double rod[], int segments,
2                        double length, double low, double high)
3    {
4      int i;
5      FILE *outFile;
6
7      outFile = fopen("heat-output.txt", "w");
8      if (!outFile) {
9        printf("Unable to open file heat-output.txt\n");
10       exit(1);
11     }
12
13     fprintf(outFile, "ListPlot[{");
14
15     for (i = 0; i < segments; i = i+1) {
16       fprintf(outFile, "{%g,%g},",
17               (i*length/segments), rod[i]);
18     }
19
20     fprintf(outFile, "{%g,%g}}, ", length, rod[segments]);
21     fprintf(outFile, "PlotJoined->True,
22             PlotRange->{%g,%g}]\n\n", low, high);
23
24     fclose(outFile);
25   }
```

Figure 18.12   Incorrectly modifying `createFrame` to write animation commands to an output file (`heat10.c`)

responsibility for opening and closing the file from `createFrame` and perform these steps inside of `simulate` instead. This will in turn require passing the pointer to `heat-output.txt` as a parameter to `createFrame`.

Figure 18.13 shows a version of `createFrame` that receives the file pointer `outFile` as a parameter. We have removed from `createFrame` the declaration of the local variable `outFile`, the code that initializes and tests `outFile`, and the code that closes `outFile`. This final version of our function will send its output to whatever file is pointed to by the parameter `outFile`, which will be determined by `simulate`.

18.3.4 Passing a File-Pointer Parameter

Now that we have an acceptable version of `createFrame`, we can turn our attention to modifying `simulate` to exploit it. We will need to modify `simulate` to open the output file, pass the file pointer each time it calls `createFrame`, and finally close the output file. A version of `simulate` that does all of this appears in Figure 18.14.

```
1    void createFrame (double rod[], int segments,
2                      double length, double low, double high,
3                      FILE *outFile)
4    {
5      int i;
6
7      fprintf(outFile, "ListPlot[{");
8
9      for (i = 0; i < segments; i = i+1) {
10       fprintf(outFile, "{%g,%g},",
11              (i*length/segments), rod[i]);
12     }
13
14     fprintf(outFile, "{%g,%g}}, ", length, rod[segments]);
15     fprintf(outFile, "PlotJoined->True,
16            PlotRange->{%g,%g}]\n\n", low, high);
17   }
```

Figure 18.13   Modifying `createFrame` to receive a file pointer as a parameter (continued in Figure 18.14)

This version of `simulate` uses the new variable `output` to hold the pointer to the output file. We open and verify the output file on lines 32–36 of Figure 18.14. We have added `output` as an extra parameter in the calls to `createFrame` on lines 38 and 48. Finally, we close the output file on line 54.

18.4 Assessment

With the program that we have developed in this chapter we have come full circle. We began this book by studying *Mathematica*, and partway through we turned our attention to C. We have now developed a solution that makes use of both *Mathematica and* C.

Mathematica and C have different strengths. *Mathematica* is especially good for symbolic mathematics and graphical output, while C is especially good for large-scale numerical computations. By exploiting the strengths of both in this chapter, we developed a solution more easily than we could have with either language alone.

This is entirely typical of solutions to problems in computational science. Obtaining a good solution often involves combining the capabilities of two or more languages or packages. Breadth in your computational background is more important than depth in any one area.

18.4.1 *Mathematica* Capability: *MathLink*

The program that we have developed in this chapter communicates with *Mathematica* in a rather circuitous fashion. We must run `heat11.c` to obtain a file containing

```
18    double simulate (double diffusivity, double length,
19                     double left, double right, double rod,
20                     double duration, int segs)
21    {
22      double delta, clock, low, high;
23      double cur[MAXSIZE];
24      FILE *output;
25
26      initRod(cur, segs, left, right, rod);
27      delta = (length*length/(segs*segs)) / (2*diffusivity);
28
29      low = min(rod, left, right);
30      high = max(rod, left, right);
31
32      output = fopen("heat-output.txt", "w");
33      if (!output) {
34        printf("Unable to open heat-output.txt\n");
35        exit(1);
36      }
37
38      createFrame(cur, segs, length, low, high, output);
39      clock = 0;
40
41      while (duration > 0) {
42
43        nextRod(cur, segs);
44        duration = duration - delta;
45
46        clock = clock + delta;
47        if (clock >= 1 - delta/2) {
48          createFrame(cur, segs, length, low, high, output);
49          clock = clock - 1;
50        }
51
52      }
53
54      fclose(output);
55      return(cur[segs/2]);
56
57    }
```

Figure 18.14   Modifying `simulate` to pass a file pointer to `createFrame` (`heat11.c`)

the *Mathematica* commands that are required to animate heat transfer. We must then start *Mathematica* and use it to read this file to actually produce the animation.

    Mathematica provides a capability, called *MathLink*, that makes it possible to more tightly couple C and *Mathematica* programs. By using *MathLink*, we can write a C program that directly calls on *Mathematica* to produce the desired animation. In effect, *MathLink* allows us to combine the capabilities of C and *Mathematica* within a single program.

In Appendix A we will discuss how *MathLink* works and use it to produce an improved version of `heat11.c`.

18.4.2 Size of Finite-Element Model

The final version of our program, `heat11.c`, gives us a good way to assess the effect of increasing the number of elements in our finite-element heat transfer model. If we create a file `heat-input.txt` containing the numbers from Figure 18.6, the program saves a six-frame animation in `heat-output.txt`. We can then read this file into *Mathematica*

$$
\begin{array}{l}
\texttt{In[1]:= Get["heat-output.txt"]} \\[2ex]
\texttt{Out[1]= (See Figure 18.15)}
\end{array}
\tag{18.1}
$$

to view the frames of the animation.

The first frame (Figure 18.15a) shows the abrupt jump between the left and right ends, which are at 100°C and 50°C respectively, and the rest of the rod, which is at 0°C. The final frame (Figure 18.15b), which characterizes the temperature distribution in the rod after five seconds, shows a much different but still quite ragged curve. If we had generated a long enough animation, the final frame would have been a straight line characterizing the rod at steady state.

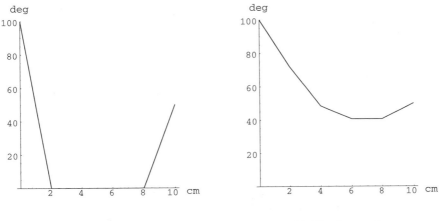

(a) Initial state   (b) After 5 seconds

Figure 18.15 Two frames from animation of heat transfer in a 10-centimeter silver rod using a five-element model

If we change the last number in heat-input.txt to specify 100 segments instead of five and run heat11.c again, it saves a different six-frame animation in heat-output.txt.

In[2]:= Get["heat-output.txt"] Out[2]= (See Figure 18.1)	(18.2)

It is evident from the first frame (Figure 18.1a) that we are using many more segments, as the sloped lines that appeared on either end of the graph in Figure 18.15a have become almost vertical. The graph in the final frame (Figure 18.1b), with 100 elements used to calculate it, is markedly smoother than the graph in Figure 18.15b.

18.5 Key Concepts

Array parameters.   Arrays can be passed as parameters to functions just like other types of values, but there are some important differences. A function has no automatic way to determine the size of an array passed to it, so the size is typically passed as a separate parameter. When a function modifies its array parameter, the modification is visible at the point of call. This is in contrast to all other types of values that we have studied.

File input and output.   Data can be read from and written to files by using the fscanf and fprintf library functions. These functions take one extra parameter relative to scanf and printf. This parameter is a pointer to the file involved in the input/output operation. Pointers to files are obtained via calls to the fopen library function; files are closed via calls to the fclose library function.

18.6 Exercises

18.1   Write a program that interacts with the user to create a file called heat-input.txt, formatted exactly as in Figure 18.6. Your program should be sure that it has received sensible values from the user before it creates the file.

18.2   Repeat Exercise 17.10, but modify heat11.c instead of heat6.c.

18.3   Modify heat11.c so that the axes of the animation are labeled.

18.4   Modify heat11.c so that the temperatures written to the output file contain only three-digit mantissas. Does this affect the quality of the animation?

18.5   As currently written, heat11.c exits immediately on determining that the file heat-input.txt cannot be opened. Modify heat11.c so that, instead of exiting, it notifies the user and then prompts for and reads the values from standard input.

18.6   As currently written, `heat11.c` will overwrite `heat-output.txt` if a file by that name already exists. Modify `heat11.c` so that it warns the user and then exits if `heat-output.txt` exists. One way to do this is to first attempt to open `heat-output.txt` for reading. Only if this attempt succeeds does the file already exist.

18.7   Modify `heat11.c` so that, instead of producing an animation, it produces a two-dimensional plot showing the temperatures in the rod at the end of the simulation.

18.8   Modify `heat11.c` so that it animates the behavior of the heat transfer problem from Exercise 17.12.

18.9   Modify `heat11.c` so that it animates the behavior of the heat transfer problem from Exercise 17.14.

18.10   Suppose that a 10-centimeter silver rod is initially at $0°$C, and that at time zero its left end is put in contact with a $100°$C heat source and its right end is put in contact with a $d°$C heat source. Modify `heat11.c` so that it creates a file containing the *Mathematica* commands required to produce an animation of how the temperature distribution in the rod after five seconds depends on d, as d ranges between 0 and 100.

18.11   Modify `heat11.c` so that it produces one frame of the animation for every s seconds of simulated time. Your program should prompt the user for the value of s.

18.12   As `heat11.c` is currently written, each frame of the animation contains one point for each boundary segment. Modify `heat11.c` so that each frame contains only half as many points. The number of different temperatures that your program calculates at each step should not change.

18.13   Is there any difference between the animations produced by the following two strategies for calculating and animating the change in temperature distribution in a rod?

(a) Model the rod with $k/2$ segment boundaries and use all $k/2$ temperatures in each frame of the animation as in `heat11.c`.

(b) Model the rod with k segment boundaries and use only half of them in each frame of the animation as in Exercise 18.12.

18.14   Extend your solution to Exercise 16.12 so that it writes to a file a *Mathematica* command that will produce the kind of plot that appears in Figure 16.2. Your program will no longer need to prompt for t.

18.15   Extend your solution to Exercise 17.6 so that it writes to a file a *Mathematica* command that will produce a plot showing how the time required for the center of the rod to come within 1% of the average of the temperatures of the two ends depends on the number of rod segments.

18.16   Repeat Exercise 17.17, but this time read the numbers from a file, stopping at the end of the file. If there are any numbers outside of the range 0–100, ignore them.

18.17   Repeat Exercise 18.16. Instead of displaying the frequency of the most commonly appearing number, write to the output file the frequency of every in-range number that occurs in the input file.

18.18   Modify your solution to Exercise 18.17 so that it writes the numbers to the output file in descending order of frequency. One way to do this is to search repeatedly through the array for the number with the highest frequency, then write it to the file and zero out its frequency. When there are no more non-zero entries in the array, the job is done.

18.19   Modify your solution to Exercise 17.20 so that it is organized around programmer-defined functions. One function should be used to read in the numbers to be added, a second function should be used to do the addition, and a third function should be used to display the sum.

18.20   Extend your solution to Exercise 18.19 to obtain a program that can multiply two 25-digit numbers.

A

Mathematica Capabilities

In Chapters 1–18 we included five sections that briefly discussed *Mathematica* capabilities and then referred the interested reader to this appendix for more details. Below, we discuss *Mathematica*'s support for units, typeset mathematics, floating-point simulation, arbitrary precision numbers, and communication with C programs.

A.1 Units

In Section 2.4.3 we saw that *Mathematica* provides support for attaching the proper units to the calculations that it performs. (This is called *dimensional analysis*.) We saw in Example 2.8, which we repeat below,

```
In[1]:= 72 in * 2.54 cm/in

Out[1]= 182.88 cm
```
(A.1)

that if we ask *Mathematica* to multiply a measurement in inches by a conversion factor in centimeters per inch, it can determine that the result is in centimeters.

    Mathematica does this by combining numerical calculations with the kinds of symbolic calculations that we first encountered in Chapter 6. It also exploits the fact that, in *Mathematica*, the use of the multiplication operator is optional.

    We have been careful throughout this book to use the multiplication operator ($*$) to denote multiplication in *Mathematica* expressions. In standard mathematical notation, as you are probably aware, multiplication operators are not typically used. Instead, the convention is that two expressions, when written next to each other, are treated as factors in a product. Thus, we write $2 + x$ to indicate the sum of 2 and x, but $2x$ to indicate the product of 2 and x.

The same rule applies to *Mathematica*. Thus, Example A.1 is actually equivalent to the calculation

```
In[2]:= 72 * in * 2.54 * cm / in

Out[2]= 182.88 cm
```

(A.2)

When *Mathematica* is presented with Example A.1 or A.2, it combines the numbers, simplifies the symbols as much as possible, and displays the result. The two occurrences of the symbol in cancel, leaving the symbol cm behind to appear in the answer.

The fact that *Mathematica* doesn't really recognize symbols like in and cm as units of measurement becomes obvious if we consider the following calculation, which attempts to convert 6 feet (as opposed to the equivalent 72 inches) into centimeters.

```
In[3]:= 6 ft * 2.54 cm/in
```

$$Out[3]= \frac{15.24 \text{ cm ft}}{\text{in}}$$

(A.3)

As you can see, *Mathematica* doesn't recognize that a foot is 12 inches. As there was no way to algebraically simplify the symbolic components of the calculation, it expresses the result using an odd combination of units.

Notice, however, that if we first tell *Mathematica* that a foot is 12 inches

```
In[4]:= ft = 12 in

Out[4]= 12 in
```

(A.4)

and then repeat Example A.3,

```
In[5]:= 6 ft * 2.54 cm/in

Out[5]= 182.88 cm
```

(A.5)

Mathematica is able to do the proper cancellation.

A.1.1 Examples

As you can see, it is quite easy to arrange for *Mathematica* to do dimensional analysis. Problems sometimes arise, however, and it is important to be able to anticipate them. Let's look at some examples from elsewhere in the book and see what happens when we incorporate units.

In Example 3.4 we determined the circumference of the earth in kilometers, based on measurements made in degrees and stadia. When we include units in the calculation

```
In[6]:= 360 deg / 7.2 deg * 5000. stade *
            0.1575 km/stade

Out[6]= 39375. deg² km
```
(A.6)

we get an unexpected answer. Because of *Mathematica*'s precedence rules, the symbol deg that we think of as labeling 7.2 actually appears in the *numerator* of the calculation. We can fix the problem by using parentheses to group 7.2 and deg together.

```
In[7]:= 360 deg / (7.2 deg) * 5000. stade *
            0.1575 km/stade

Out[7]= 39375. km
```
(A.7)

In Example 5.3, we determined the distance to the horizon from the top of a 66-foot-high hill. If we attach the appropriate units

```
In[8]:= Sqrt[(2.09*^7 ft + 66 ft)^2 - (2.09*^7 ft)^2]

Out[8]= 52524.3√ ft²
```
(A.8)

we see that the answer is indeed reported in feet, although certainly not in the simplest possible form. (The problem here is that the square root of ft^2 can be either ft or $-ft$, and *Mathematica* is unwilling to make an arbitrary choice.)

Units can be used in combination with functions. For example, in Chapter 7, we defined a function horizontal

```
In[9]:= horizontal[v_, theta_, t_] :=
            v * t * Cos[theta]
```
(A.9)

and a function vertical.

```
In[10]:= vertical[v_, theta_, t_] :=
            v * t * Sin[theta] - 1/2 * g * t^2
```
(A.10)

If we also define the gravitational constant g using appropriate units,

```
In[11]:= g = 9.8 m/sec^2
```

$$\text{Out[11]}= \frac{9.8\,\text{m}}{\text{sec}^2}$$

(A.11)

we can perform calculations using `horizontal` and `vertical` that produce properly dimensioned results.

For example, if a projectile is fired at 100 m/sec, at an angle of $\frac{\pi}{4}$ radians, then after 10 seconds it will have traveled

```
In[12]:= horizontal[100 m/sec, Pi/4., 10 sec]
```

Out[12]= 707.107 m

(A.12)

approximately 707 meters down range and will be

```
In[13]:= vertical[100 m/sec, Pi/4., 10 sec]
```

Out[13]= 217.107 m

(A.13)

approximately 217 meters above the ground.

Notice, however, that we must be careful not to label the angle measurement as being in radians. As this example illustrates, the built-in function `Sin` cannot make sense of a dimensioned parameter.

```
In[14]:= horizontal[100 m/sec, Pi/4. rad, 10 sec]
```

Out[14]= 1000 m Cos[0.785398 rad]

(A.14)

The preceding examples show that *Mathematica* can do dimensional analysis if we are a bit judicious in preparing our calculations and interpreting *Mathematica*'s results. Labeling measurements with units does not always work so smoothly, however. For example, suppose that we wish to recreate the plot from Example 7.28 using dimensioned parameters.

```
In[15]:= Plot[vertical[100 m/sec, Pi/4, t],
                {t, 0 sec, 20 sec}]

         Plot::plln : Limiting value 20 sec in
            {t, 0 sec, 20 sec} is not a
            machine-sized real number
```

(A.15)

Instead of a plot, we get an error message complaining about the presence of the symbol `sec`.

Even if we leave out the units from the `Plot` command

```
In[16]:= Plot[vertical[100, Pi/4, t], {t, 0, 20}]

        Plot::plnr : vertical[100, π/4, t] is not a
            machine-sized real number at
            t = 8.33333333333333214`*^-7.
```
(A.16)

we still get an error message instead of a plot. The problem here is that the free variable g, which `vertical` uses, has a dimensioned value.

A.1.2 `Units` Package

In addition to its support for dimensional analysis, *Mathematica* also provides a function `Convert` that provides conversions between different units. This function is part of the `Units` package

```
In[17]:= Needs["Miscellaneous`Units`"]
```
(A.17)

This example shows how to convert from miles/hour to kilometers/second.

```
In[18]:= Convert[60 Mile/Hour, Kilo Meter/Second]

Out[18]= 0.0268224 Kilo Meter / Second
```
(A.18)

You should consult *Mathematica*'s online documentation to learn more about the `Units` package, including how to spell the names of the various units.

A.2 Typeset Mathematics

In Section 2.4.6 we saw that you can give the calculations that you enter into a *Mathematica* notebook the appearance of typeset mathematics. Doing this can make your notebooks more succinct and easier to read, and is especially valuable if you plan to include copies of your notebooks in printed documents. In this section, we will present typeset versions of examples from earlier in the book. To create the examples that follow, we used the "palettes" provided by *Mathematica*'s notebook interface to insert mathematical symbols and Greek letters.

Example 4.2 is an assignment expression that adds up the first ten even terms of the harmonic series, each of which is a small fraction. Using typeset mathematics, we can write the fractions in a more conventional form.

$$
\text{In[19]:= rat10} = \tfrac{1}{2} + \tfrac{1}{4} + \tfrac{1}{6} + \tfrac{1}{8} + \tfrac{1}{10} + \tfrac{1}{12} + \tfrac{1}{14} +
$$
$$
\tfrac{1}{16} + \tfrac{1}{18} + \tfrac{1}{20}
$$
$$
\text{Out[19]} = \frac{7381}{5040}
$$

(A.19)

Example 5.16 is an expression that calculates the distance to the horizon from the top of a hill by taking the square root of an expression that involves an exponent. We can replace the function name `Sqrt` with a radical symbol and display the exponent in its customary raised position.

$$
\text{In[20]:= } \sqrt{2} \ * \ 2.09*{\char`\^}7 \ * \ 100 \ + \ 100^2
$$
$$
\text{Out[20]= 64653.}
$$

(A.20)

Example 5.17 is a programmer-defined function named `horizon` that is based on the expression from Example 5.16. Below, we not only use a radical symbol and a raised exponent, but we omit the multiplication operators in favor of juxtaposing the factors of $2Rh$, separated by spaces. As a result, the body of the function looks like something from a math book.

$$
\text{In[21]:= horizon[R\_, h\_] := } \sqrt{2\,R\,h + h^2}
$$

(A.21)

In a similar fashion, we can improve the appearance of the `compound` function from Example 6.23.

$$
\text{In[22]:= compound[p\_,r\_,n\_,m\_] := p } (1 + \tfrac{r}{m})^{mn}
$$

(A.22)

At several places in the book, we spelled out the names of Greek letters (for example, `Pi` and `theta`). By exploiting *Mathematica*'s notebook interface, we can enter these letters in their symbolic form (for example, π and θ). Here, for example, is an improved version of Example 7.13.

$$
\text{In[23]:= horizontal[100, } \tfrac{\pi}{4}, \text{ 10]}
$$
$$
\text{Out[23]= } 500\sqrt{2}
$$

(A.23)

In a similar vein, here is an improved version of the `horizontal` function itself.

$$
\text{In[24]:= horizontal[v\_, } \theta\_, \text{ t\_] := v t Cos[}\theta]
$$

(A.24)

Mathematica will also typeset expressions from calculus. For example, we can rewrite `makePower` from Example 8.20, which involves differentiation, as

$$
\boxed{\text{In[25]:= makePower[pos\_, mass\_] := mass } \partial_t \partial_t \text{ pos } \partial_t \text{ pos}} \tag{A.25}
$$

We can also rewrite Example 15.12, which involves integration, as

$$
\boxed{\begin{array}{l} \text{In[26]:= } \int_{0.0}^{8.0} \text{ (Sin[x]+2) dx} \\[2mm] \text{Out[26]= 17.1455} \end{array}} \tag{A.26}
$$

A.3 Floating-Point Simulation

We pointed out in Section 2.7.5 that *Mathematica* provides a capability that you can use to create and experiment with your own simplified floating-point number systems. This capability is a good way for you to gain a better understanding of the issues surrounding floating-point numbers that we raised in Chapters 2–5. In this section we will tell you enough about this capability so that you can experiment on your own.

The first step is to load the `ComputerArithmetic` package.

$$
\boxed{\text{In[27]:= Needs[\"NumericalMath\`ComputerArithmetic\`\"]}} \tag{A.27}
$$

Doing this makes three functions available: `SetArithmetic`, `Computer-Number`, and `Arithmetic`. We will discuss each in turn.

A.3.1 Specifying a Simulated System

You can use the `SetArithmetic` function to specify the details of the floating-point number system that you wish to simulate. To specify a base-b floating-point system with p-digit mantissas and exponents that range from e_{min} to e_{max}, you must evaluate `SetArithmetic[`p`, `b`, ExponentRange->{`e_{min}`, `e_{max}` + 1}]`. The first parameter is the desired mantissa size, the second parameter is the base in which the mantissa is to be represented, and the optional parameter `ExponentRange` gives the minimum and maximum exponents. (Notice that if you want the maximum exponent to be e_{max}, you must actually specify a value of $e_{max} + 1$.)

For example, suppose that you would like to simulate the base ten floating-point system with two-digit mantissas and exponents between -1 and 1 that we discussed in Section 2.7.3. To do this, you must evaluate

```
In[28]:= SetArithmetic[2, 10, ExponentRange->{-1, 2}]

Out[28]= {2, 10, RoundingRule→RoundToEven,
            ExponentRange→{-1,2}, MixedMode→False,
            IdealDivide→False}
```
(A.28)

Once you have specified a floating-point system with a call to SetArithmetic, subsequent simulated floating-point calculations, as discussed in the next section, will be carried out in this system. If you wish to change the system being simulated, you must make another call to SetArithmetic.

A.3.2 Using a Simulated System

Suppose that a system with precision p is being simulated. ComputerNumber takes a number as its parameter and rounds it to the nearest number with a p-digit mantissa. If that number is a valid simulated floating-point number, it is displayed as the result. Otherwise, an overflow or underflow is reported, as appropriate.

Thus, we see that

```
In[29]:= ComputerNumber[11.2]

Out[29]= 11.00000000000000000
```
(A.29)

rounds 11.2 to the nearest simulated floating-point number (11.),

```
In[30]:= ComputerNumber[.555]

Out[30]= 0.560000000000000000
```
(A.30)

rounds 0.555 to the nearest simulated floating-point number (0.56),

```
In[31]:= ComputerNumber[112.]

        ComputerNumber::ovrflw : Overflow occurred
            in computation.  The exponent is 3.
```
(A.31)

112. causes an overflow error, and

```
In[32]:= ComputerNumber[.0555]

        ComputerNumber:undflw :  Underflow occurred
            in computation.  The exponent is -1.
```
(A.32)

0.0555 causes an underflow error. (Notice that *Mathematica* doesn't do a very good job of reporting the exponents that cause underflow and overflow error.)

Besides creating simulated floating-point numbers, we can also do arithmetic with them. For example, the two-digit version of $\frac{1}{8}$ is

```
In[33]:= ComputerNumber[1] / ComputerNumber[8]

Out[33]= 0.1200000000000000000
```
(A.33)

and the two-digit product of 12 and 1.2 is

```
In[34]:= ComputerNumber[12] * ComputerNumber[1.2]

Out[34]= 14.00000000000000000
```
(A.34)

We can easily use simulated floating-point arithmetic to work through the example from Section 2.9.1 that we used to argue that floating-point multiplication is not associative. The first step is to change the simulation so that exponents between -2 and 2 are permitted.

```
In[35]:= SetArithmetic[2, 10, ExponentRange->{-2, 3}]

Out[35]= {2, 10, RoundingRule→RoundToEven,
           ExponentRange→{-2,3}, MixedMode→False,
           IdealDivide→False}
```
(A.35)

We can then see that $(14 \times 11) \times 2$

```
In[36]:= quot1 =
             (ComputerNumber[14] * ComputerNumber[11]) *
                ComputerNumber[2]

Out[36]= 300.0000000000000000
```
(A.36)

is 300, but that $14 \times (11 \times 2)$

```
In[37]:= quot2 =
             ComputerNumber[14] *
                (ComputerNumber[11] * ComputerNumber[2])

Out[37]= 310.0000000000000000
```
(A.37)

is 310.

We can use `InputForm` to learn much more about a simulated floating-point number. For example, we can more closely examine `quot1`, which is the result from Example A.36 above.

```
In[38]:= InputForm[quot1]

Out[38]= ComputerNumber[1, 30, 1, 300, 308.`21.5229]
```
(A.38)

The first three numbers above tell us that the simulated floating-point number is $1 \times 30 \times 10^1$, the fourth number tells us the value of the number when written as a rational number is 300, and the fifth number gives us the value that the number would have if it had been calculated to 21 digits. (The fifth number is one of *Mathematica*'s arbitrary-precision numbers, which we will discuss in Section A.4.)

By comparing the arbitrary-precision value to the actual value, you can see how much roundoff error has accumulated. Thus, a close examination of `quot2`, which is the result from Example A.37 above,

```
In[39]:= InputForm[quot2]

Out[39]= ComputerNumber[1, 31, 1, 310, 308`21.5229]
```
(A.39)

reveals that its arbitrary-precision value is the same as that of `quot1`.

A.3.3 Other Options

The function `Arithmetic` displays the characteristics of the simulated floating-point system.

```
In[40]:= Arithmetic[]

Out[40]= {2, 10, RoundingRule→RoundToEven,
           ExponentRange→{-2,3}, MixedMode→False,
           IdealDivide→False}
```
(A.40)

The result that is displayed suggests that it is possible to control more about the simulated floating-point system than just the precision, base, and exponent range.

`ComputerNumber` takes three additional optional parameters. The optional parameter `RoundingRule` can be given a value of `RoundToEven` (which is the default), `RoundToInfinity`, or `Truncation`. If the precision is p,

- `RoundToEven` rounds to the nearest p-digit number, and breaks ties by rounding to the number with an even mantissa;
- `RoundToInfinity` rounds to the nearest p-digit number, and breaks ties by rounding away from zero; and
- `Truncation` discards all but the first p digits from the mantissa.

The optional parameter `MixedMode` can be set to `True` or `False`. Only if it is `True` can simulated floating-point numbers be combined with ordinary *Mathematica* numbers in arithmetic expressions.

The optional parameter `IdealDivide` can also be set to `True` or `False`. If it is `False`, simulated division is done by multiplying the numerator by the reciprocal of the denominator, which provides two opportunities for roundoff error. If `IdealDivide` is `True`, division is done in one step in a manner similar to long division.

A.4 Arbitrary-Precision Numbers

In this book, we have discussed *Mathematica*'s support for rational and floating-point numbers and C's support for integers and floating-point numbers. Most programming languages provide both exact numbers (such as rational numbers or integers) and floating-point numbers. *Mathematica* goes beyond the norm by also providing arbitrary-precision numbers. Arbitrary-precision numbers are related to floating-point numbers, but, as we will discuss below, there are several important differences.

A.4.1 Precision

Every floating-point number has the same precision, or mantissa length. In the version of *Mathematica* that we have used while writing this book, for example, all floating-point numbers have 16-digit mantissas. The precision of an arbitrary-precision number, on the other hand, can be specified by the programmer. This means that two different arbitrary-precision numbers can have different mantissa lengths.

The built-in function `N`, which we have previously used to convert an exact number into a floating-point number, can also be used to convert an exact number into an arbitrary-precision number. We simply provide the desired precision as its second parameter. For example,

```
In[41]:= radius = N[1/7, 30]

Out[41]= 0.142857142857142857142857142857
```
(A.41)

creates a 31-digit representation of $\frac{1}{7}$, while

```
In[42]:= pi = N[Pi, 25]

Out[42]= 3.141592653589793238462643
```
(A.42)

creates a 25-digit representation of π, and

```
In[43]:= area = N[Pi*(1/7)^2, 25]

Out[43]= 0.064114135787546800784951 9
```

(A.43)

creates a 25-digit representation of $\pi \frac{1}{7}^2$.

When *Mathematica* creates a p-digit arbitrary-precision approximation to an exact number, it guarantees that *all* of the digits are correct. We can determine the precision of an arbitrary-precision number by using the built-in function `Precision`. Thus, `radius` has a

```
In[44]:= Precision[radius]

Out[44]= 30
```

(A.44)

30-digit mantissa, `pi` has a

```
In[45]:= Precision[pi]

Out[45]= 25
```

(A.45)

25-digit mantissa, and `area` also has a

```
In[46]:= Precision[area]

Out[46]= 25
```

(A.46)

25-digit mantissa.

A.4.2 Calculations

Mathematica can do calculations with arbitrary-precision numbers just as it does with floating-point numbers. In Example A.43, we converted the exact number $\pi \frac{1}{7}^2$ directly into a 25-digit arbitrary-precision number. We can calculate the same value with the 25-digit `pi` and the 30-digit `radius`.

```
In[47]:= area = pi * radius^2

Out[47]= 0.0641141357875468007849519
```

(A.47)

Mathematica assumes that the unknown digits of an arbitrary-precision number can be anything whatsoever. When it does calculations, it keeps track of the influence that these unknown digits could have on the answer and adjusts the precision of the result accordingly. This means that *all* of the digits of an arbitrary-precision result can be trusted; *Mathematica* discards the low-order digits that have been compromised by roundoff error. Thus, the precision of `area`, which we calculated above, is 25.

```
In[48]:= Precision[area]

Out[48]= 25
```
(A.48)

We could not have hoped for a precision *greater* than 25, since `pi` was only known to 25 digits.

We have seen that the effects of roundoff error can accumulate during a sequence of floating-point calculations. The same is true with arbitrary-precision numbers. As we pointed out above, however, *Mathematica* keeps track of the roundoff error and adjusts the precision of the result accordingly. Thus,

```
In[49]:= Precision[pi^2]

Out[49]= 25
```
(A.49)

produces a 25-digit result, but

```
In[50]:= Precision[pi^4]

Out[50]= 24
```
(A.50)

produces 24-digit result, and

```
In[51]:= Precision[pi^100000]

Out[51]= 20
```
(A.51)

produces only a 20-digit result.

A.4.3 Conversion from Floating-Point

We saw in Section A.4.1 that exact numbers can be converted into arbitrary-precision numbers by using the built-in function N and specifying the desired precision. Floating-point numbers, however, cannot be converted into arbitrary-precision numbers by using N. No matter what precision we suggest,

```
In[52]:= Precision[N[1.234, 50]]

Out[52]= 16
```
(A.52)

the result has the usual floating-point precision.

If we write what looks like a floating-point number but include more than the requisite number of digits, *Mathematica* will treat it as an arbitrary-precision number. Thus, the 20-digit

```
In[53]:= x = .12345678901234567890*^20
```

$$\text{Out[53]} = 1.2345678901234567890 \times 10^{19}$$

(A.53)

is treated as a 20-digit arbitrary-precision number.

```
In[54]:= Precision[x]
```

$$\text{Out[54]} = 20$$

(A.54)

The largest possible positive arbitrary-precision number can be determined by evaluating the special symbol $MaxNumber,

```
In[55]:= $MaxNumber
```

$$\text{Out[55]} = 1.44039712 \times 10^{323228010}$$

(A.55)

and the smallest possible positive arbitrary-precision number can be determined by evaluating the special symbol $MinNumber.

```
In[56]:= $MinNumber
```

$$\text{Out[56]} = 1.05934595 \times 10^{-323228015}$$

(A.56)

The exponents in these numbers dwarf those in the largest possible positive floating-point number,

```
In[57]:= $MaxMachineNumber
```

$$\text{Out[57]} = 1.79769 \times 10^{308}$$

(A.57)

and the smallest possible positive floating-point number.

```
In[58]:= $MinMachineNumber
```

$$\text{Out[58]} = 2.22507 \times 10^{-308}$$

(A.58)

As is the case with $MaxMachineNumber and $MinMachineNumber, the exact values of $MaxNumber and $MinNumber vary among different versions of *Mathematica*.

If the result of a floating-point calculation is too large to represent as a floating-point number, *Mathematica* will automatically represent it as an arbitrary-precision number, if possible, rather than generate an overflow error. For example,

```
In[59]:= $MaxMachineNumber^2

Out[59]= 3.23170060713110 × 10^616
```
(A.59)

produces an arbitrary-precision number instead of overflowing. On the other hand,

```
In[60]:= $MaxMachineNumber ^ $MaxMachineNumber

        General::ovfl : Overflow occurred in
           computation.
```
(A.60)

is too large to represent even as an arbitrary-precision number. The same principle applies to floating-point underflow as well.

A.4.4 Floating Point versus Arbitrary Precision

Arbitrary-precision numbers, with their support for arbitrary-length mantissas and their automatic accounting for roundoff error, are an attractive alternative to floating-point numbers. We did not feature them more prominently in this book because we wanted to cover widely applicable concepts that would carry over to other programming languages. Whereas arbitrary-precision numbers are unique to *Mathematica*, almost all languages support floating-point numbers.

Another potential drawback of arbitrary-precision numbers has to do with performance. Arbitrary-precision calculations take longer than corresponding floating-point calculations, and the time required to do an arbitrary-precision calculation increases with the precision of the numbers involved. Table A.1 shows the length of time required to sum different numbers of terms of Series 4.1 when using *Mathematica*'s floating-point numbers and arbitrary-precision numbers with 20 and 40 digits. (Our function `BlockPrecision` uses arbitrary-precision numbers when a precision in excess of regular floating-point precision is specified.)

Arbitrary-precision numbers are particularly useful when you are trying to understand the numerical properties of a new kind of calculation. By looking at the precision of the result, you can get a feel for whether the calculation in question suffers from a catastrophic loss of significance.

A.5 Using *Mathematica* and C Together

As you are certainly aware, we used *Mathematica* in the first half of this book and C in the second half. We have also discussed the advantages and disadvantages of the

Table A.1   Time required (in seconds) to sum series of different lengths using `BlockFloat` for the floating-point computation and `BlockPrecision` for the arbitrary-precision computations. (A 200 MHz Pentium Pro processor was used to obtain these timing measurements.)

	Floating	Arbitrary Precision	
Terms	Point	20 digits	40 digits
50	0.007	0.022	0.025
100	0.015	0.044	0.050
500	0.094	0.194	0.241
1000	0.171	0.384	0.478
5000	0.891	1.938	2.403
10000	1.781	3.860	4.805
50000	8.953	19.305	24.141

two languages. You should appreciate the fact that there are large collections of both *Mathematica* and C programs, written by thousands of other programmers, that you can exploit in your own work. What we have not emphasized until now, however, is that you can use existing C functions in a *Mathematica* program, and you can use existing *Mathematica* functions in a C program. *Mathematica*'s *MathLink* capability is what makes this possible.

Fully understanding how to use *MathLink* would require that you know more about both *Mathematica* and C than we have covered in this book. Our goal in this section is more modest. We will illustrate, via two examples, the sorts of things that *MathLink* can do. We will focus on the problem of calling a C function from a *Mathematica* notebook, because it is both the simplest capability to explain and the easiest to illustrate.

A.5.1 Calling a C Function from *Mathematica*

MathLink makes it possible to call a C function from within *Mathematica*, exactly as if that function had been built into *Mathematica*. For example, once we have set things up properly, we can invoke the `simulate` function from the program `heat6.c` of Chapter 17 directly from *Mathematica*.

```
In[61]:= simulate[1.752, 10., 100., 50., 0., 5., 5]

Out[61]= 48.4375
```

(A.61)

It is far more convenient to do this than it would be to rewrite `simulate` as a *Mathematica* function.

A C program such as `heat6.c` must be modified slightly before it can be called from within *Mathematica*. The `main` function must be replaced, and each function that *Mathematica* is to call must be annotated with a *MathLink* template. The annotated program—which will not be a C program because of the *MathLink* templates—must then be translated into a C program by using the `mprep` translator that is provided as part of *Mathematica*. The resulting C program must then be compiled, and the executable that is produced by the compilation must be installed into *Mathematica*. Only then will Example A.61 behave as illustrated.

Let's look at what is involved in carrying out these steps with `heat6.c`. Recall that `heat6.c` consists of two functions: `main`, which prompts the user for input values and displays the final result, and `simulate`, which does all of the calculations. Because *Mathematica* will be calling `simulate` directly, we no longer have any need for the original version of `main`. Instead, we will replace it with the simple `main` function from Figure A.1, which is required if *MathLink* is to work properly. (Don't be put off by the fact that this `main` function has a different header from all of the others in this book, or by the fact that it uses aspects of C that we have not discussed.)

```
1    int main (int argc, char *argv[]) {
2       return(MLMain(argc, argv));
3    }
```

Figure A.1   The `main` function required by *MathLink* (continued in Figure A.2)

The next step is to annotate the `simulate` function with a *MathLink* template, as shown in Figure A.2. The function itself has not changed from Figure 17.11; all we have done is add the template on lines 4–14. The template is designed to tell *Mathematica* everything that it needs to know in order to call `simulate`: the function's name, its return type, and the name and type of each of its parameters. (Some of the information in the template is redundant, but it is all required.) Notice that the type name `Real` stands for `double` and the type name `Integer` stands for `int`.

At this point our program consists of `main`, `simulate`, and the *MathLink* template for `simulate`. Because our program contains a template, it is no longer a valid C program, which is why the complete version of the program on the diskette is named `heat12.tm` instead of `heat12.c`. (Think of "tm" as standing for "template.")

The rest of the process is purely mechanical. We must first use `mprep`—a program that is packaged with *Mathematica*—to translate `heat12.tm` into a C program `heat12.c`. We must next use a C compiler to convert `heat12.c` into

```
4    :Begin:
5    :Function:       simulate
6    :Pattern:        simulate[diffusivity_Real, length_Real,
7                               left_Real, right_Real, rod_Real,
8                               duration_Real, segs_Integer]
9    :Arguments:      {diffusivity, length, left, right, rod,
10                      duration, segs}
11   :ArgumentTypes: {Real, Real, Real, Real, Real,
12                      Real, Integer}
13   :ReturnType:     Real
14   :End:
15
16   #define MAXSIZE 1001
17
18   double simulate (double diffusivity, double length,
19                    double left, double right, double rod,
20                    double duration, int segs)
21   {
22     double delta;
23     double cur[MAXSIZE], nxt[MAXSIZE];
24     int i;
25
26     delta = (length*length/(segs*segs)) / (2*diffusivity);
27
28     cur[0] = left;
29     for (i = 1; i < segs; i = i+1) {
30       cur[i] = rod;
31     }
32     cur[segs] = right;
33
34     while (duration > 0) {
35
36       for (i = 1; i < segs; i = i+1) {
37         nxt[i] = (cur[i-1] + cur[i+1]) / 2;
38       }
39
40       for (i = 1; i < segs; i = i+1) {
41         cur[i] = nxt[i];
42       }
43
44       duration = duration - delta;
45     }
46
47     return(cur[segs/2]);
48   }
```

Figure A.2   The simulate function from heat6.c annotated with a *MathLink* template (heat12.tm)

an executable heat12. Finally, we must install the executable into *Mathematica* by using the Install function.

```
In[62]:= Install["heat12"]

Out[62]= LinkObject[./heat12, 2, 2]
```
(A.62)

When we have done all this, we will be able to call simulate from *Mathematica* as illustrated in Example A.61

If you try to follow the procedure that we have just outlined, you will doubtless run into some of the following problems.

- The main function that is required by *MathLink*, as illustrated in Figure A.1, is different on some types of computers.
- To translate heat12.tm into heat12.c, you must learn how to use mprep.
- The C program heat12.c that mprep produces depends on an include file (mathlink.h) and a C library file that you must locate and make visible to your C compiler.
- The executable heat that the C compiler produces must be made visible to *Mathematica*.

It is difficult to go into detail about how to solve these potential problems, since the solutions differ from computer to computer. *Mathematica*'s online help will prove useful, however.

A.5.2 Communication from C to *Mathematica*

The C program that we developed in Chapter 18 produced as its output a *Mathematica* animation, but in a rather circuitous fashion. To obtain an animation, we had to run the C program, capture its output in a text file, and then read the text file into *Mathematica*. We can use *MathLink* to streamline this process by arranging for the C program to communicate directly to *Mathematica* the commands required to produce the animation.

In this section we will modify heat9.c to obtain heat13.tm. When heat13.tm is converted by mprep into heat13.c, and heat13.c is compiled to produce an executable heat13, it will be possible to install heat13 into *Mathematica*

```
In[63]:= Install["heat13"]

Out[63]= LinkObject[./heat13, 2, 2]
```
(A.63)

and then invoke simulate

```
In[64]:= simulate[1.752, 10., 100., 50., 0., 5., 5]

Out[64]= (See Figure 18.15)
```

(A.64)

to produce directly the animation that we previously produced indirectly with Example 18.1.

Recall that `heat9.c` consists of

- `max` and `min`, which compute the maximum and minimum, respectively, of three floating-point numbers;
- `initRod`, which initializes the segment temperatures in the simulated rod;
- `nextRod`, which computes the segment temperatures in the simulated rod at the next instant in simulated time;
- `createFrame`, which writes to the display the *Mathematica* command required to plot the segment temperatures of the simulated rod at a given instant in simulated time;
- `simulate`, which controls the simulation by first calling `initRod` and then repeatedly calling `nextRod` and `createFrame`; and
- `main`, which reads the simulation parameters from the user and passes them along to `simulate`.

We will modify `createFrame` extensively so that it sends animation commands directly to *Mathematica* instead of writing them to the display. We will make a minor change to `simulate` and also add a *MathLink* template so that *Mathematica* can call `simulate` directly. Finally, we will replace `main` with the version from Figure A.1, as required by *MathLink*. None of the remaining functions—`max`, `min`, `initRod`, and `nextRod`—need to be modified.

Figure A.3 shows the modified version of `createFrame`; the original version appears in Figure 18.10. Both versions produce a call to the *Mathematica* function `ListPlot`. Six examples of such calls appear in Figure 18.9.

The original version of `createFrame` writes a function call to the display. This function call must then be typed (or cut-and-pasted) into a *Mathematica* notebook before it can be evaluated. The modified version doesn't produce any output at all. Instead, it describes the function call directly to *Mathematica*, where it is immediately evaluated. It does this by using the three C functions `MLPutFunction`, `MLPutReal`, and `MLPutSymbol`, which are defined by *MathLink*.

A good way to understand how these three functions work is to imagine that a friend is using *Mathematica* and you are trying to dictate to him the *Mathematica* expression

```
ListPlot[{{0,100},{2,0},{4,0},{6,0},{8,0},{10,50}},
         PlotJoined->True, PlotRange->{0,100}]
```

To do this, you go through the following steps.

1. You say, "`ListPlot` takes three parameters," and he enters

```
1    void createFrame (double rod[], int segments,
2                          double length, double low, double high)
3    {
4      int i;
5
6      MLPutFunction(stdlink, "ListPlot", 3);
7      MLPutFunction(stdlink, "List", segments+1);
8
9      for (i = 0; i <= segments; i = i+1) {
10       MLPutFunction(stdlink, "List", 2);
11       MLPutReal(stdlink, i*length/segments);
12       MLPutReal(stdlink, rod[i]);
13     }
14
15     MLPutFunction(stdlink, "Rule", 2);
16     MLPutSymbol(stdlink, "PlotJoined");
17     MLPutSymbol(stdlink, "True");
18
19     MLPutFunction(stdlink, "Rule", 2);
20     MLPutSymbol(stdlink, "PlotRange");
21     MLPutFunction(stdlink, "List", 2);
22     MLPutReal(stdlink, low);
23     MLPutReal(stdlink, high);
24
25   }
```

Figure A.3   The `createFrame` function from `heat9.c`, modified to transmit a `ListPlot` command directly to *Mathematica* (continued in Figure A.4)

```
ListPlot[?, ?, ?]
```

putting question marks where the three as yet unknown parameters are to go.

2. You say, "`List` with six elements," and he replaces the first question mark

```
ListPlot[{?, ?, ?, ?, ?, ?}, ?, ?]
```

with a list containing question marks where the six list elements will eventually go.

3. You say, "`List` with two elements," and he replaces the first question mark

```
ListPlot[{{?, ?}, ?, ?, ?, ?, ?}, ?, ?]
```

with a list containing two question marks.

4. You say, "`Real` number 0," and he replaces the first question mark with 0.

```
ListPlot[{{0, ?}, ?, ?, ?, ?, ?}, ?, ?]
```

5. You say, "`Real` number 100," and he replaces the first question mark with 100.

```
ListPlot[{{0, 100}, ?, ?, ?, ?, ?}, ?, ?]
```

6. You repeat steps 3, 4, and 5 (using different real numbers in steps 4 and 5) five more times, at which point your friend has entered

```
ListPlot[{{0,100},{2,0},{4,0},{6,0},{8,0},{10,50}},
          ?, ?]
```

7. You say "Rule with two parts," and he replaces the first question mark

```
ListPlot[{{0,100},{2,0},{4,0},{6,0},{8,0},{10,50}},
          ?->?, ?]
```

with a rule composed of two question marks.
8. You say "Symbol PlotJoined," and he replaces the first question mark with PlotJoined.

```
ListPlot[{{0,100},{2,0},{4,0},{6,0},{8,0},{10,50}},
          PlotJoined->?, ?]
```

9. You say "Symbol True," and he replaces the first question mark with True.

```
ListPlot[{{0,100},{2,0},{4,0},{6,0},{8,0},{10,50}},
          PlotJoined->True, ?]
```

10. You say "Rule with two parts," and he replaces the first question mark

```
ListPlot[{{0,100},{2,0},{4,0},{6,0},{8,0},{10,50}},
          PlotJoined->True, ?->?]
```

with a rule composed of two question marks.
11. You say "Symbol PlotRange," and he replaces the first question mark with PlotRange.

```
ListPlot[{{0,100},{2,0},{4,0},{6,0},{8,0},{10,50}},
          PlotJoined->True, PlotRange->?]
```

12. You say "List with two elements," and he replaces the first question mark

```
ListPlot[{{0,100},{2,0},{4,0},{6,0},{8,0},{10,50}},
          PlotJoined->True, PlotRange->{?, ?}]
```

with a list containing two question marks.
13. You say "Real number 0," and he replaces the first question mark with 0.

```
ListPlot[{{0,100},{2,0},{4,0},{6,0},{8,0},{10,50}},
          PlotJoined->True, PlotRange->{0, ?}]
```

14. You say "Real number 100," and he replaces the first question mark with 100.

```
ListPlot[{{0,100},{2,0},{4,0},{6,0},{8,0},{10,50}},
          PlotJoined->True, PlotRange->{0, 100}]
```

At this point, your friend notices that there are no remaining question marks, realizes that you have described the entire expression to him, and asks *Mathematica* to evaluate the expression.

If you trace through the implementation of `createFrame` in Figure A.3, you will discover that it works in a way that is completely analogous to our hypothetical example of dictating a *Mathematica* expression to a friend. `MLPutFunction` tells *Mathematica* to enter a function with a particular number of parameters, `MLPutReal` tells *Mathematica* to enter a floating-point number, and `MLPutSymbol` tells *Mathematica* to enter a symbol. (The first parameter to each function is always `stdlink`.)

There are 14 steps in our hypothetical example. Table A.2 shows how each step corresponds to one or more lines of `createFrame`.

Table A.2   Correspondence between steps of hypothetical conversation with friend and lines of `createFrame` from Figure A.3

Step	Corresponding portion of `createFrame`
1	Line 6
2	Line 7
3	First repetition of line 10
4	First repetition of line 11
5	First repetition of line 12
6	Remaining five repetitions of lines 10–12
7	Line 15
8	Line 16
9	Line 17
10	Line 19
11	Line 20
12	Line 21
13	Line 22
14	Line 22

Figure A.4 shows the modified and annotated version of `simulate`. The template is exactly the same as the one from Figure A.2. We have made only one change to the version of `simulate` from `heat9.c`. On line 51 we have put a call to `MLPutFunction` that tells *Mathematica* to create a list of duration+1 elements. This list will be made up of the individual `ListPlot` commands `simulate` produces by repeatedly calling `createFrame`. This is necessary because otherwise *Mathematica* will ignore all but the first of the `ListPlot` commands. By packaging them into a list, we arrange for *Mathematica* to evaluate them all in order.

```
26    :Begin:
27    :Function:        simulate
28    :Pattern:         simulate[diffusivity_Real, length_Real,
29                                 left_Real, right_Real, rod_Real,
30                                 duration_Real, segs_Integer]
31    :Arguments:       {diffusivity, length, left, right, rod,
32                        duration, segs}
33    :ArgumentTypes: {Real, Real, Real, Real, Real,
34                        Real, Integer}
35    :ReturnType:      Real
36    :End:
37
38    double simulate (double diffusivity, double length,
39                         double left, double right, double rod,
40                         double duration, int segs)
41    {
42      double delta, clock, low, high;
43      double cur[MAXSIZE];
44
45      initRod(cur, segs, left, right, rod);
46      delta = (length*length/(segs*segs)) / (2*diffusivity);
47
48      low = min(rod, left, right);
49      high = max(rod, left, right);
50
51      MLPutFunction(stdlink, "List", duration+1);
52
53      createFrame(cur, segs, length, low, high);
54      clock = 0;
55
56      while (duration > 0) {
57
58        nextRod(cur, segs);
59        duration = duration - delta;
60
61        clock = clock + delta;
62        if (clock >= 1 - delta/2) {
63          createFrame(cur, segs, length, low, high);
64          clock = clock - 1;
65        }
66
67      }
68
69      return(cur[segs/2]);
70
71    }
```

Figure A.4   The `simulate` function from `heat9.c`, slightly modified and annotated with a *MathLink* template (`heat13.tm`)

B

Mathematica Functions and Constants

Much of the power of *Mathematica* is provided by its extensive collection of built-in functions and constants. This appendix contains a brief description of each *Mathematica* function and constant that we have used in this book. Refer to the index to find the places in the text where examples appear. You can consult *Mathematica*'s online help facility to learn more about these functions and the hundreds of others that *Mathematica* provides.

Abs[x] returns $|x|$.

AnimateNewton[function, guess, nsteps, {a,b}] produces an animation of how Newton's method finds an approximate root of *function* beginning with an initial *guess*. The animation consists of *nsteps*+1 frames, and each frame displays the *x*-axis between *a* and *b*. To use AnimateNewton, the Newton package must be accessed via Needs["ISP`Newton`"]. This package is not a standard part of *Mathematica*, but is on the diskette included with this book.

AnimateRectangular[f, a, b, nframes] produces an animation showing how the rectangular method approximates $\int_a^b f(x)\,dx$. The animation consists of *nframes* frames. The first frame shows a two-rectangle approximation, and the number of rectangles doubles in each subsequent frame. To use this function, the Integ package must be accessed via Needs["ISP`Integ`"]. This package is not a standard part of *Mathematica*, but is on the diskette included with this book.

AnimateTrapezoidal[f, a, b, nframes] produces an animation showing how the trapezoidal method approximates $\int_a^b f(x)\,dx$. The animation consists of *nframes* frames. The first frame shows a two-trapezoid approximation, and the number of trapezoids doubles in each subsequent frame. To use this function, the Integ package must be accessed via Needs["ISP`Integ`"]. This package is not a standard part of *Mathematica*, but is on the diskette included with this book.

ArcCos[x] returns arccos(*x*), with the result expressed in radians.

ArcSin[x] returns arcsin(*x*), with the result expressed in radians.

ArcTan[x] returns arctan(*x*), with the result expressed in radians.

Arithmetic[] returns the current characteristics of the simulated floating-point numbers created by ComputerNumber, as initialized by SetArithmetic. To use Arithmetic, the ComputerArithmetic package must be accessed via Needs["NumericalMath`ComputerArithmetic`"].

BlockFast[n] solves the block-stacking problem of Chapter 4 by returning the maximum extension possible with *n* blocks. It uses PolyGamma and PolyEuler to quickly sum the first *n* even terms of the harmonic series. To use BlockFast, the Blocks package must be accessed via Needs["ISP`Blocks`"]. This package is not a standard part of *Mathematica*, but is on the diskette included with this book.

BlockFloat[n] solves the block-stacking problem of Chapter 4 by returning the maximum extension possible with *n* blocks. It uses floating-point division and addition to sum the first *n* even terms of the harmonic series. To use BlockFloat, the Blocks package must be accessed via Needs["ISP`Blocks`"]. This package is not a standard part of *Mathematica*, but is on the diskette included with this book.

BlockPrecision[n, p] solves the block-stacking problem of Chapter 4 by returning the maximum extension possible with |*n*| blocks. If *n* is positive, it sums the first *n* even terms of the harmonic series from left to right; if *n* is negative, it sums the terms from right to left. If $p \leq 10$, it uses simulated floating-point arithmetic with a precision of *p*. If *p* is larger than $MachinePrecision, it uses arbitrary-precision arithmetic with a precision of *p*. Otherwise, it uses floating-point arithmetic. To use BlockPrecision, the Blocks package must be accessed via Needs["ISP`Blocks`"]. This package is not a standard part of *Mathematica*, but is on the diskette included with this book.

BlockRat[n] solves the block-stacking problem of Chapter 4 by returning the maximum extension possible with *n* blocks. It uses rational division and addition to sum the first *n* even terms of the harmonic series. To use BlockRat, the Blocks package must be accessed via Needs["ISP`Blocks`"]. This package is not a standard part of *Mathematica*, but is on the diskette included with this book.

Clear[x] removes any value that has been assigned to *x*.

`ComputerNumber[x]` converts *x* into a simulated floating-point number, with a base and a precision as specified by `SetArithmetic`. To use `ComputerNumber`, the `ComputerArithmetic` package must first be accessed by evaluating `Needs["NumericalMath`ComputerArithmetic`"]`.

`Convert[n, u]` converts the dimensioned number *n* into the type of unit specified by *u*. To use `Convert`, the `Units` package must be accessed via `Needs["Miscellaneous`Units`"]`.

`Cos[x]` returns $\cos(x)$, where *x* is expressed in radians.

`D[expr, x]` returns $\frac{d}{dx}expr$.

`E` is *e*.

`Exp[x]` returns e^x.

`Expand[expr]` returns the result of expanding products and positive integer products in *expr*.

`EulerGamma` is Euler's constant, which is used in summing the harmonic series.

`Factor[expr]` returns the factorization of *expr*.

`FindRoot[eqn, var]` requires that *eqn* be an equation in the single unknown *var*. It returns a solution of *eqn* that is obtained numerically.

`Get["filename"]` reads and executes the *Mathematica* commands contained in the file `filename`.

`I` is *i*, the square root of -1.

`If[b,e1,e2]` returns the value of e_1 if *b* evaluates to `True`, and returns the value of e_2 if *b* evaluates to `False`.

`Infinity` is ∞.

`InputForm[expr]` displays what must be typed at the keyboard to reconstruct the value of *expr*. It is useful for seeing all of the digits of a floating-point number.

`Install["file"]` makes the *MathLink*-templated functions contained in the executable file `file` available for use from *Mathematica*.

`Integrate[expr, x]` returns $\int expr\, dx$.

`Integrate[expr, {x,a,b}]` returns $\int_a^b expr\, dx$. The integral is evaluated symbolically.

`Interval[{x, y}]` represents a number that can lie anywhere in the interval $[x \ldots y]$. It is used in combination with interval arithmetic.

`Limit[expr, x->a]` returns $\lim_{x \to a} expr$. Limits at infinity can be obtained by using `Infinity` as the value of a.

`ListPlot[points]` creates a plot of the points contained in the list *points*.

`Log[x]` returns $\log_e(x)$.

`Log[b, x]` returns $\log_b(x)$.

`$MachinePrecision` is the number of digits in a *Mathematica* floating-point number.

`$MaxMachineNumber` is *Mathematica*'s largest positive floating-point number.

`$MaxNumber` is *Mathematica*'s largest positive arbitrary-precision number.

`$MinMachineNumber` is *Mathematica*'s smallest positive floating-point number.

`$MinNumber` is *Mathematica*'s smallest positive arbitrary-precision number.

`Module[vars, expr]` returns the result of evaluating *expr* in a context in which the variables in the list *vars* are considered local to the evaluation.

`N[x]` converts x into a floating-point number.

`N[x, p]` converts x into a floating-point number if p is no greater than `$MachinePrecision`, and converts x into an arbitrary-precision number with precision p otherwise.

`Names["Library`Package`*"]` returns a list of the functions contained in the package `Package` of the library `Library`.

`Needs["Library`Package`"]` makes the functions contained in the package `Package` of the library `Library` available for use.

`NIntegrate[expr, {x,a,b}]` evaluates $\int_a^b expr\,dx$ using numerical techniques.

`ParametricPlot[{expr1, expr2}, {x,a,b}]` requires that *expr1* and *expr2* be symbolic expressions in x. It produces a parametric plot of *expr1* and *expr2* as x ranges from a to b. If a list of parametric specifications is supplied as the first parameter, a plot containing multiple parametric curves will be produced.

`Pi` is π.

`Plot[expr, {x,a,b}]` requires that *expr* be a symbolic expression in *x*. It produces a plot of *expr* as *x* ranges from *a* to *b*. If a list of expressions is supplied as the first parameter, a plot containing multiple curves will be produced.

`PolyGamma[x]` returns the value of the polygamma function at *x*. It is used for summing the harmonic series.

`Precision[x]` returns the precision of the arbitrary-precision number *x*.

`Product[expr, {x,a,b}]` returns $\prod_{x=a}^{b} expr$.

`ReplaceAll[expr, subs]` returns the result of applying the replacements *subs* to the expression *expr*.

`SetArithmetic[p, b]` tells *Mathematica* that simulated floating-point numbers created with `ComputerNumber` are to be represented in base *b* with *p*-digit mantissas. The four optional parameters and their possible values are

`RoundingRule->rule`	If `rule` is `RoundToEven`, round to nearest floating-point number; break ties by rounding to an even number. If `rule` is `RoundToInfinity`, round to nearest floating-point number; break ties by rounding away from zero. If `rule` is `Truncation`, discard excess digits instead of rounding.
`ExponentRange->{l, h}`	Restrict exponents to a minimum of *l* and a maximum of $h - 1$.
`MixedMode->b`	If *b* is `True`, allow mixed arithmetic on simulated floating-point numbers and integers. If *b* is `False`, do not allow mixed arithmetic.
`IdealDivide->b`	If *b* is `False`, do division by first computing the reciprocal of the denominator and then multiplying by the denominator. If *b* is `True`, do long division instead.

To use `SetArithmetic`, the `ComputerArithmetic` package must be accessed via `Needs["NumericalMath`ComputerArithmetic`"]`.

`Show[{plot1,...,plotn}]` creates a plot that combines *plot1* through *plotn*.

`Simplify[expr]` returns the result of applying algebraic simplifications to *expr*.

`Sin[x]` returns $\sin(x)$, where *x* is expressed in radians.

`Solve[eqn, var]` returns values for *var* that solve *eqn*. The solutions are obtained symbolically. Lists of equations and variables can also be specified as parameters.

`Sqrt[x]` returns $\sqrt{x}$.

Sum[expr, {x,a,b}] returns $\sum_{x=a}^{b} expr$.

Table[expr, {x,a,b,c}] returns the list of values obtained by evaluating *expr* under the assumption that the value of x is a, then under the assumption that x is $a + c$, and so on up to the assumption that x is b.

Tan[x] returns $\tan(x)$, where x is expressed in radians.

Timing[expr] returns a list of the time required to evaluate *expr* and the value of *expr*.

Which[b1,e1, ... ,bn,en] evaluates the Boolean expressions b_i in order until it finds one whose value is True, then returns the value of the corresponding expression e_i.

While[b,expr] evaluates *expr* repeatedly so long as the value of b remains True.

C

C Library Functions

This appendix contains a brief description of the C library functions that we have used in this book. Refer to the index to find the places in the text where examples appear. We have documented only those aspects of the functions that are relevant to this book. Most C systems have online information about the library functions, which you can consult for more complete information about these and other library functions.

The math functions are declared in <math.h>; the input/output functions are declared in <stdio.h>; and exit is declared in <stdlib.h>.

double acos (double x) returns arccos(x), where the result is expressed in radians.

double asin (double x) returns arcsin(x), where the result is expressed in radians.

double atan (double x) returns arctan(x), where the result is expressed in radians.

double cos (double x) returns cos(x), where x is expressed in radians.

void exit (int status) immediately terminates execution of the program and specifies *status* as the termination status to the operating system.

double exp (double x) returns e^x.

double fabs (double x) returns $|x|$.

`int fclose (FILE *file)` closes `file`, which should be a file pointer obtained with a call to `fopen`.

`FILE *fopen (char *filename, char *mode)` opens and returns a pointer to the file named `filename`. If `mode` is `"w"`, the file is opened for writing. If `mode` is `"r"`, the file is opened for reading. If the file cannot be opened, a null pointer is returned. The returned pointer can be tested to see if it is null by using the `!` operator.

`int fprintf (FILE *file, char *format, ...)` writes information to `file`, which should be either `stdout` or a file pointer obtained with a call to `fopen`. The `format` string and the parameters that follow it are treated exactly as in `printf`.

`int fscanf (FILE *file, char *format, ...)` reads information from `file`, which should be either `stdin` or a file pointer obtained with a call to `fopen`. The `format` string and the parameters that follow it are treated exactly as in `scanf`.

`double log (double x)` returns $\ln(x)$.

`double log10 (double x)` returns $\log_{10}(x)$.

`double pow (double x, double y)` returns x^y.

`int printf (char *format, ...)` writes information to the display. The `format` string should contain zero or more conversion specifications and should be followed by the same number of integer and floating-point expressions. The `format` string is written to the display, but each conversion specification that it contains is replaced with the value of the corresponding integer or floating-point expression.

All conversion specifications begin with a percent sign. The possible conversion specifications include

`%d`	which displays all of the digits of an integer.
`%f`	which displays a floating-point number. The number is displayed without an exponent and with six digits after the decimal point.
`%e`	which displays a floating-point number. The number is displayed with an exponent and with six digits after the decimal point.
`%g`	which displays a floating-point number. The number is displayed with a six-digit mantissa and, if necessary, an exponent.

A column width w may be included in a conversion specification, as in `%8d` or `%12g`. This specifies that the value is to be padded out with spaces if necessary so that it occupies a column of width w.

A precision p may be included in a floating-point conversion specification, as in `%.8f`, `%.7e`, or `%.6g`. This specifies that the value is to be displayed with p digits after the decimal point (in the case of the `%f` and `%e` formats), and with `p` digits in the mantissa (in the case of the `%g` format). Do not ask for more digits than are contained in the number being displayed.

Floating-point conversion specifications may include both a column width and a precision, as in `%12.8f`.

If a + symbol follows the percent sign, as in `%+d` or `%+.8f`, and the corresponding value is positive, a leading + symbol is displayed.

`int scanf (char *format, ...)` reads information from the keyboard. The `format` string should contain zero or more conversion specifications and should be followed by the same number of `int` and `double` variables, each preceded by an ampersand character (`&`). The `format` string is processed from left to right, and as each conversion specification is encountered a value of the appropriate type is read and stored into the corresponding variable.

All conversion specifications begin with a percent sign. The possible conversion specifications include

  `%d`   which reads a value into an `int` variable.
  `%lf`   which reads a value into a `double` variable.

If any characters other than conversion specifications, spaces, or tabs appear in the `format` string, those characters must also appear in the input entered at the keyboard.

`double sin (double x)` returns $\sin(x)$, where x is expressed in radians.

`double sqrt (double x)` returns $\sqrt{x}$.

`double tan (double x)` returns $\tan(x)$, where x is expressed in radians.

D

Using *Mathematica* 2.2

All of the *Mathematica* programs in this book are compatible with *Mathematica* 3.0. In this appendix we will detail a handful of changes that you must make if you wish to use the programs with the older *Mathematica* 2.2.

D.1 Floating-Point Syntax

Beginning in Chapter 2, and continuing through all of the *Mathematica* chapters, we exploited *Mathematica* 3.0's special syntax for writing floating-point numbers in scientific notation. Thus, in Example 2.6,

```
In[1]:= 57.8*^6 / 5.713*^9

Out[1]= .0101173 mi²
```
(D.1)

we wrote 57.8×10^6 as `57.8*^6` and 5.713×10^9 as `5.713*^9`.

This floating-point syntax is not supported by *Mathematica* 2.2. Instead, you must write 57.8×10^6 as the slightly longer `57.8*10^6`. You must take care when using the longer form because it is actually the product of two numbers. If we naively translate Example D.1 as

```
In[2]:= 5.78*10^6 / 5.713*10^9

Out[2]= 1.01173 × 10¹⁵ mi²
```
(D.2)

for example, we get the wrong answer because *Mathematica*'s precedence rules cause the 10^9 factor to appear in the numerator. We get the right answer only if we use parentheses in the translation of Example D.1.

```
In[3]:= 5.78*10^6 / (5.713*10^9)

Out[3]= .0101173 mi²
```
(D.3)

D.2 Typeset Mathematics

In Appendix A, we showed how *Mathematica* 3.0 permits the entry of expressions in a typeset format, and throughout the book we showed how the results produced by *Mathematica* 3.0 are displayed in a typeset format. These are features of the *Mathematica* 3.0 notebook interface, and are not supported by *Mathematica* 2.2.

For example, in *Mathematica* 3.0 we can produce the interaction

```
In[4]:= ∫ 1/(2√x) dx

Out[4]= √x
```
(D.4)

whereas in *Mathematica* 2.2 the same interaction must be written as

```
In[5]:= Integrate[1/(2*Sqrt[x]), x]

Out[5]= Sqrt[x]
```
(D.5)

D.3 Special Constants

In this book we used several *Mathematica* constants, including Pi, E, and EulerGamma, that are treated as exact symbolic representations in rational contexts and as numerical approximations in floating-point contexts. Notice the difference in *Mathematica* 3.0, for example, between

```
In[6]:= 2 * Pi

Out[6]= 2π
```
(D.6)

and

```
In[7]:= 2.0 * Pi

Out[7]= 6.28319
```
(D.7)

Mathematica 2.2, in contrast, does not automatically convert the symbolic constants into numerical approximations. In *Mathematica* 2.2, Example D.7 would evaluate as

```
In[8]:= 2.0 * Pi

Out[8]= 2. Pi
```
(D.8)

There are several places in the book where this aspect of *Mathematica* 2.2 is apparent. The first example appears in Chapter 4, where the result of Example 4.18

```
In[9]:= 0.5 * (PolyGamma[1.*10^9 + 1] + EulerGamma)

Out[9]= 0.5(20.7233 + EulerGamma)
```
(D.9)

is expressed by *Mathematica* 2.2 with a mixture of numbers and symbols. To obtain a numerical solution, we must use the built-in function N.

```
In[10]:= N[0.5 * (PolyGamma[1.*10^9 + 1] + EulerGamma)]

Out[10]= 10.6502
```
(D.10)

A more serious problem arises in Chapter 9, where the function cow is defined in Example 9.24 as

```
In[11]:= cow[theta_] := Sin[theta] -
                        theta*Cos[theta] - Pi/2
```
(D.11)

Because this definition involves the constant Pi, the value of

```
In[12]:= cow[2.5]

Out[12]= 2.60133 - Pi/2
```
(D.12)

in *Mathematica* 2.2 is expressed in terms of Pi.

The problem with this behavior is that in *Mathematica* 2.2, expressions involving symbols such as Pi interact badly with relational operators such as >=. For example, the expression

```
In[13]:= Pi >= 3

Out[13]= Pi >= 3
```

(D.13)

evaluates to neither True nor False. As a result, if we use this inequality in an If expression,

```
In[14]:= If[Pi >= 3, 1, 2]
```

(D.14)

neither the True value nor the False value is delivered as the value of the conditional.

Let's see how this behavior of *Mathematica* 2.2 causes trouble with the implementation of the bisection method in Chapter 9. We explored the idea behind the bisection method in Figure 9.6, which we evaluate in Figure D.1 using *Mathematica* 2.2.

The value of the expression sequence in Figure D.1 is 2.0, as compared to the 1.75 we obtained with *Mathematica* 3.0 in Figure 9.6. The problem is that the expressions f[avg] on lines 6 and 11 produce results that contain Pi, which means that neither the True parts nor the False parts of the conditionals on lines 6–8 and 11–13 are evaluated.

```
In[15]:=

1      (f = cow;
2       pos = 2.5;
3       neg = 1.5;
4
5       avg = (pos + neg) / 2.0;
6       If[f[avg] >= 0,
7          pos = avg,
8          neg = avg];
9
10      avg = (pos + neg) / 2.0;
11      If[f[avg] >= 0,
12         pos = avg,
13         neg = avg];
14
15      avg)

Out[15]= 2.0
```

Figure D.1   The bisection process with conditionals in *Mathematica* 2.2

There are a variety of ways to solve this problem, but the simplest is to modify the program in Figure D.1 by using N to ensure that f[avg] produces floating-point results. This change appears in Figure D.2.

```
In[16]:=

1      (f = cow;
2       pos = 2.5;
3       neg = 1.5;
4
5       avg = (pos + neg) / 2.0;
6       If[N[f[avg]] >= 0,
7           pos = avg,
8           neg = avg];
9
10      avg = (pos + neg) / 2.0;
11      If[N[f[avg]] >= 0,
12          pos = avg,
13          neg = avg];
14
15      avg)

Out[16]= 1.75
```

Figure D.2 Corrected bisection process with conditionals in *Mathematica* 2.2

We have replaced f[avg] on lines 6 and 11 with N[f[avg]]. As a result, *Mathematica* 2.2 produces the correct 1.75 as the value of the expression. All of the bisection programs in Chapter 9 (Figures 9.5—9.10) must be changed in the same fashion if they are to run correctly in *Mathematica* 2.2.

D.4 Symbolic Capabilities

With each new release, the symbolic capabilities of *Mathematica* become more powerful. This is illustrated by three symbolic calculations in this book which *Mathematica* 3.0 can evaluate but which *Mathematica* 2.2 cannot.

The first occurs in Example 6.47,

$$\text{In[17]:= Sum[1/2^n, \{n,1,Infinity\}]}$$
$$\text{Out[17]= Sum[2}^{-n}\text{, \{n, 1, Infinity\}]} \tag{D.17}$$

an infinite sum which *Mathematica* 2.2 cannot reduce to 1.

The second calculation occurs in Example 8.14,

```
In[18]:= threshold = Solve[Abs[pow] == maxPower, t]

Out[18]= {{}}
```

(D.18)

an equation with two roots. *Mathematica* 2.2 cannot solve it because it contains the Abs function. The workaround is to rewrite the original equation into two equations that do not involve Abs. We use one equation for the case where pow is positive,

```
In[19]:= Solve[pow == maxPower, t]

Out[19]= {{t->271.267}}
```

(D.19)

and a second equation for the case where pow is negative.

```
In[20]:= Solve[-pow == maxPower, t]

Out[20]= {{t-> - 271.267}}
```

(D.20)

The third symbolic calculation occurs in Example 9.12,

```
In[21]:= Solve[Cos[alpha] == Sin[alpha], alpha]

        Solve::tdep: The equations appear to involve
            transcendental functions of the variables
            in an essentially non-algebraic way.
```

(D.21)

which *Mathematica* 2.2 is unable to solve.

Bibliography

Mathematica is a large system, and it has many capabilities beyond those we covered in this text. The first book below is a huge volume that was published as a companion to Version 3 of *Mathematica*. It is comprehensive, but it is also 1400 pages long. The second book below discusses *Mathematica* from a programming perspective.

- Stephen Wolfram. *The Mathematica Book*, third edition. Wolfram Media/ Cambridge University Press, 1996.
- Richard J. Gaylord, Samuel N. Kamin, and Paul R. Wellin. *An Introduction to Programming with Mathematica*, second edition. TELOS/Springer-Verlag, 1996.

We covered a greater percentage of C in this text than we did of *Mathematica*. Now that you have a working knowledge of how C can be used to do scientific programming, you may find the following three books useful for extending your knowledge. Dozens of books on C have been published, but Kernighan and Ritchie's book sets the standard for conciseness and completeness. Winston's book, which is equally concise, takes a different approach that may be more to your taste. Plauger's book is an encyclopedic treatment of all of the nuances of the standard C library.

- Brian Kernighan and Dennis Ritchie. *The C Programming Language*. Prentice Hall, Englewood Cliffs, N.J., second edition, 1989.
- P.J. Plauger. *The Standard C Library*. Prentice Hall, Englewood Cliffs, N.J., 1992.
- Patrick Henry Winston. *On to C*. Addison-Wesley, Reading, Mass., 1994.

We touched on a variety of numerical methods in this text, including significant digits, interval arithmetic, numerical root finding, numerical integration, and the finite-element method. These two books treat these and related topics in more detail.

- Richard L. Burden and J. Douglas Faires. *Numerical Analysis*. PWS–Kent Publishing Company, Boston, fifth edition, 1993.
- E. Ward Cheney and David R. Kincaid. *Numerical Mathematics and Computing*. Brooks/Cole, Pacific Grove, Cal., third edition, 1994.

Most of the models that we developed in this text are based on concepts from geometry, trigonometry, and calculus. These two books provide a rich source of mathematical background.

- Erwin Kreyszig. *Advanced Engineering Mathematics*. John Wiley and Sons, Inc., New York, seventh edition, 1993.
- George B. Thomas, Jr. and Ross L. Finney. *Calculus and Analytic Geometry*. Addison-Wesley, Reading, Mass., ninth edition, 1996.

Many of the problems in this text are drawn from classical physics. The following two books provide a solid background in physics and physics-based problem solving.

- Murray R. Spiegel. *Schaum's Outline of Theoretical Mechanics*. McGraw-Hill, New York, 1967.
- Hugh D. Young and Roger A. Freedman. *University Physics*. Addison-Wesley, Reading, Mass., ninth edition, 1996.

We drew on these books for the historical and biographical background with which we framed many of the chapters in this text.

- Gwynne Evans. *Practical Numerical Integration*. John Wiley & Sons, Chichester, U.K., 1993.
- Martin Gardner. Limits of infinite series. In *Martin Gardner's Sixth Book of Mathematical Games from Scientific American*, pages 163–172. Charles Scribner's Sons, New York, 1971.
- Owen Gingerich. *The Eye of Heaven: Ptolemy, Copernicus, Kepler*. American Institute of Physics, New York, 1993.
- Herman H. Goldstine. *The Computer: From Pascal to von Neumann*. Princeton University Press, Princeton, N.J., 1972.
- John Herivel. *Joseph Fourier: The Man and the Physicist*. Oxford University Press, London, 1975.
- Frank Kreith and Mark S. Bohn. *Principles of Heat Transfer*. West Publishing Company, St. Paul, Minn., fifth edition, 1993.
- Alexandre Koyré. *The Astronomical Revolution: Copernicus — Kepler — Borelli*. Cornell University Press, Ithaca, N.Y., 1973. Translated by R.E.W. Maddison from the original French text *La révolution astronomique*, 1961, Hermann, Paris.
- G.E.R. Lloyd. *Greek Science After Aristotle*. W.W. Norton & Company, New York, 1973.
- Samuel Eliot Morison. *Leyte: June 1944–January 1945*, volume XII of *History of United States Naval Operations in World War II*. Little, Brown and Company, Boston, 1958.
- L. Rosemblum, R.A. Earnshaw, J. Encarnacao, H. Hagen, A. Kaufman, S. Klimenko, G. Nielson, F. Post, and D. Thalmann, editors. *Scientific Visualization: Advances and Challenges*. Academic Press, San Diego, 1994.
- George Sarton. *A History of Science: Hellenistic Science and Culture in the Last Three Centuries B.C.* Harvard University Press, Cambridge, Mass., 1959.
- Victor E. Thoren. *The Lord of Uraniborg: A Biography of Tycho Brahe*. Cambridge University Press, Cambridge, U.K., 1990.
- Michael S. Turner. In the beginning. In *Scientific American: Triumph of Discovery: A Chronicle of Great Adventures in Science*, pages 32–37. Henry Holt and Company, Inc., New York, 1995.
- Michael R. Williams. *A History of Computing Technology*. Prentice-Hall, Inc., Englewood Cliffs, N.J., 1985.
- Stanley P. Wyatt. *Principles of Astronomy*. Allyn and Bacon, Inc., Boston, 1964.

Index

Since this field is fast-moving, we expect updates and changes to occur that might necessitate sending you the most current pertinent information by paper, electronic media, or both, regarding *Introduction to Scientific Programming* (Mathematica and C edition). Therefore, in order to not miss out on receiving important update information, please fill out this card and return it to us promptly. Thank you.

Name: _____

Title: _____

Company: _____

Address: _____

City: _____ State: _____ Zip: _____

E-mail: _____

Areas of Interest/Technical Expertise: _____

Comments on this Publication: _____

_____

_____

❑ Please check this box to indicate that we may use your comments in our promotion and advertising for this publication.

Purchased from: _____

Date of Purchase: _____

❑ Please add me to your mailing list to receive updated information on *Introduction to Scientific Programming* and other TELOS publications.

I have a(n)   ❑ IBM compatible   ❑ Macintosh   ❑ Unix   ❑ Other
Designate specific model: _____

BUSINESS REPLY MAIL

FIRST CLASS MAIL PERMIT NO. 1314 SANTA CLARA, CA

POSTAGE WILL BE PAID BY ADDRESSEE

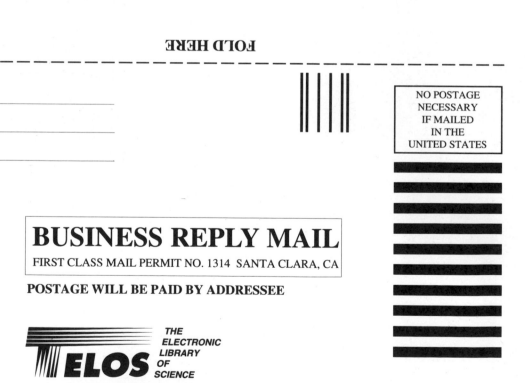

TELOS
THE
ELECTRONIC
LIBRARY
OF
SCIENCE

3600 PRUNERIDGE AVE STE 200
SANTA CLARA CA 95051-9835